INTERNATIONAL CENTRE FOR MECHANICAL SCIENCES

COURSES AND LECTURES No. 274

DYNAMICS OF HIGH-SPEED VEHICLES

EDITED BY

W.O. SCHIEHLEN

UNIVERSITY OF STUTTGART

SPRINGER-VERLAG WIEN GMBH

This volume contains 191 figures

Originally published by Springer-Verlag Wien New York in 1982

ISBN 978-3-211-81719-3 ISBN 978-3-7091-2926-5 (eBook)
DOI 10.1007/978-3-7091-2926-5

PREFACE

High-speed ground transportation requires the development of vehicle systems with better and better dynamical behaviour. Such vehicle systems include the guideway, the vehicle itself and the passenger or freight, respectively.

The present volume contains the courses and lectures presented at the International Centre for Mechanical Sciences at Udine, in fall, 1981. All the main aspects of high-speed vehicle engineering are summarized with respect to the dynamics of automobiles, railways and magnetic levitated vehicles. Analytical, experimental and numerical methods are applied using the deterministic and stochastic approach of dynamics.

The contributions to the vertical motion of vehicles (K. Popp, W.O. Schiehlen and P.C. Müller) are followed by more specific papers on the horizontal motion. Automobiles (P. Lugner, A. Zomotor and W.O. Schiehlen), railways (P. Meinke, A.D. de Pater and P. Meinke) and maglev vehicles (G. Bohn, W. Crämer and K. Popp) are treated in detail.

I hope the contributions presented will be of interest to engineers and research workers in companies and univerties who want first-hand information on the present trends and problems in this important field of technology.

Finally, I would like to thank the authors for their efforts in presenting the lectures and preparing the manuscripts for publication. My thanks are also due to Professor G. Bianchi, Secretary General of CISM and to Professor H.B. Pacejka, Secretary General of IAVSD, for advice and help during the preparation of the course. I am also grateful to my secretary, U. Wachendorff, for her engaged service with respect to the editorial work.

W. Werner O. Schiehlen
Stuttgart, June 1982

CONTENTS

Page

Introduction to Vehicle Dynamics
by Werner O. Schiehlen 1

Stochastic and Elastic Guideway Models
by K. Popp 13

Modeling by Multibody Systems
by Werner O. Schiehlen 39

Mathematical Methods in Vehicle Dynamics
by Peter C. Müller 51

Horizontal Motion of Automobiles, Theoretical and Practical Investigations
by P. Lugner 83

Horizontal Motion of Automobiles, Vehicle Handling, Measurement Methods and Experimental Experimental Results
by A. Zomotor 147

Complex Nonlinear Vehicles under Stochastic Excitation
by Werner O. Schiehlen 185

Principal Concepts of High-Speed Traffic using Wheel/Rail Technology
by P. Meinke, H. Örley 199

The Lateral Behaviour of Railway Vehicles
by A.D. de Pater 223

Design and Evaluation of Trucks for High-Speed Wheel/Rail Application
by P. Meinke, A. Mielcarek 279

Mathematical Modeling and Control System Design of Maglev Vehicles
by K. Popp 333

Some Design Criteria for the Layout of Maglev-Vehicle-Systems
by W. Crämer 365

Magnetics and Experimental Results of Maglev-Vehicles
by G. Bohn 381

INTRODUCTION TO VEHICLE DYNAMICS

Werner O. Schiehlen
Institut B für Mechanik
Universität Stuttgart
Pfaffenwaldring 9, Stuttgart 80, F.R.G.

KINDS AND MOTIONS OF VEHICLES

Today's and tomorrow's vehicles are based on various principles and travel with very different speeds. For a first classification of all kinds of vehicles the support and propulsion principles are used. The support mechanism has to balance the gravity acting on vehicles and the propulsion generates the forward speed, Fig. 1.

Ground vehicles are supported by reaction forces generated by wheels, air cushion or magnets. They are driven by friction, flow or magnetic forces. *Fluid vehicles* are supported by static or dynamic lift forces generated by water or air, and they are propelled generally by flow forces. *Inertia vehicles* are supported by dynamic lift or inertia forces generated by air, jet propulsion or orbital motion, and they are accelerated by inertia forces only. Due to the applied support and propulsion principles very different speeds are obtained. Ground vehicles come up to 450 km/h traveling speed and 650 km/h maximum speed. Fluid vehicles reach 1000 km/h traveling speed and 3300 km/h maximum speed while inertia vehicles may

Types of vehicles	Traveling speed km/h	Maximum speed km/h
GROUND VEHICLES		
Guided ground vehicles		
Railway vehicles	250	380
Tracked air cushion vehicles	-	400
Magnetically levitated vehicles	-	450
Nonguided ground vehicles		
Road vehicles	220	650
Wheeled off-road vehicles	60	-
Air cushion vehicles	-	140
FLUID VEHICLES		
Marine crafts		
Ships	40	300
Hydrofoils	150	-
Aircrafts		
Airships	80	140
Helicopters	300	-
Airplanes	1 000	3 300
INERTIA VEHICLES		
Aircrafts		
Airplanes	-	7 300
Spacecrafts		
Launch vehicles	30 000	-
Satellites	50 000	-

Table 1. Traveling and maximum speed of vehicles.

have 50 000 km/h traveling speed. More details are given in Table 1.

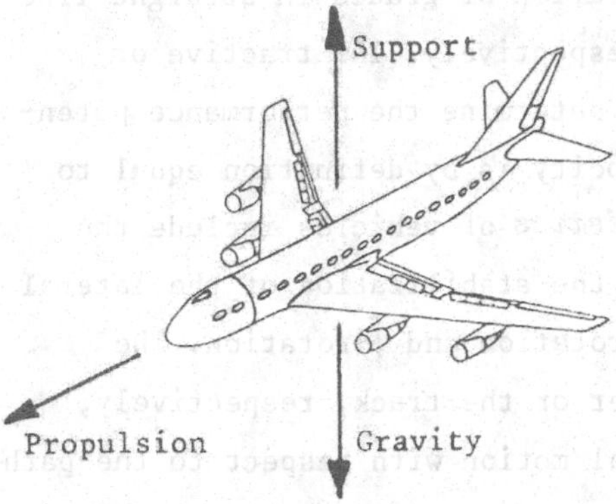

Fig. 1. Propulsion and support of an airplane

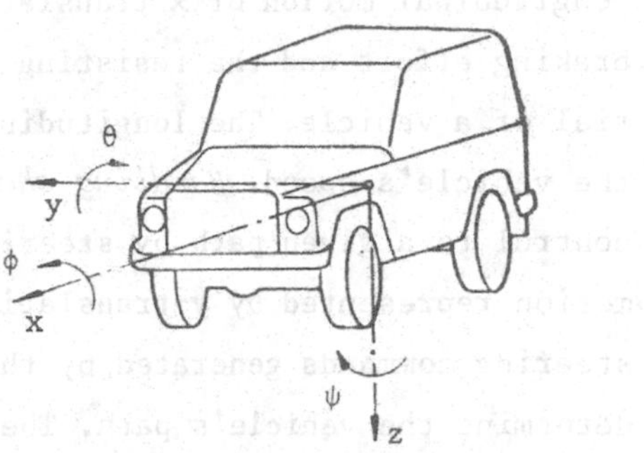

Fig. 2. Direction of motion

High-speed vehicles discussed in this course are ground vehicles with traveling speeds of 200 to 400 km/h. There are included railway vehicles, road vehicles and tracked magnetically levitated vehicles.

The motions of vehicles are generally rated by their directions and velocities with respect to the speed. The directions of translational and rotational displacements are defined with respect to the x, y, z-axis fixed in the vehicle's body, Fig. 2. The x-axis represents the longitudinal displacement in forward direction, the y-axis describes the lateral displacement to the right and the z-axis is adjusted to the vertical displacement in gravity direction. The rotations around the x, y, z-axis are called roll, yaw and pitch displacements, respectively. The ratio of the translational or the corresponding rotational velocities to the speed may be one or less than one. Small ratios often result in essential simplications by linearization of the equations. The following symbols are usually used for the motions:

x	longitudinal	ϕ	roll
y	lateral	θ	pitch
z	vertical	ψ	yaw .

The motions are related to the characteristic features of vehicles.

Performance characteristics of vehicles are concerned with the acceleration, the deceleration and the negotiation of grades in straight-line longitudinal motion or x-translation, respectively. The tractive or braking effort and the resisting forces determine the performance potential of a vehicle. The longitudinal velocity is by definition equal to the vehicle's speed. *Handling characteristics* of vehicles include the control to a given path by steering and the stabilization of the lateral motion represented by y-translation, ψ-rotation and ϕ-rotation. The steering commands generated by the driver or the track, respectively, determine the vehicle's path. The lateral motion with respect to the path is affected by inputs from the environment and the support.
Ride characteristics are related to the vehicle's vibrations in vertical direction excited by support irregularities, engine forces and the environment. Vertical vibrations are due to z-translation, θ-rotation and ϕ-rotation, they affect passengers and goods. Therefore, the understanding of human response is also very essential for the ride characteristics.

The *dynamics* of high-speed vehicles treated in this course cover the longitudinal motion, the lateral or horizontal motion, respectively, and the vertical motion. In particular, there are considered handling and ride characteristics.

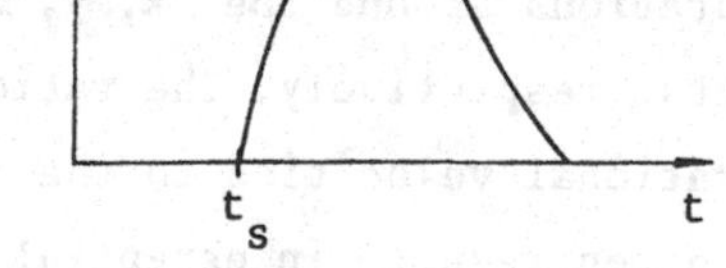

Fig. 3. Step excitation

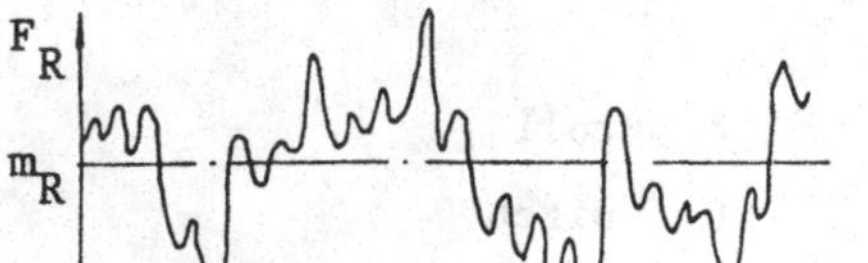

Fig. 4. Stochastic excitation

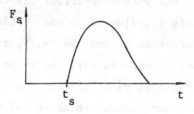

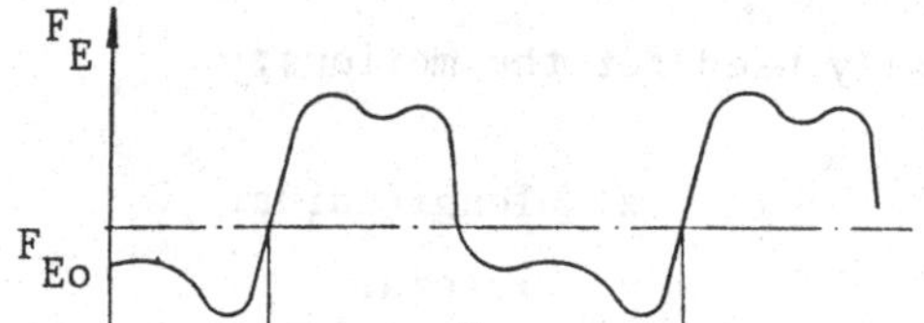

Fig. 5. Periodic excitation

MATHEMATICAL MODELING FOR HANDLING AND RIDE INVESTIGATIONS

For the dynamical analysis of the handling and ride characteristics of ground vehicles mathematical models of the excitations, the vehicle itself and the rating of the resulting motions are required. The excitations follow from aerodynamics, engine, wheels and guideway irregularities. The vehicle has to follow the path and is modeled by appropriate mechanical systems including the driver. The motions are expected to be stable, men and goods require low acceleration.

Aerodynamic forces and torques due to cross winds affect particularly the lateral motion. The time history of the wind may be gusty or random, resulting in different mathematical descriptions. Fig. 3 shows a typical excitation by a gust. The step excitation is then characterized by a polynominal

$$F_s(t) = \sum_{k=0}^{\infty} F_k\{t-t_s\}^k \qquad (1)$$

where F_k are constants, $\{t-t_s\}^k$ the Heaviside functions, $\{t-t_s\}^k = (t-t_s)^k$ for $t>t_s$ and $\{t-t_s\}^k = 0$ for $t<t_s$, and t_s is the step time. A steady-state random excitation is presented in Fig. 4. The random wind can be characterized by a stationary, Gaussian, ergodic stochastic process

$$F_R(t) \sim (m_R, N_R(s)) \qquad (2)$$

where m_R is the mean value, N_R the correlation function and s the correlation time. The aerodynamic forces act usually in the y-axis, the torques in the z-axis.

The rotary motion of the *engine* and the *wheels*, respectively, results at constant speed in a periodic force and torque excitation, Fig. 5. The periodic excitation is represented by Fourier expansion as

$$F_E(t) = F_E(t+T) = F_{Eo} + \sum_{k=1}^{\infty} (F'_{Ek} \cos k\Omega t + F''_{Ek} \sin k\Omega t) \tag{3}$$

where F_{Eo} , F'_{Ek} and F''_{Ek} are Fourier coefficients and

$$\Omega = \frac{2\pi}{T} \tag{4}$$

the excitation frequency. Thus, only the superposition of harmonic functions remain. Engine and wheel forces and torques may affect the lateral and vertical motion.

The surface irregularities of the *guideway* affect the vertical motion, and in the case of guided vehicles also the lateral motion. The guideway may have a rigid or a flexible surface. Rigid surfaces result in excitation functions of the vehicles while flexible surfaces have to be modeled as mechanical systems. The different types of guideways are summarized in Table 2 and Table 3.

The *path* is given for guided vehicles, in contrary to nonguided vehicles where the driver controls the path. In addition to the path usually also the tangential plane is given representing the superelevation of the road or track. Typical paths are straight lines, plane circles, plane and spatial curves.

The choice of the mathematical model for the *vehicle* depends on the technical problem under consideration. There are three mechanical systems available for different geometry and stiffness properties, Table 4. The final decision for one or more of these systems can be made with respect to the technical problem, Table 5. The equations of motion read for nonlinear ordinary multibody systems as

$$M(y,t)\,\ddot{y} + k(y,\dot{y},t) = q(y,\dot{y},t) \ , \tag{5}$$

for linear finite element systems as

$$M\ddot{y} + D\dot{y} + K y = h(t) \tag{6}$$

and for linear continuous systems in modal representation

$$\ddot{y} + 2\,\mathrm{diag}\,(\delta_j)\,\dot{y} + \mathrm{diag}\,(\omega^2_j)\,y = f(t) \tag{7}$$

where $y(t)$ is the corresponding position vector and $q(t)$, $h(t)$ and $f(t)$ are excitation vectors. The equations are completed by the coefficient matrices M, D, K representing inertia, damping and stiffness.

The *driver* controls the path of nonguided vehicles by adequate steering inputs. The steering inputs are collected from the visual observation of the path and the physical sensation of the vehicle's absolute and relative motion. Therefore, the steering problem of a nonguided vehicle can be characterized by a closed-loop control system, Fig. 6. The dynamical behavior of the driver has to be found by measurements in simulator or onboard experiments.

The lateral motion has to be directional stable with respect to the path. This means that particularly the differential equations for the y-translation, ψ-rotation and ϕ-rotation have to be asymptotically stable. Usually the stability boundary depends on the vehicle speed and, in the nonlinear case, on the characteristic amplitude, Fig. 7. A complete analysis requires linear and nonlinear stability theory.

In vehicles *men and goods* are subject to vertical mechanical vibrations. The human response to vibrations or the human sensation, respectively, has been investigated in medical and technical sciences for many years. It was found that there exists an open control loop, Fig. 8. The human sensation is correlated to the mechanical vibrations by a frequency response. Due to the guideway irregularities the vertical vehicle motion is random and the sensation has to be characterized by its standard deviation.

Surface	Figure	Excitation
Plane		Vanishing
Obstacle	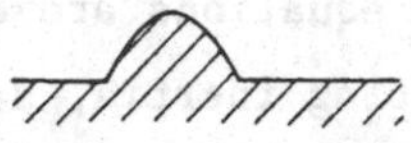	Step
Wave	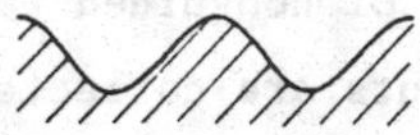	Harmonic
Rough		Stochastic

Table 2. Excitation functions by rigid surfaces

Surface	Figure	Mechanical System
Flexible Support	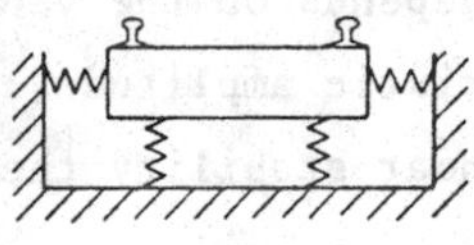	Multibody System (MBS)
Flexible Beams	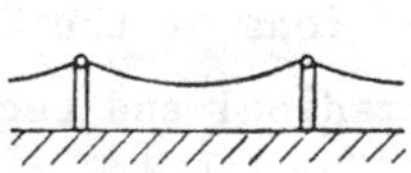	Finite Element System (FES)
Elastic Half - Space	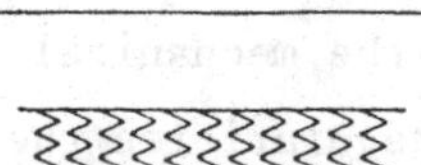	Continuous System (COS)

Table 3. Mechanical systems for flexible surfaces

Mechanical System	Geometry	Stiffness
Multibody System (MBS)	Complex	Inhomogen
Finite Element System (FES)	Complex	Homogen
Continuous System (COS)	Simple	Homogen

Table 4. Mathematical models for vehicles

Technical Problem	Figure	System
Maglev vehicle with secondary suspension, vertical motion	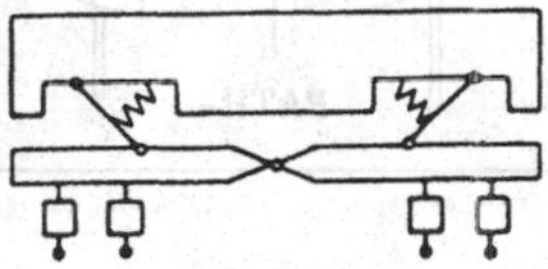	MBS
Maglev vehicle with primary suspension, vertical motion	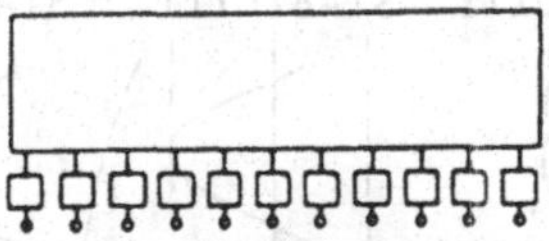	COS
Vehicle body, bending motion	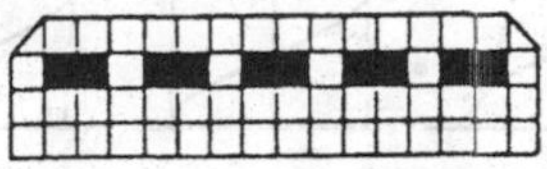	FES

Table 5. Mechanical systems and technical problems

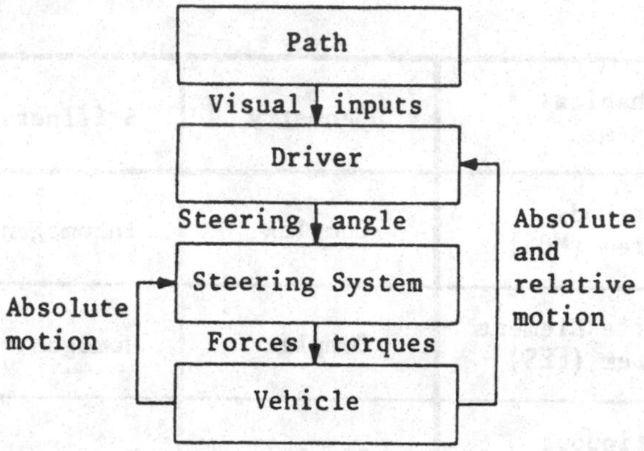

Fig. 6. Steering problem

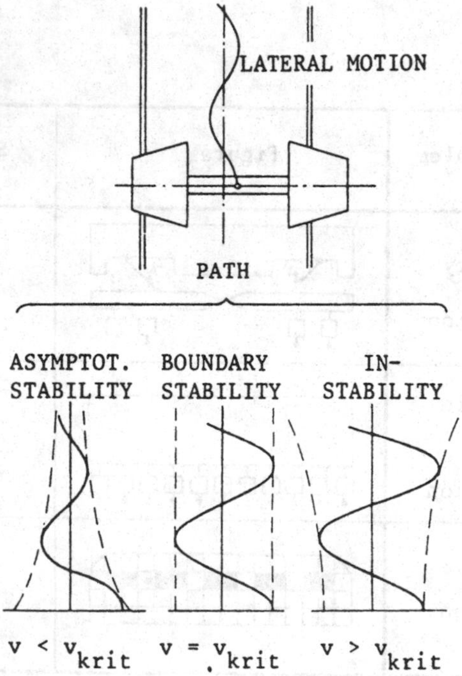

Fig. 7. Lateral stability of a wheelset

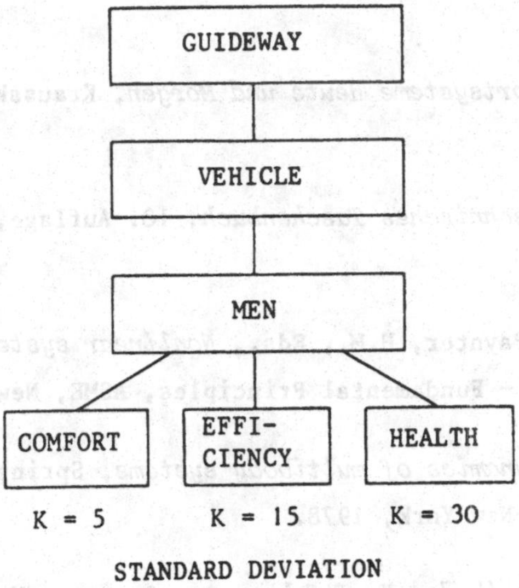

Fig. 8. Human sensation of vibration

METHODS OF INVESTIGATION

The dynamical research of high-speed vehicles requires a broad spectrum of theoretical and experimental methods. From *dynamics* the multibody systems, the finite element systems and the continuous systems are applied. *Control theory* contributes with closed-loop and open-loop systems including human response. *Vibration analysis* includes stability theory of free vibrations as well as forced vibrations excited by step, harmonic and stochastic forces. *Numerical methods* like algorithms for linear equations, eigenvalue procedures and integration methods of differential equations have always to be used in connection with the digital computer. Sophisticated *measurements* and experienced *design* have to be conducted during the development of each kind of vehicle. Thus, vehicle dynamics are a very challenging field for an engaged engineer.

BIBLIOGRAPHY

1 Bahke, E., *Transportsysteme Heute und Morgen*, Krausskopf-Verlag, Mainz, 1973.

2 Bosch, *Kraftfahrtechnisches Taschenbuch*, 18. Auflage, VDI-Verlag, Düsseldorf, 1976.

3 Hedrick, J.K and Paynter, H.M., Eds., *Nonlinear system analysis and synthesis*, Vol. 1 - Fundamental Principles, ASME, New York, 1978.

4 Magnus, K. Ed., *Dynamics of multibody systems*, Springer-Verlag, Berlin-Heidelberg-New York, 1978.

5 Mitschke, M., *Dynamik der Kraftfahrzeuge*, Springer-Verlag, Berlin-Heidelberg-New York, 1972.

6 Müller, P.C. and Schiehlen, W.O., *Forced linear vibrations*, CISM Courses and Lectures No. 172, Springer-Verlag, Wien-New York, 1977.

7 Pacejka, H.B., Ed., *The dynamics of vehicles on roads and railway tracks*, Swets & Zeitlinger, Amsterdam, 1976.

8 Slibar, A. and Springer, H., Eds., *The dynamics of vehicles on roads and tracks*, Swets & Zeitlinger, Amsterdam, 1978.

9 Stoer, J., *Einführung in die numerische Mathematik I, II*, Springer-Verlag, Berlin-Heidelberg-New York, 1976, 1978.

10 Willumeit, H.-P., Ed., *The dynamics of vehicles on roads and on rail tracks*, Swets & Zeitlinger, Lisse, 1980.

STOCHASTIC AND ELASTIC GUIDEWAY MODELS

K. Popp
Universität Hannover, FRG

1. Introduction

The theoretical investigations in vehicle system dynamics are based upon a suitable mathematical system description, called mathematical model. The mathematical model can be gained either by application of the fundamental laws of physics to a physical model of the real technical system or by evaluation of measurements performed on the real technical system itself or on parts or experimental models of it. Which way is taken depends on the problem, purpose of investigation, knowledge of the system, desired accuray, and last not least on equipment, time and money available.

The quality of the theoretical results is only as good as the underlying mathematical model. Thus, the mathematical model must be as complete and accurate as necessary. On the other hand, from the computational point of view the mathematical model must be as simple and easy to handle as possible. It is obvious that the modelling process is a tough engineering problem.

The aim of the contributions in this Chapter is to show the different steps in the process of mathematical modelling which are common for different types of vehicles like automobiles, magnetically levitated vehicles (Maglev vehicles) and railway vehicles. The general vehicle setup is shown in Fig. 1 by means of the block diagram. Subsequent blocks are in dynamical interaction with each other. In the following the mathematical models of the subsystems are developed. From this the mathematical model of the entire vehicle-guideway system can be composed.

We start with the mathematical description of the disturbances and

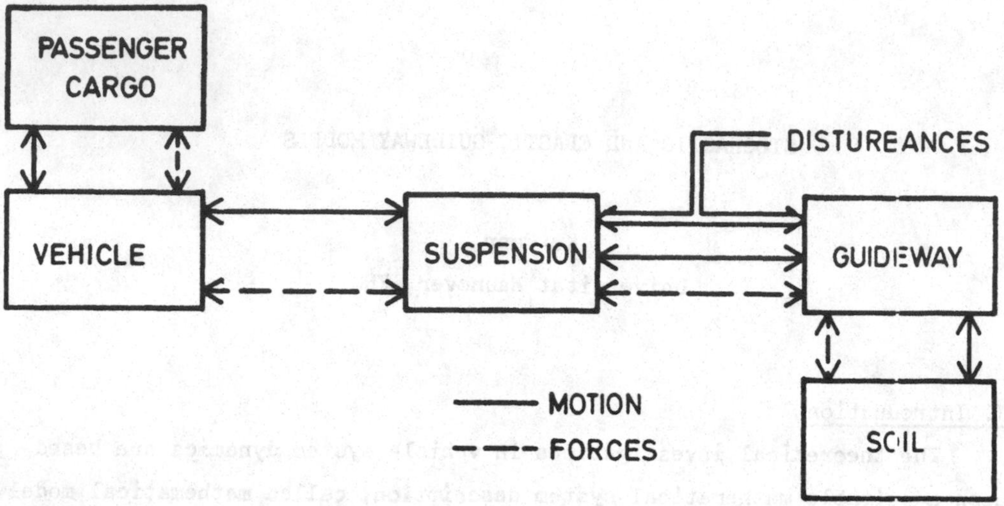

Fig. 1: Vehicle-guideway-interaction.

the guideway dynamics. We restrict ourselves to the two most important cases in applications:

i) stochastic excitation models for rigid guideways and

ii) deterministic models for elastic guideways.

Since the mathematical description turns out to be linear in either case, both models can be superposed, if necessary.

2. Stochastic excitation models for rigid guideways

Research in the field of guideway roughness models is going on for a long time. Numerous measurements of road roughness profiles have been performed, cf. the classical book by Mitschke[1], or publications by Braun[2], Wendeborn[3], Voy[4]. With respect to railway tracks four roughness profiles have to be distinguished; lateral alignment, vertical profile, cross-level and gauge, respectively. Here, measurements are summarized in ORE[5]. For elevated guideways as used for Maglev vehicles the vertical irregularities comprise the vertical offset and random walk of the piers, camber of the spans and surface roughness of the tracks as pointed out by Snyder III and Wormley[6]. The evaluation of measurements as well as theo -

retical investigations concerning the superposition of random irregularities have shown the common fact that the different roughness profiles can be modelled as

- stationary ergodic Gaussian random processes.

Bevor the guideway roughness models are described in detail, some general remarks on random processes may be in order, cf. also Newland[7], Crandall, Mark[8].

2.1 Mathematical description of random processes

Suppose an infinite ensemble of roughness profile measurements $\zeta^{(r)}(x)$ for a special guideway typ, let say for highways in Europe, is given, see Fig. 2. Here, the independent variable x describes the distance from an arbitrary starting point. Each profile sample differs from all others, $\zeta^{(r)}(x) \neq \zeta^{(s)}(x)$ for $r \neq s$. The family of profiles forms a random process $\zeta(x)$. The Profile values $\zeta_j = \zeta(x_j)$ at discrete distances x_j are random variables. The probability distribution of the random variable ζ_j is characterized by the probability density function $p(\zeta_j)$, see Fig. 2, which yields the probability Pr that the profile value ζ_j lies between certain limits a and b,

$$Pr\,(a \leq \zeta_j \leq b) = \int_a^b p(\zeta_j)d\zeta_j \quad \left(\int_{-\infty}^{\infty} p(\zeta_j)d\zeta_j = 1 \right) . \tag{1}$$

The random variable ζ_j, $j = 1, 2, \ldots$, can be characterized by ensemble averages. The most important ones are the mean $m_\zeta(x_1)$ (first order moment) and the mean square value $m_{\zeta^2}(x_1)$ (second order moment),

$$m_\zeta(x_1) = E\{\zeta(x_1)\} \equiv \int_{\infty}^{\infty} \zeta_1\, p(\zeta_1)\, d\zeta_1 \quad , \tag{2}$$

$$m_{\zeta^2}(x_1) = E\{\zeta^2(x_1)\} \equiv \int_{\infty}^{\infty} \zeta_1^2\, p(\zeta_1)d\zeta_1 \quad , \tag{3}$$

where the operator $E\{\ \}$ is called mathematical expectation of $\{\ \}$. The square root of (3) is called the root mean square value of ζ_1 or rms value. An important statistical parameter ist the variance $\sigma_\zeta^2(x_1)$,

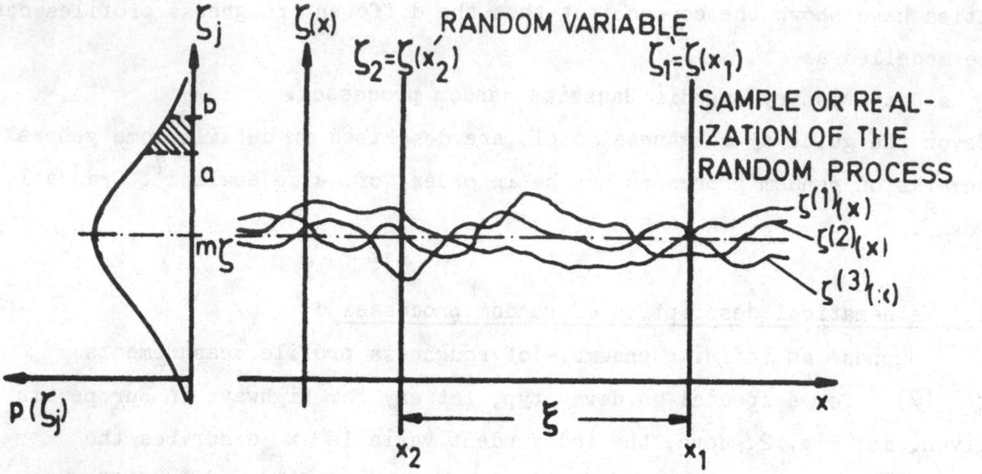

Fig. 2: Guideway roughness as random process.

$$\sigma_\zeta^2(x_1) = E\{(\zeta_1 - E\{\zeta_1\})^2\} \equiv \int_{-\infty}^{\infty} (\zeta_1 - E\{\zeta_1\})^2 p(\zeta_1) d\zeta_1$$
$$= E\{\zeta_1^2\} - (E\{\zeta_1\})^2. \tag{4}$$

If the mean is zero then the variance is identical with the mean square. The square root of (4) is called the standard deviation $\sigma_\zeta(x_1)$. The correlation between any two random variables, let say ζ_1 and ζ_2, gives insight into the random process. The joint ensemble average is called (auto)correlation function $R_\zeta(x_1, x_2)$,

$$R_\zeta(x_1, x_2) = E\{\zeta_1, \zeta_2\} \equiv \int_{-\infty}^{\infty} \zeta_1 \zeta_2 p(\zeta_1, \zeta_2) d\zeta_1 d\zeta_2 , \tag{5}$$

where $p(\zeta_1, \zeta_2)$ denotes the joint probability density function. Analogous to (4) the covariance $P_\zeta(x_1, x_2)$ can be defined as

$$P_\zeta(x_1, x_2) = E\{(\zeta_1 - E\{\zeta_1\})(\zeta_2 - E\{\zeta_2\})\} = E\{\zeta_1, \zeta_2\} - E\{\zeta_1\}E\{\zeta_2\} . \tag{6}$$

If ζ_1 and ζ_2 have zero means, then the covariance $P_\zeta(x_1, x_2)$

is identical with the (auto)correlation function $R_\zeta(x_1, x_2)$
When $x_1 = x_2$, the covariance (6) becomes identical with the variance (4). Now we can describe mathematically the properties: stationarity, gaussian or normal distribution and ergodicity.

Stationary means that the probability distributions are invariant under a shift of the x-axis, i.e. $p(\zeta_1) = p(\zeta_2) = p(\zeta)$ and $p(\zeta_1, \zeta_2)$ depends only on the distance $\xi = x_1 - x_2$. As a consequence all random variables and thus the entire random process have one and the same mean, mean square and variance, respectively. The random process can always be centered resulting in a zero mean. Thus

$$m_\zeta(x) = 0 \,, \tag{7}$$

$$R_\zeta(x_1, x_2) = R_\zeta(x_1 - x_2 = \xi) = P_\zeta(x_1, x_2) \,, \tag{8}$$

$$\sigma_\zeta^2(x) = m_{\zeta^2}(x) = R_\zeta(0) = \text{const.} \tag{9}$$

Gaussian or normal distribution means that the probability density functions read

$$p(\zeta_1) = \frac{1}{\sqrt{2\pi}\,\sigma_1} \exp\left[- (\zeta_1 - m_1)^2 / 2\,\sigma_1^2 \right] \,, \tag{10}$$

$$p(\zeta_1, \zeta_2) = \frac{1}{2\pi\sqrt{\sigma_1^2\,\sigma_2^2 - P_{12}^2}} \exp\left\{ \frac{-\,\sigma_1^2\,\sigma_2^2}{2(\sigma_1^2\sigma_2^2 - P_{12}^2)} \left[\frac{(\zeta_1 - m_1)^2}{\sigma_1^2} - \right.\right. \tag{11}$$

$$\left.\left. - \frac{2P_{12}(\zeta_1 - m_1)(\zeta_2 - m_2)}{\sigma_1^2\,\sigma_2^2} + \frac{(\zeta_2 - m_2)^2}{\sigma_2^2} \right] \right\} \,,$$

where the abbreviations $m_i = m_\zeta(x_i)$, $i = 1,2$, $P_{12} = P_\zeta(x_1, x_2)$, have been used. These parameters are sufficient for the complete mathematical description. Eqs. (10) and (11) can be simplified for stationary random processes regarding (7), (8) and (9). From (10) can be seen, that the maximum value of the function $p(\zeta_1)$ is proportional to $1/\sigma_1$

Another way to signify σ_1 is given by considering confidence intervals. Solving (1) e.g. for

$a = m_1 - \sigma_1$, $b = m_1 + \sigma_1$ yields

$$Pr(m_1 - \sigma_1 \le \zeta_1 \le m_1 + \sigma_1) = 0.6827 \quad , \tag{12}$$

which means that 68,27 % of all values of the random variable ζ_1 lie in the $2\sigma_1$-band centered at the mean m_1.

Within the subclass of stationary random processes there exists a further subclass known as ergodic processes, for which the ensemble averages are equal to the sample averages of a single sample $\zeta^{(r)}(x)$ of infinite length,

$$m_\zeta = \lim_{X\to\infty} \frac{1}{2X} \int_{-X}^{X} \zeta^{(r)}(x)\, dx \quad , \tag{13}$$

$$R_\zeta(\xi) = \lim_{X\to\infty} \frac{1}{2X} \int_{-X}^{X} \zeta^{(r)}(x)\, \zeta^{(r)}(x-\xi)\, dx \quad . \tag{14}$$

A sufficient condition for ergodicity reads, cf. Drenick[9],

$$\int_{-\infty}^{\infty} |R_\zeta(\xi)|\, d\xi < \infty \quad . \tag{15}$$

In technical applications stationary random process are very often represented by power spectral density (PSD) functions $S_\zeta(\Omega)$ depending on the (spacial) circular frequency Ω. The connection with the correlation function $R_\zeta(\xi)$ is simply given by the Fourier transform,

$$S_\zeta(\Omega) = \frac{1}{2\pi} \int_{-\infty}^{\infty} R_\zeta(\xi)\, e^{-i\Omega\xi} d\xi \quad , \tag{16}$$

$$R_\zeta(\xi) = \int_{-\infty}^{\infty} S_\zeta(\Omega)\, e^{i\Omega\xi} d\Omega \quad , \tag{17}$$

$$\sigma_\zeta^2 = R_\zeta(0) = \int_{-\infty}^{\infty} S_\zeta(\Omega)\, d\Omega \tag{18}$$

Since the PSD $S_\zeta(\Omega)$ is an even function of Ω, $S_\zeta(\Omega) = S_\zeta(-\Omega)$, nearly always single sided PSDs $\Phi_\zeta(\Omega)$ are used, $\Phi_\zeta(\Omega) = 2\, S_\zeta(\Omega)$,

$0 \leq \Omega < \infty$, with non-negative (special) circular frequencies Ω .

Usually, in technical applications neither an infinite ensemble nor a single sample infinite in length is available. Thus, only approximations of $R_\zeta(\xi)$ or $S_\zeta(\Omega)$ can be given.

2.2. Guideway roughness models

In recent publications there is a trend towards standardization of guideway roughness models. A simple but useful road roughness model reads, cf. Mitschke[1], Voy[4],

$$\phi_\zeta(\Omega) = \phi_o\left(\frac{\Omega_o}{\Omega}\right)^w \quad , \tag{19}$$

where Ω_o[rad/m] , $\phi_o = \phi_\zeta(\Omega_o)$[$m^2$/(rad/m)] and w are constants describing the reference (spacial) circular frequency, unevennes and waviness, respectively. Usually, the waviness ranges between $1{,}75 \leq w \leq 2{,}25$. The roughness model (19) is often plotted in a Ω, ϕ-diagramm with logarithmic scales as a sloping straight line. A similar but more sophisticated model is suitable for roads and for tracks, see Dodds, Robson[10], Hedrick, Anis[11],

$$\phi_\zeta(\Omega) = \begin{cases} \phi_o\left(\frac{\Omega_o}{\Omega}\right)^{w_1} & \Omega \leq \Omega_o \quad , \\ & \text{for} \\ \phi_o\left(\frac{\Omega_o}{\Omega}\right)^{w_2} & \Omega \geq \Omega_o \quad , \end{cases} \tag{20}$$

where different exponents w_1 and w_2 are introduced. In case of track irregularities (20) is used to describe vertical profile, lateral alignment, gauge and cross-level as well. However, no crosscorrelation between these four profiles are knwon. The standardized models (19) and (20) are approximations to measured PSDs in a distinct frequency range $0 < \Omega_1 \leq \Omega \leq \Omega_2$. In either case, the limit $\Omega \to 0$ results in $\phi_\zeta(\Omega \to 0) \to \infty$ and thus an infinite variance would follow which is not realistic. To avoid these difficulties two other roughness models are used, cf. Dincă, Theodosiu[12], Fábián[13], Sussman[14],

$$\phi_\zeta(\Omega) = \frac{2\alpha\sigma^2}{\pi} \frac{1}{\alpha^2 \Omega^2} \tag{21a}$$

$$\phi_\zeta(\Omega) = \frac{2\alpha\sigma^2}{\pi} \frac{\Omega^2+\alpha^2+\beta^2}{(\Omega^2-\alpha^2-\beta^2)^2+4\alpha^2\Omega^2} , \tag{22a}$$

where α, β and σ^2 are constant. Since eqs. (21a), (22a) are valid in the entire frequency range, the corresponding (auto)correlation function $R_\zeta(\xi)$ can be calculated utilizing (17) which reads for single sided PSDs,

$$R_\zeta(\xi) = \int_0^\infty \phi_\zeta(\Omega) \cos \Omega \xi \, d\Omega \quad . \tag{17a}$$

Eq. (17) applied to (21a) and (22a) yields

$$R_\zeta(\xi) = \sigma^2 e^{-\alpha|\xi|} , \tag{21b}$$

$$R_\zeta(\xi) = \sigma^2 e^{-\alpha|\xi|} \cos \beta \xi \quad . \tag{22b}$$

Here, $\sigma^2 = R_\zeta(0)$ characterizes the (finite) variance of the random roughness process $\zeta(x)$.

2.3. Vehicle excitation models

From the guideway roughness models $\zeta(x)$ given in the space domain the corresponding vehicle excitation models $\zeta(t)$ in time domain can be obtained using

$$x = vt \ , \ \xi = v\tau, \quad \omega = v\Omega , \tag{23}$$

where v = const [m/s] is the vehicle speed, τ denotes the correlation time and ω [rad/s] the (time) circular frequency. Since the roughness profil $\zeta(x)$ and the vehicle excitation $\zeta(t)$ have the same variance $R\zeta(0)$, from eq. (17a) it follows

$$\Phi_\zeta(\omega)d\omega = \phi_\zeta(\Omega)d\Omega \ . \tag{24}$$

Thus, using (23) the single sided PSD $\phi_\zeta(\omega)$ reads,

$$\phi_\zeta(\omega) = \frac{1}{v}\,\phi_\zeta(\Omega = \frac{\omega}{v}) \quad . \tag{25}$$

Considering model (19) vor example, where the waviness $w = 2$ is chosen and $\tilde{\phi}_o$ is used to denote the corresponding unevenness, ohne gets

$$\phi_\zeta(\omega) = \frac{1}{v}\,\tilde{\phi}_o(\frac{v\,\Omega_o}{\omega})^2 = v\,\tilde{\phi}_o(\frac{\Omega_o}{\omega})^2 \quad . \tag{26}$$

For $w \neq 2$ eq.(25) has to be applied. However, as a frist approximation the simple model (26) can still be used if the unevenness $\tilde{\phi}_o$ is chosen properly. Equating the variance of model (19) $(w \neq 2)$ and the variance of the approximation with $w = 2$ in the frequency range $\Omega_1 \leq \Omega \leq \Omega_2$,

$$\phi_o \int_{\Omega_1}^{\Omega_2} (\frac{\Omega_o}{\Omega})^w \, d\Omega = \tilde{\phi}_o \int_{\Omega_1}^{\Omega_2} (\frac{\Omega_o}{\Omega})^2 \, d\Omega \quad , \tag{27}$$

yields the equivalent unevennes $\tilde{\phi}_o$,

$$\tilde{\phi}_o = \frac{\phi_o}{w-1}(\frac{\Omega_o}{\Omega_1})^{w-2} \frac{1-(\Omega_1/\Omega_2)^{w-1}}{1-(\Omega_1\ \Omega_2)} \qquad (w \neq 1) \quad . \tag{28}$$

Often in applications $\Omega_1/\Omega_2 \ll 1$ is given which simplifies (28).

Up to now only the excitation profile $\zeta(t)$ has been considered. But also the time derivatives $\dot{\zeta}(t)$ and $\ddot{\zeta}(t)$ are of interest. Due to the classical theory, cf. Newland[7], the PSDs of the derivated random process can easily be calculated,

$$\phi_{\dot{\zeta}}(\omega) = \omega^2\,\phi_\zeta(\omega) \quad , \quad \phi_{\ddot{\zeta}}(\omega) = \omega^4\,\phi_\zeta(\omega) \quad . \tag{29}$$

Eq. (29) applied to (26) yields a white noise process, i.e. a constant PDS for the random excitation velocity process,

$$\tilde{\phi}_{\dot{\zeta}}(\omega) = \omega^2\, v\,\tilde{\phi}_o\,(\frac{\Omega_o}{\omega})^2 = v\,\tilde{\phi}_o\,\Omega_o^{\,2} \quad , \tag{30}$$

corresponding to the (auto)correlation function

$$R_{\dot\zeta}(\tau) = q_{\dot\zeta}\,\delta(\tau) \quad , \quad q_{\dot\zeta} = \pi\, v\, \phi_o\, \Omega_o^{\,2} \quad , \tag{31}$$

where $q_{\dot\zeta}$ is the noise intensity and $\delta(\cdot)$ denotes the Dirac distribution. The white noise process results clearly in an infinite variance which is by no means realistic. On the other hand, this simple model reduces considerably the computation work and can serve as a first approximation, cf. Karnopp[15], Müller et al.[16].

Better vehicle excitation models $\zeta(t)$ are given by stationary Gaussian colored noise processes which can be obtained from a white noise process $w(t)$ by means of a shape filter. The shape filter is an asymptotically stable linear dynamical system, see Fig. 3, which - roughly speaking - changes the shape of the correlation function but does not influence the stationarity, normal distribution and ergodicity. The mathematical shape filter description reads

$$\zeta(t) = \underline{h}^T\,\underline{v}(t) \quad , \tag{32}$$

$$\dot{\underline{v}}(t) = \underline{F}\,\underline{v}(t) + \underline{g}\,w(t) \quad , \ \mathrm{Re}\,\lambda(\underline{F})<0 \quad , \ w(t) \sim N(0,q_q) \quad ,$$

where the state vector $\underline{v}(t)$ and the quantities $\underline{F}$, $\underline{g}$, and $\underline{h}$ determine uniquely the shape filter. The input process $w(t)$ is assumed to be Gaussian white noise with zero mean and intensity q_w. For colored noise characterized by (21a), (21b) or (22a), (22b) the corresponding shape filter quantities read

$$\underline{F} = -\,\alpha v \quad , \quad \underline{g} = g \quad , \quad \underline{h} = 1 \quad , \tag{21c}$$

$$\underline{F} = \begin{bmatrix} 0 & 1 \\ -(\alpha^2+\beta^2)v^2 & -2\alpha v \end{bmatrix} , \quad \underline{g} = g\begin{bmatrix} 0 \\ 1 \end{bmatrix} , \quad \underline{h} = \begin{bmatrix} v\sqrt{\alpha^2+\beta^2} \\ 1 \end{bmatrix} . \tag{22c}$$

In either case g and q_w can be chosen arbitrarily regarding

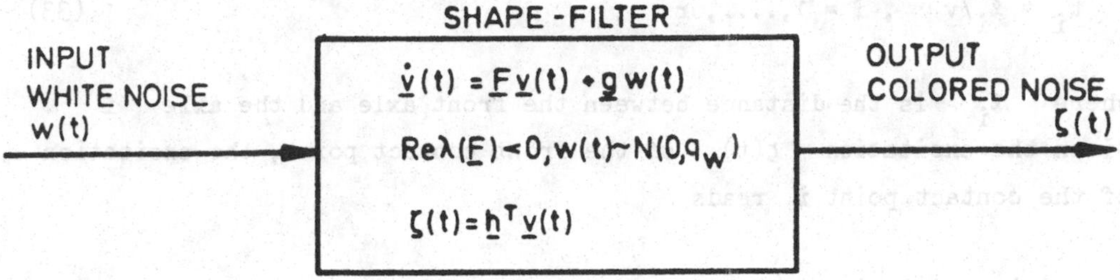

Fig. 3: Shape filter.

$g^2 q_w = 2\alpha v \sigma^2$. In applications one is interested in computing the shape filter quantities directly from measured data rather than from analytical approximations. This can be done by parameter identification procedures as described e.g. in Müller et.al.[17].

All vehicle excitation model up to now are models where only single contact is taken into account. However, real multi-axle vehicles have multiple contact with the guideway. Thus, the time delays between successive contact points have to be regarded, see Fig. 4. For r contact points in a line the time delays read

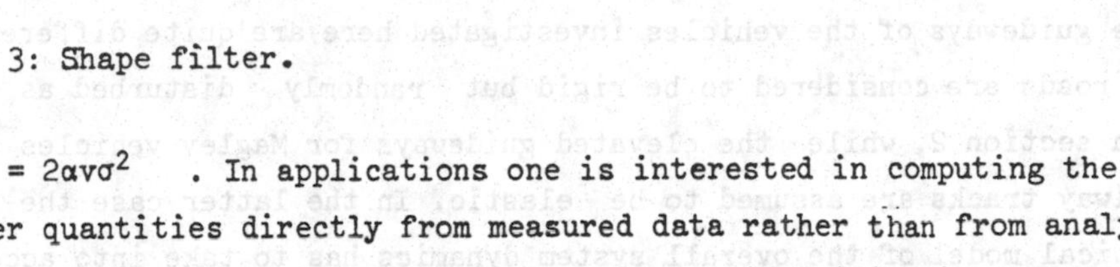

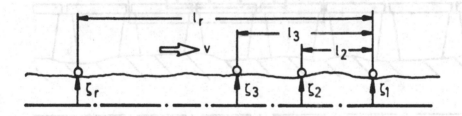

Fig. 4: Multiple vehicle random excitation.

$$t_i = \ell_i/v \quad , \; i = 1,\ldots\ldots, r \, , \tag{33}$$

where ℓ_i is the distance between the front axle and the axle i. Given the excitation $\zeta(t)$ of the front contact point, the excitation of the contact point i reads

$$\zeta_i(t) = \zeta(t-t_i) \quad , \quad 0 = t_1 < t_2 < \ldots\ldots < t_r \quad . \tag{34}$$

3. Deterministic models for elastic guideways

The guideways of the vehicles investigated here are quite different. Usually roads are considered to be rigid but randomly disturbed as shown in section 2, while the elevated guideways for Maglev vehicles and the railway tracks are assumed to be elastic. In the latter case the mathematical model of the overall system dynamics has to take into account the elastic guideway deflections. Since the guideways stretch over long distances, only parts of them can be included in the corresponding system models. If we separate the vehicles from their guideways and introduce the forces of interaction, then we receive the models shown in Fig. 5.

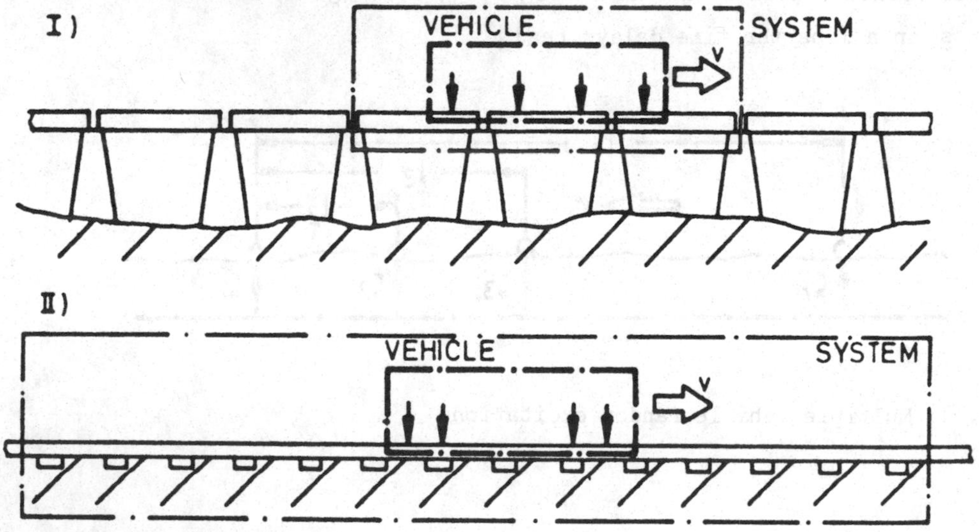

Fig. 5: Vehicle-guideway models for I) Maglev vehicles, II) railway vehicles.

The guideways for Maglev vehicles, Fig. 5 I), are elevated periodic structures, modelled as a sequence of identical prefabricated structural elements on rigid supports. The system model has to comprise at least as much guideway elements as coupled by the moving vehicle. To the contrary, railway tracks are modelled as continuous beams on elastic foundation, Fig. 5 II). Here, the system model has to include a guideway section of sufficient length, so that the boundaries are free of deflections as in reality. In either case the system bounds have to follow the moving vehicle.

The mathematical description of the elastic guideway deflections is generally achieved in three steps:

i) Analysis of a single guideway element.

ii) Setup of the guideway model within the system bounds.

iii) Calculation of the vehicle-guideway interaction regarding the shift of the system bounds due to the moving vehicle.

In the following sections, step i) will be considered in more detail for elevated guideways, where the elastic deformations are essential (steps ii) and iii), then, are carried out in Chapter 3). While details on the deformation of railway tracks may be found in the literature, cf. e.g. Timoshenko[18], Dörr[19], Korb[20], Popp[21].

3.1. Mathematical description of guideway deflections under moving forces

Suppose an elevated guideway for Maglev vehicles is given with guideway elements as shown in Fig. 6. Each guideway element of lenght L consists of s uniform beam segments of length L_i, with constant bending stiffness $(EI)_i$, and constant mass per unit lenght $(\rho A)_i$, $i = 1,\ldots, s$. Due to the moving forces $F_\mu(t)$, $\mu = 1,\ldots, m$, where the travelling speed is v, guideway deflections $w_i(\xi_i,t)$, $0 \leq \xi_i \leq L_i$, occur in each segment. These deflections can be calculated using Bernoulli-Euler-beam theory which implies small deformations, Bernoulli's hypothesis of linear stress distribution along the cross section, Hooke's law of linear stress-strain relation and neglects shear effects and rotational inertia effects. The wellknown beam equation reads for segment i,

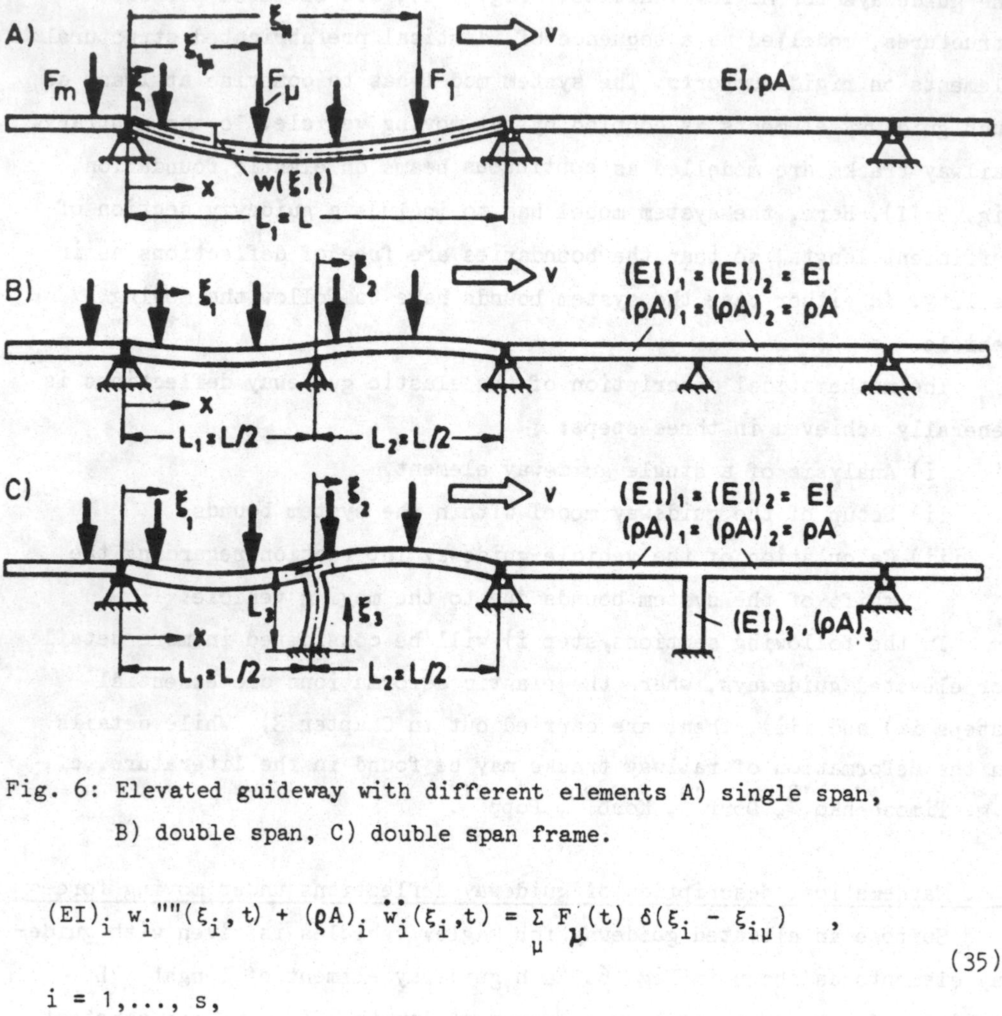

Fig. 6: Elevated guideway with different elements A) single span, B) double span, C) double span frame.

$$(EI)_i\, w_i''''(\xi_i, t) + (\rho A)_i\, \ddot{w}_i(\xi_i, t) = \sum_{\mu} F_{\mu}(t)\, \delta(\xi_i - \xi_{i\mu}) \quad , \qquad i = 1, \ldots, s, \tag{35}$$

where $(\ )'$ and $\dot{(\ )}$ denote spacial and time derivatives, respectively, $\delta(\ )$ is the Dirac distribution, and $\xi_{i\mu}$ characterizes the distance where the force $F_{\mu}(t)$ is acting on segment i. The solution of (35) can be obtained regarding initial and boundary conditions using modal expansion,

$$w_i(\xi_i, t) = \sum_{j=1}^{\infty} \varphi_{ij}(\xi_i)\, z_j(t) \quad , \quad i = 1, \ldots, s \quad . \tag{36}$$

Here, $\varphi_{ij}(\xi_i)$ are normal modes and $z_j(t)$ generalized coordinates. The normal modes show the following properties:

I) They are solution of the eigenvalue problem

$$\varphi_i''''(\xi_i) - \lambda_i^4 \varphi_i(\xi_i) = 0 \quad , \quad \lambda_i^4 = \omega^2 (\varrho A/EI)_i \quad , \tag{37}$$

where the eigenvalue λ_i has been introduced. Eq. (37) yields the eigenfrequencies ω_j and the corresponding eigenmodes $\varphi_{ij}(\xi_i)$

II) They fulfill the boundary conditions.

III) They are orthogonal with respect to the entire guideway element,

$$\sum_{i=1}^{s} \int_0^{Li} (\varrho A)_i \, \varphi_{ij}(\xi_i) \, \varphi_{ik}(\xi_i) \, d\,\xi_i = \begin{cases} 0 & \text{for } j \neq k \\ M_j = \sum_{i=1}^{s} M_{ij} & \text{for } j = k \end{cases} \tag{38}$$

Utilizing these properties, from (35) follows the equation governing the generalized coordinates $z_j(t)$,

$$\ddot{z}_j(t) + \omega_j^2 \, z_j(t) = \frac{1}{M_j} \sum_{\mu} \varphi_{ij} \, (\xi_{i\mu}) \, F_\mu(t) \quad , \tag{39}$$

which is subject to the initial conditions. In technical applications the procedure is modified in the following way:

1) A finite number f of modes is regarded only, $j = 1, \ldots, f$

2) Modal damping $2\zeta_j \, \omega_j \, \dot{z}_j(t)$ is added to the right hand side of (39), where ζ_j denotes the modal damping coefficient. Usually, from measurement only ζ_1 is known, thus, assumptions have to be made for ζ_j , $j = 2, \ldots, f$. Often $\zeta_j = \zeta_1 \, \omega_j/\omega_1$ is assumed for convenience.

3) Vector notation is used instead of (36), (39):

$$w_i(\xi_i, t) = \underline{\varphi}^T(\xi_i) \, \underline{z}(t) \quad , \tag{40}$$

$$\underline{\ddot{z}}(t) + \underline{\Delta} \, \underline{\dot{z}}(t) + \underline{\Omega} \, \underline{z}(t) = \underline{M}^{-1} \sum_{\mu} \underline{\varphi}(\xi_{i\mu}) \, F_\mu(t) \quad , \tag{41}$$

$$\underline{\varphi} = [\varphi_1, \ldots, \varphi_f]^T, \quad \underline{z} = [z_1, \ldots, z_f]^T ,$$

$$\underline{M} = \underline{\text{diag}}\,(M_j),\ \underline{\Delta} = \underline{\text{diag}}\,(2\zeta_j\,\omega_j)\ ,\ \underline{\Omega} = \underline{\text{diag}}\,(\omega_j^{\,2}),\ j = 1, \ldots, f \quad . \tag{42}$$

As can be seen, the entire analysis boils down to the calculation of the eigenfrequencies ω_j and the corresponding normal modes $\varphi_{ij}(\xi_i)$. Once these quantities are available, the setup of the guideway model within the system bounds (step ii) can easily be performed composing eqs. (40), (41) for those guideway elements which are coupled by the vehicle. The remaining step iii) requires knowledge about the interactive forces $F_\mu(t)$, cf. Chapter 3.

3.2. Modal analysis of beam structures

Prior to more general considerations the eigenfrequencies ω_j and corresponding normal modes φ_j of the single span element, see Fig. 6A), shall be calculated. Here, the index i can be dropped.

The solution of the eigenvalue problem (37) reads

$$\varphi(\xi) = C_1 \cosh \lambda\xi/L + C_2 \sinh \lambda\xi/L + C_3 \cos \lambda\,\xi/L + C_4 \sin \lambda\xi/L \quad , \tag{43}$$

or in equivalent vector notation,

$$\underline{\varphi}(\xi) = \underline{a}^T(\lambda\xi/L)\,\underline{c} \quad , \tag{44}$$

$$\underline{a}(\cdot) = [C(\cdot)\ ,\ S(\cdot),\ c(\cdot),\ s(\cdot)]^T \quad , \tag{45}$$

$$\underline{c} \ = [C_1, C_2, C_3, C_4]^T \quad , \tag{46}$$

$$C(\cdot) = \cosh(\cdot),\ S(\cdot) = \sinh(\cdot),\ c(\cdot) = \cos(\cdot),\ s(\cdot) = \sin(\cdot), \tag{47}$$

where some abbreviations have been introduced for convenience. The boundary conditions are $\varphi(0)=0$, $\varphi''(0)=0$, $\varphi(L) = 0$, $\varphi''(L) = 0$,

which yields $C_1 = C_2 = C_3 = 0$ and provides the frequency equation

$$C_4 \sin \lambda = 0 \quad . \tag{48}$$

The nontrivial solution reads

$$\lambda_j = j\pi , \quad \omega_j^2 = \left(\frac{j\pi}{L}\right)^4 \frac{EI}{\varrho A} , \quad j = 1,2,\ldots \quad ; \tag{49}$$

introducing $C_4 = 1$ gives finally

$$\varphi_j = \sin j\pi , \quad M_j = \int_0^L \varrho A \ \varphi_j^2(\xi)\, d\xi = \varrho AL/2. \tag{50}$$

The modal analysis of more complicated beam structures cannot be performed analytically any more. One has to rely on the computer and thus appropriate methods are required. From the various methods, see e.g. Knothe[22], only three shall be mentioned:

α) Transfer-matrix method (TMM), cf. e.g. Pestel, Leckie[23],

β) Deformation method (DEM), cf. Kolou̇šek[24], which is also known as dynamic-stiffness method, cf. Clough, Penzien[25],

γ) Finete-element method (FEM), cf. e.g. Gallagher[26].

Here, the deformation method shall briefly be described, which leads to the same results as the finite-element method after a linearization process. The modal analysis can be carried out in five steps, which are the same as in the finite-element approach.

Step 1: The guideway element is subdivided into s uniform beam segments which are connected in nodes; global and local coordinates are defined.

Step 2: A single beam segment is considered. On the left end (index l) and right end (index r) forces Q and moments M resulting from end displacements w and end rotations ζ are introduced, see Fig. 7. Nodal loads and nodal deformations are arranged in vectors,

$$\underline{f} = [\ M_l \ Q_l \ M_r \ Q_r \ ]^T \quad , \quad \underline{v} = [\ \zeta_l \ w_l \ \zeta_r \ w_r \ ]^T . \tag{51}$$

Since the beam segment is assumed to be uniform and subjected to no loading within the span, the associated mode shapes can be described by (37) resulting in the solution (44). Thus, the vectors $\underline{f}$ and $\underline{v}$ can be expressed by $\varphi(\xi) = \underline{a}^T(\lambda\xi/L)\underline{c}$ and its derivatives, where the left end (l) corresponds to $\xi = 0$ and the right end (r) is reached for

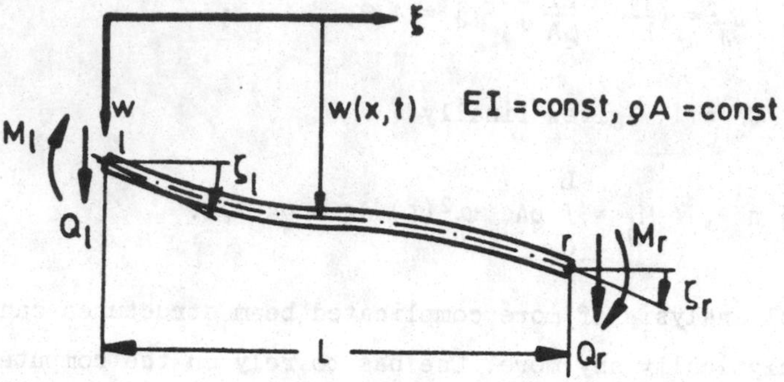

Fig. 7: Boundary forces and deformations for a uniform beam segment.

$\xi = L$.

$$\underline{f} = EI\,[-\varphi''(0)\;,\;\varphi'''(0)\;,\;\varphi''(L)\;,\;-\varphi'''(L)]^T = \underline{D}(\lambda)\,\underline{c}\quad, \tag{52}$$

$$\underline{v} = [\;\varphi'(0)\;,\;\varphi(0)\;,\;\varphi'(L)\;,\;\varphi(L)]^T = \underline{C}^{-1}(\lambda)\,\underline{c}\quad. \tag{53}$$

Obviously both vectors $\underline{f}$ and $\underline{v}$ can be expressed by the vector $\underline{c}$ (46) and approriate matrices $\underline{D}(\lambda)$, $\underline{C}^{-1}(\lambda)$. Eliminating $\underline{c}$ yields

$$\underline{f} = \underline{D}(\lambda)\;\underline{C}\;(\lambda)\;\underline{v} = \underline{F}(\lambda)\;\underline{v} \tag{54}$$

where

$$\underline{F}(\lambda) = E\,I \begin{bmatrix} F_2/L & -F_4/L^2 & F_1/L & -F_3/L^2 \\ -F_4/L^2 & F_6/L^3 & F_3/L^2 & F_5/L^3 \\ F_1/L & F_3/L^2 & F_2/L & F_4/L^2 \\ -F_3/L^2 & F_5/L^3 & F_4/L^2 & F_6/L^3 \end{bmatrix}, \tag{55}$$

$$
\begin{aligned}
F_1 &= -\lambda\,(S(\lambda) - s(\lambda)) / N \quad , \\
F_2 &= -\lambda\,(C(\lambda)\, s(\lambda) - S(\lambda)\, c(\lambda)) / N \quad , \\
F_3 &= -\lambda^2(C(\lambda) - c(\lambda)) / N \quad , \\
F_4 &= \lambda^2(S(\lambda)\, s(\lambda)) / N \quad , \\
F_5 &= \lambda^3(S(\lambda) + s(\lambda)) / N \quad , \\
F_6 &= -\lambda^3(C(\lambda)\, s(\lambda) + S(\lambda)\, c(\lambda)) / N \quad , \\
N &= C(\lambda)\, c(\lambda) - 1 \quad .
\end{aligned}
\tag{56}
$$

The matrix $\underline{F}(\lambda)$ is called dynamic stiffness matrix of the beam segment, because it expresses nodal loads in terms of nodal deformations. It depends on the frequency functions $F_\nu = F_\nu(\lambda)$, $\nu = 1,\ldots,6$, introduced by Koloušek[24].

Step 3: The entire guideway element comprises s segments and k nods. Each segment is characterized by the following quantities,

$$
\begin{aligned}
&\lambda_i = \lambda_i(\omega) = L\sqrt[4]{\omega^2(\varrho A/EI)_i} \quad , \\
&\varphi_i = \varphi_i(\xi_i) = \underline{a}_i^T\, \underline{c}_i \quad , \\
&\underline{c}_i = \underline{C}_i\, \underline{v}_i \quad , \\
&\underline{f}_i = \underline{F}_i\, \underline{v}_i \quad , \\
&\underline{a}_i = \underline{a}\,(\lambda_i\ \xi_i/L_i)\ ,\ \underline{C}_i = \underline{C}(\lambda_i)\ ,\ \underline{F}_i = \underline{F}(\lambda_i)\ , \quad i = 1,\ldots, s \quad .
\end{aligned}
\tag{57}
$$

The boundary conditions may allow n nodal deformations, called the nodal degrees of freedom. They are described by n generalized coordinates q_ν , $\nu = 1,\ldots, n$. The following global quantities (index g) are introduced

$$
\begin{aligned}
&\underline{q} = [\ q_1,\ldots,\ q_n\ ]^T \quad , \\
&\underline{v}_g = [\ \underline{v}_1^T,\ldots,\ \underline{v}_s^T]^T \quad ,
\end{aligned}
\tag{58}
$$

$$\underline{F}_g = \underline{\text{diag}}\,(\underline{F}_i) \quad , \quad i = 1,\dots, s \quad .$$

The connection between $\underline{v}_i$, $\underline{v}_g$ an $\underline{q}$ is given by incidence matrices $\underline{I}_i$ and $\underline{I}_g$, respectively, where the boundary and transition conditions are regarded,

$$\underline{v}_i = \underline{I}_i\, \underline{q} \quad , \quad \underline{v}_g = \underline{I}_g\, \underline{q} \quad . \tag{59}$$

Applying the principle of virtual work, $\delta V_g = 0$, yields

$$\begin{aligned} \delta V_g &= \sum_{i=1}^{s} \delta\, \underline{v}_i^T\, \underline{f}_i = \sum_{i=1}^{s} \delta\, \underline{v}_i^T \underline{F}_i\, \underline{v}_i = \delta\, \underline{v}_g^T\, \underline{F}_g\, \underline{v}_g = \\ &= \delta\, \underline{q}^T\, \underline{I}_g^T\, \underline{F}_g\, \underline{I}_g\, \underline{q} = 0 \quad , \end{aligned} \tag{60}$$

where (57) - (59) has been used. Since $\delta\underline{q}$ is arbitrary, we receive

$$\underline{I}_g^T\, \underline{F}_g(\omega)\, \underline{I}_g\, \underline{q} = \bar{\underline{F}}\,(\omega)\, \underline{q} = \underline{0} \quad . \tag{61}$$

Here, $\bar{\underline{F}}(\omega)$ denotes the dynamical stiffness matrix of the structure. Eq. (61) represents an implicit eigenvalue problem.

The influence of additional springs and masses located in node p , see Fig. 8, can be characterized by the virtual work, δW , of the applied forces. If k_p , c_p denote the lateral and torsional spring constant, and m_p , J_p the mass and moment of inertia, respectively, then the total virtual work δW_g is given by

$$\begin{aligned} \delta W_g &= - \sum_{p=1}^{k} [\delta\zeta_p\, (c_p - \omega^2 J_p)\, \zeta_p + \delta w_p\, (k_p - \omega^2 m) w_p\,] = \\ &= - \delta\underline{q}^T\, (\, \underline{K}^z - \omega^2 \underline{M}^z\,)\, \underline{q} \quad , \end{aligned} \tag{62}$$

where the stiffness matrix $\underline{K}^z$ and inertia matrix $\underline{M}^z$ of the appendages have been introduced. Equating $\delta V_g = \delta W_g$ yields finally

$$[\bar{\underline{F}}(\omega) + \underline{K}^z - \omega^2\, \underline{M}^z]\, \underline{q} = \hat{\underline{F}}\,(\omega)\, \underline{q} = \underline{0} \quad . \tag{63}$$

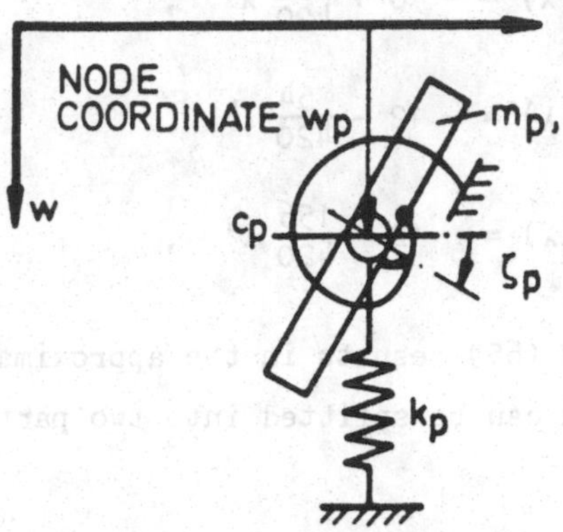

Fig. 8: Characterization of additional springs and masses.

Here, the overall dynamic stiffness matrix $\hat{\underline{F}}(\omega)$ has been introduced.
Step 4: The numerical solution of the implicit eigenvalue problem (63) give the eigenfrequencies ω_j and the eigenvectors $\underline{q}_j$, j = 1,...,f

Step 5: Backward computation according to the following scheme provide the eigenfunctions $\varphi_{ij}(\xi)$,

$$\lambda_{ij} = \lambda_i(\omega_j) = L_i \sqrt[4]{\omega_j^2 \, (\rho A/EI)_i} \quad ,$$

$$\underline{v}_{ij} = \underline{I}_i \, \underline{q}_j \quad , \tag{64}$$

$$\underline{c}_{ij} = \underline{C}_{ij} \, \underline{v}_{ij} \, , \quad \underline{C}_{ij} = \underline{C}_i \, (\lambda_{ij}) \quad ,$$

$$\varphi_{ij} = \underline{a}^T_{ij} \, \underline{c}_{ij} = \underline{a}^T_{ij} \, \underline{C}_{ij} \, \underline{v}_{ij} \quad , \quad \underline{a}_{ij} = \underline{a}_i \, (\lambda_{ij} \, \xi_i \, / \, L_i) \quad ,$$

which completes the modal analysis by the deformation method.

The finite-element approach requires the same steps as the deformation method. The FEM relations follow easily if the frequency functions $F_\nu(\lambda)$ are expanded into a power series. The first two terms read,

$$F_1(\lambda) = 2 + \frac{3}{420}\lambda^4 \quad , \quad F_4(\lambda) = -\ 6 + \frac{22}{420}\lambda^4 \quad ,$$

$$F_2(\lambda) = 4 - \frac{4}{420}\lambda^4 \quad , \quad F_5(\lambda) = -\ 12 - \frac{54}{420}\lambda^4 \quad , \tag{65}$$

$$F_3(\lambda) = 6 + \frac{13}{420}\lambda^4 \quad , \quad F_6(\lambda) = \ 12 - \frac{156}{420}\lambda^4 \quad ,$$

Substituting F_ν in (55) by (65) results in the approximate dynamic stiffness matrix $\underline{\tilde{F}}$, which can be splitted into two parts,

$$\underline{\tilde{F}} = \underline{K} - \tilde{\omega}^2\underline{M} \quad , \tag{66}$$

$$\underline{K} = \frac{EI}{L^3}\begin{bmatrix} 4\,L^2 & 6\,L & 2\,L^2 & -6\,L \\ 6\,L & 12 & 6\,L & -12 \\ 2\,L^2 & 6\,L & 4\,L^2 & -6\,L \\ -6\,L & -12 & -6\,L & 12 \end{bmatrix}, \quad \underline{M} = \frac{\varrho AL}{420}\begin{bmatrix} 4\,L^2 & 22\,L & -3\,L^2 & 13\,L \\ 22\,L & 156 & -13\,L & 54 \\ -3\,L^2 & -13\,L & 4\,L^2 & -22\,L \\ 13\,L & 54 & -22\,L & 156 \end{bmatrix}.$$

The stiffness matrix $\underline{K}$ and the inertia matrix $\underline{M}$ are identical with the corresponding matrices in the FEM. Utilizing (66) instead of (55) in the modal analysis results in

$$[\ \underline{\bar{K}} + \underline{K}^z - \tilde{\omega}^2(\ \underline{\bar{M}} + \underline{M}^z)\ ]\ \underline{\tilde{q}} = \underline{0} \quad ,$$
$$\underline{\bar{K}} = \underline{I}_g^T\ \underline{\mathrm{diag}}\ (\underline{K}_i)\ \ \underline{I}_g \quad , \quad \underline{\bar{M}} = \underline{I}_g^T\ \underline{\mathrm{diag}}\ (\underline{M}_i)\ \ \underline{I}_g\ ,\ i = 1,\ldots,\ s \quad , \tag{67}$$

which represents an explicit eigenvalue problem.

A comparison of DEM and FEM is given in Table 1

	DEM	FEM
Result	exact	approximate
Eigenvalue problem	implicit	explicit
Node number	minimum	large for high accuracy

Table 1: Comparison of the deformation method DEM with the finite-element method FEM

In the DEM only the minimum node number is required, this allows it to enter in competition with the FEM in applications.

As an example the results of the modal analysis for the double span frame, cf. Fig. 6 C), are shown in Table 2.

SYMMETRIC 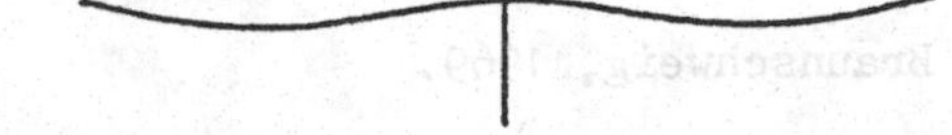	$f = 12{,}13$ Hz $\omega = 76{,}22\ s^{-1}$ $\lambda_1 = 3{,}926$
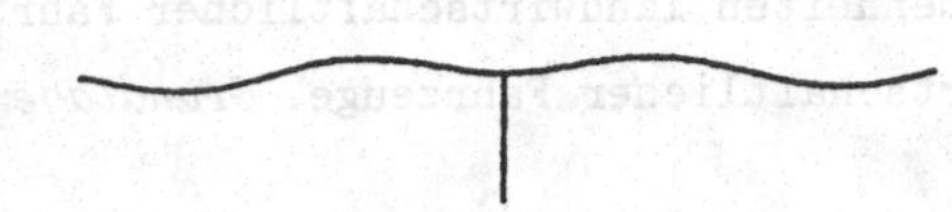	$f = 39{,}33$ Hz $\omega = 247{,}1\ s^{-1}$ $\lambda_1 = 7{,}068$
ANTIMETRIC, T-NODE FIXED 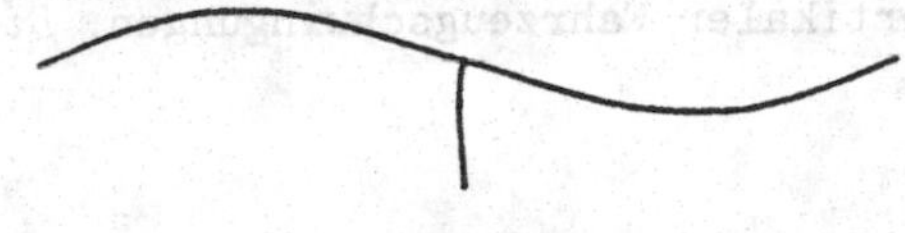	$f = 9{,}46$ Hz $\omega = 59{,}42\ s^{-1}$ $\lambda_1 = 3{,}467$ $\lambda_3 = 1{,}086$
	$f = 33{,}12$ Hz $\omega = 208{,}1\ s^{-1}$ $\lambda_1 = 6{,}487$ $\lambda_3 = 2{,}031$
ANTIMETRIC, T-NODE FREE 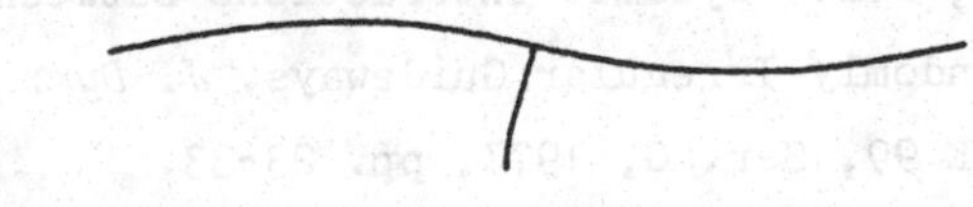	$f = 6{,}14$ Hz $\omega = 38{,}55\ s^{-1}$ $\lambda_1 = 2{,}792$ $\lambda_3 = 0{,}874$
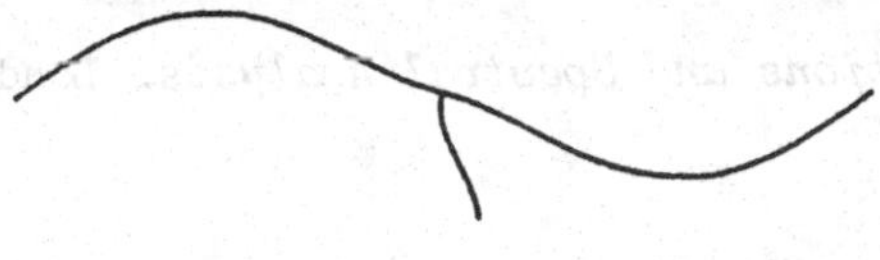	$f = 10{,}54$ Hz $\omega = 66{,}25\ s^{-1}$ $\lambda_1 = 3{,}660$ $\lambda_3 = 1{,}146$
	$f = 33{,}18$ Hz $\omega = 208{,}5\ s^{-1}$ $\lambda_1 = 6{,}494$ $\lambda_3 = 2{,}034$
$L_1 = 20$ m $(\rho A)_1 = 2{,}3\cdot 10^3 \frac{kg}{m}$; $(EI)_1 = 0{,}9\cdot 10^{10}$ Nm^2 $L_3 = 5{,}66$ m $(\rho A)_3 = 1{,}725\cdot 10^3 \frac{kg}{m}$; $(EI)_3 = 0{,}45\cdot 10^{10}$ Nm^2	

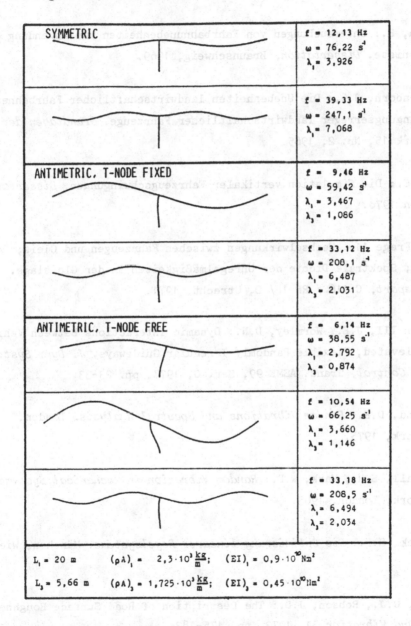

Table 2: Modal analysis of a double span frame.

REFERENCES

1. Mitschke, M.: *Dynamik der Kraftfahrzeuge*. Berlin, Heidelberg, New York, 1972.

2. Braun, H.: Untersuchungen von Fahrbahnunebenheiten und Anwendung der Ergebnisse. *Dissertation,* Braunschweig, 1969.

3. Wendenborn, J.O.: Die Unebenheiten landwirtschaftlicher Fahrbahnen als Schwingungserreger landwirtschaftlicher Fahrzeuge. *Grundlagen der Landtechnik* 15, Nr. 2, 1965.

4. Voy, C.: Die Simulation vertikaler Fahrzeugschwingungen. *Dissertation,* Berlin 1976.

5. ORE, Frage 116: Wechselwirkungen zwischen Fahrzeugen und Gleis; Bericht Nr. 1; Spektrale Dichte der Unregelmäßigkeiten in der Gleislage. *ORE-Report,* C 116 / RP 1 / D, Utrecht, 1971.

6. Snyder III, J.E., Wormley, D.N.: Dynamic Interactions Between Vehicles and Elevated, Flexible Randomly Irregular Guideways. *J. Dyn. Syst. Meas. Control,* Trans. ASME 99, Ser. G, 1977, pp. 23-33.

7. Newland, D.E.: *Random Vibrations and Spectral Analysis*. London, New York, 1975.

8. Crandall, S.H., Mark, W.D.: *Random Vibration in Mechanical Systems*. New York, London, 1963.

9. Drenick, R.F.: *Die Optimierung linearer Regelsysteme*. München, Wien, 1967.

10. Dodds, C.J., Robson, J.D.: The Description of Road Surface Roughness. *J. Sound Vibration* 31, 1973, pp. 175-183.

11. Hedrick, J.K., Anis, Z.: Proposal for the Characterization of Rail Track Irregularities. MIT, MA 02139, *ISO-Meeting*, Berlin, 1978.

12. Dincă, F., Theodosiu, C.: *Nonlinear and Random Vibration*. New York, London, 1973.

13. Fábián, L.: *Zufallsschwingungen und ihre Behandlung*. Berlin, Heidelberg, New York, 1973.

14. Sussmann, N.E.: Statistical Ground Excitation Models for High Speed Vehicle Dynamic Analysis. *High Speed Ground Transportation* 8, 1974, pp. 145-154.

15. Karnopp, D.C.: Vehicle Response to Stochastic Roadways. *Vehicle System Dynamics* 7, 1978, pp. 97-109.

16. Müller, P.C., Popp, K., Schiehlen, W.O.: Covariance Analysis of Nonlinear Stochastic Guideway-Vehicle-Systems. *Proc. 6th IAVSD Symp. on Dynamics of Vehicles or Roads and on Tracks*, Berlin, 1979, pp. 337-349.

17. Müller, P.C., Popp, D., Schiehlen, W.O.: Berechnungsverfahren für stochastische Fahrzeugschwingungen. *Ing.-Arch.* 49, 1980, pp. 235-254.

18. Timoshenko, S.: Method of Analysis of Statical and Dynamical Stresses in Rail. *Proc. of the 2nd Int. Congr. for Applied Mechanics*, Zürich, 1926.

19. Dörr, J.: Der unendliche, federnd gebettete Balken unter dem Einfluß einer gleichförmig bewegten Last. *Ing.-Arch.* 14, 1943, pp. 167-192.

20. Korb, J.: Parametererregung beim Rad-Schiene-System. *VDI-Berichte*, Nr. 381, Düsseldorf 1981, pp. 99-104.

21. Popp, K.: Dynamik von Fahrzeug-Strukturen unter wandernden Lasten. *VDI-Bericht* Nr. 419, Düsseldorf 1981, pp. 153-161.

22. Knothe, K.: Vergleichende Darstellung verschiedener Verfahren zur Berechnung der Eigenschwingungen von Rahmentragwerken . *Fortschr.-Ber. VDI-Z,* Reihe 11, Nr. 9, Düsseldorf, 1971.

23. Pestel, E., Leckie, S.: *Matrix Methods in Elastomechanics,* New York, 1963.

24. Koloušek, V.: *Dynamik der Baukonstruktionen.* Berlin 1962.

25. Clough, R.W., Penzien, J.: *Dynamics of Structures.* New York, 1975.

26. Gallagher, R.H.: *Finite-Element-Analysis.* Berlin, Heidelberg, New York, 1976.

MODELING BY MULTIBODY SYSTEMS

Werner O. Schiehlen
Institut B für Mechanik
Universität Stuttgart
Pfaffenwaldring 9, Stuttgart 80, F.R.G.

INTRODUCTION

For the investigation of vehicle handling and ride, due to frequencies less than 50 Hz, the method of multibody systems is well qualified. The derivation of equations of motion is discussed in general without any specific application in mind. A complex automobile model is treated by this method later on.

ELEMENTS OF MULTIBODY SYSTEMS

Multibody systems are characterized by rigid bodies with inertia, and springs, dashpots, servomotors without inertia, Fig. 1. The bodies are interconnected by rigid bearings and subject to additional applied forces and torques by supports[1,2].

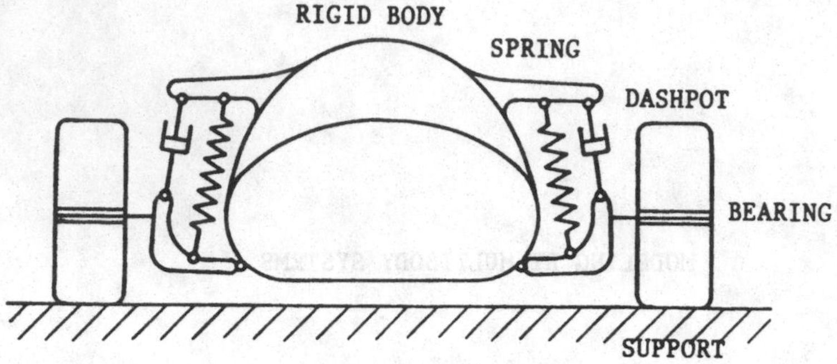

Fig. 1. Multibody system

The position of a system of p bodies is given relative to an inertial frame x_I, y_I, z_I by the 3x1-translation vector

$$r_i = [r_{xi}\ r_{yi}\ r_{zi}]^T, \quad i = 1(1)p, \tag{1}$$

of the center of mass C_i and the 3x3-rotation tensor.

$$S_i = S_i(\alpha_i, \beta_i, \gamma_i), \quad i = 1(1)p, \tag{2}$$

written down for each body, Fig. 2. The translation vector r_i and the rotation tensor S_i characterize each body K_i by a corresponding body-fixed frame x_i, y_i, z_i. The rotation tensor S_i depends on the three generalized coordinates α_i, β_i, γ_i and follows from the direction cosine matrix relating the inertial frame to each body-fixed frame. The rigid body rotations are comprehensively presented in gyro dynamics, see e.g. Magnus[3].

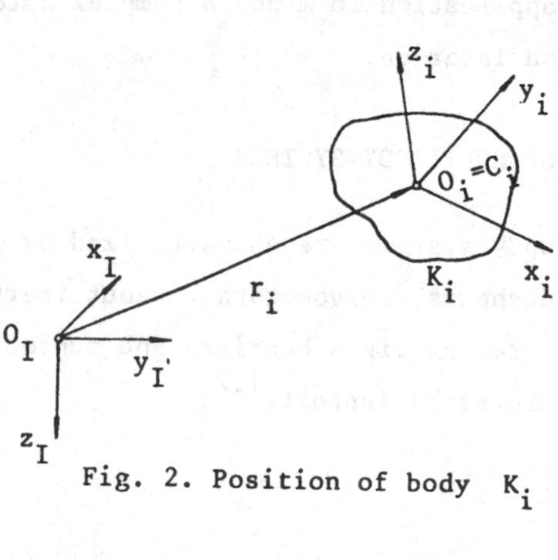

Fig. 2. Position of body K_i

In vehicle dynamics some additional frames are very convenient for the mathematical description of the path, the reference position and local bearing or support axes, Fig. 3. They are summarized in Table 1. Typical transformations between the frames will be now presented. From the path given in the inertial frame it follows the translation vector

$$ {}_I r_{IR} = R(u) \tag{3} $$

where the 3x1-path trajectory vector R is a function of the path length u. Then, the remaining vehicle translation

$$ {}_R r_{Ri} = \begin{bmatrix} x_i & y_i & z_i \end{bmatrix}^T \tag{4} $$

depends on the displacement x_i, y_i, z_i. The rotation tensor of the path frame follows also from the path trajectory,

$$ S_{IP} = \begin{bmatrix} t & n & b \end{bmatrix} , \tag{5} $$

where the 3x1-tangential vector

$$ t = \frac{d}{du} R(u) \; , \quad \left| \frac{d}{du} R(u) \right| = 1 \; , \tag{6} $$

the 3x1-normal vector

$$ n = \frac{1}{\kappa} \frac{d}{du} t(u) \; , \quad \left| \frac{d}{du} t(u) \right| = \kappa \tag{7} $$

and the 3x1-binormal vector

$$ b = t \times n \tag{8} $$

are used. The rotation tensor of the reference frame introduces the superelevation α of the tangential plane with respect to the path frame,

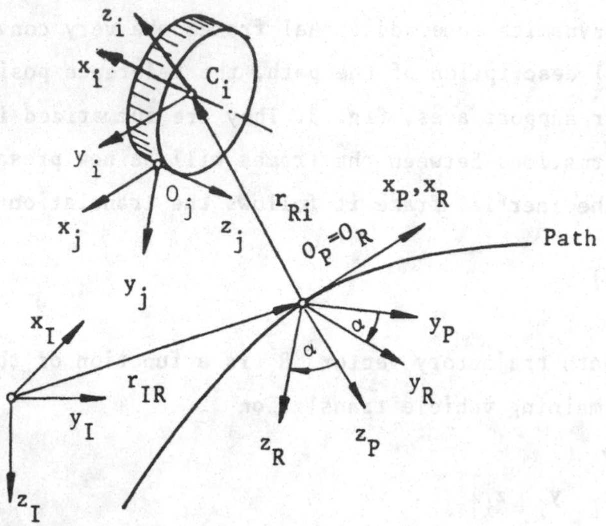

Fig. 3. Additional frames in vehicle dynamics

Frame	Initial point	Direction of axes
Inertial frame $I = \{O_I; x_I, y_I, z_I\}$	O_I fixed in space	x_I, y_I horizontal plane z_I in gravity direction
Path frame $P = \{O_P; x_P, y_P, z_P\}$	O_P moving on path with given velocity	x_P in velocity direction y_P normal, to the right z_P binormal to path
Reference frame $R = \{O_R; x_R, y_R, z_R\}$	$O_R = O_P$	$x_R = x_P$ y_R in tangential plane z_R normal to tang. plane
Body-fixed frame $i = \{O_i; x_i, y_i, z_i\}$	$O_i = C_i$ center of mass	x_i, y_i, z_i principal axes if possible
Local frame $j = \{O_j; x_j, y_j, z_j\}$	O_j fixed in bearing or support	x_j, y_j, z_j according to bearing or support axes

Table 1. Frames for modeling of vehicle systems

$$S_{PR} = \begin{bmatrix} 1 & 0 & 0 \\ 0 & \cos\alpha & -\sin\alpha \\ 0 & \sin\alpha & \cos\alpha \end{bmatrix} . \tag{9}$$

Then, the remaining small vehicle rotations are characterized by the displacement ϕ_i, θ_i, ψ_i, resulting in the rotation tensor

$$S_{Ri} = \begin{bmatrix} 1 & -\psi_i & \theta_i \\ \psi_i & 1 & -\phi_i \\ -\theta_i & \phi_i & 1 \end{bmatrix} . \tag{10}$$

Such transformations of frames are also widely used in aircraft dynamics[4].

KINEMATICS OF MULTIBODY SYSTEMS

Vehicles are holonomic multibody systems, i.e. their internal motions are constrained by bearings. Further, most of the supports result also in holonomic constraints, e.g. a wheelset of railway vehicles on a rigid track. Therefore, only holonomic systems are treated.

A holonomic system of p bodies and q holonomic, rheonomic constraints by rigid bearings holds only f degrees of freedom:

$$f = 6p - q . \tag{11}$$

Then, the position of the system can be uniquely described by f generalized coordinates summarized in a fx1-position vector $y(t)$. Position, velocity and acceleration of the system read now

$$r_i = r_i(y,t) \quad , \quad S_i = S_i(y,t) \quad , \quad i = 1(1)p \quad , \tag{12}$$

$$v_i = J_{Ti}\dot{y} + \bar{v}_i \quad , \quad \omega_i = J_{Ri}\dot{y} + \bar{\omega}_i \quad , \tag{13}$$

$$\dot{v}_i = J_{Ti}\ddot{y} + K_{Ti}\dot{y} + \dot{\bar{v}}_i \quad , \quad \dot{\omega}_i = J_{Ri}\ddot{y} + K_{Ri}\dot{y} + \dot{\bar{\omega}}_i \quad , \tag{14}$$

where the 3xf-Jacobian matrices

$$J_{Ti}(y,t) = \frac{\partial r_i}{\partial y} \quad , \qquad J_{Ri}(y,t) = \frac{\partial s_i}{\partial y} \tag{15}$$

with the infinitesimal 3x1-rotation vector s_i, the 3xf-acceleration matrices

$$K_{Ti}(y,\dot{y},t) = \frac{\partial (J_{Ti}\dot{y})}{\partial y} + 2\,\frac{\partial v_i}{\partial y} \quad ,$$
$$K_{Ri}(y,\dot{y},t) = \frac{\partial (J_{Ri}\dot{y})}{\partial y} + \frac{\partial J_{Ri}}{\partial y} + \frac{\partial \omega_i}{\partial y} \tag{16}$$

and the 3x1-vectors

$$\overline{v}_i(y,t) = \frac{\partial r_i}{\partial t} \quad , \qquad \overline{\omega}_i(y,t) = \frac{\partial s_i}{\partial t} \quad , \tag{17}$$

$$\dot{\overline{v}}_i(y,t) = \frac{\partial^2 r_i}{\partial t^2} , \qquad \dot{\overline{\omega}}(y,t) = \frac{\partial^2 s_i}{\partial t^2} \tag{18}$$

are introduced. For scleronomic constraints the vectors (17) and (18) vanish.

For special motions, e.g. a circular path, it may be more convenient to use the reference frame instead of the inertial frame. This means that the absolute velocities and accelerations are divided into reference and relative velocities and accelerations, respectively.

KINETICS OF MULTIBODY SYSTEMS

According to the free body diagram Newton's and Euler's equations have to be applied to each body of the system:

$$m_i \dot{v}_i = f^e_i + f^r_i \quad , \quad i = 1(1)p \tag{19}$$

$$I_i \dot{\omega}_i + \tilde{\omega}_i I_i \omega_i = l^e_i + l^r_i \quad . \tag{20}$$

The inertia is represented by the mass m_i and the inertia tensor I_i relative to the inertial frame with respect to the center of mass C_i. From the time-invariant inertia tensor ${}_iI_i$ in the body-fixed frame it follows

$$I_i = S_i\, {}_iI_i\, S_i^T \quad . \tag{21}$$

The external forces and torques in (19) and (20) are composed by applied forces f_i^e and torques l_i^e due to springs, dashpots, servomotors and weight and by constraint forces f_i^r and torques l_i^r due to bearings and supports. The applied forces and torques, repectively, may have proportional, differential and/or integral behavior.

The Proportional forces (P-forces) are characterized by the system's position:

$$f_i^e = f_i^e(r_i, S_i, t) \quad . \tag{22}$$

E.g., conservative spring and weight forces as well as purely time-varying servomotor forces are P-forces. The Proportional-Differential forces (PD-forces) depend on the position and the velocity:

$$f_i^e = f_i^e(r_i, S_i, v_i, \omega_i, t) \quad . \tag{23}$$

E.g., parallel spring-dashpot configurations result in PD-forces. The Proportional-Integral forces (PI-forces) are a function of the position, the velocity and integrals of position:

$$\begin{aligned} f_i^e &= f_i^e(r_i, S_i, v_i, \omega_i, w, t) \quad , \\ \dot{w} &= \dot{w}(r_i, S_i, v_i, \omega_i, w, t) \end{aligned} \tag{24}$$

where the $p \times 1$-vector w describes the position integrals. E.g., serial spring-dashpot configurations and many servomotors yield PI-forces.

Constraint forces and torques can be characterized by the qx1-vector g of the generalized constraint forces:

$$f_i^r = F_i(r_i, S_i, t)g , \quad l_i^r = L_i(r_i, S_i, t)g \tag{25}$$

where the 3xq-matrices F_i, L_i describe the distribution of the generalized constraint forces to each body.

Newton's equation (19) is valid only in the inertial frame. Euler's equation (20) yields in the inertial frame as well as in the corresponding body-fixed frame. However, in the reference frame both equations have to be extended by the reference motion[5].

EQUATIONS OF MOTION

Multibody systems may be ordinary or general depending on the constraints and the applied forces. Since in kinematics only holonomic constraints have been discussed, the applied forces remain as criterium.

A holonomic system with PD-forces results in an ordinary multibody system. From (12), (13), (14), (19), (20), (23), (25) the Newton-Euler-equations are summarized:

$$M(y,t)\ddot{y}(t) + \bar{k}(y,\dot{y},t) = \bar{q}^e(y,\dot{y},t) + \bar{Q}(y,t)g(t) \quad , \tag{26}$$

$$\bar{M} = \begin{bmatrix} m_1 J_{T1} \\ \vdots \\ m_p J_{Tp} \\ \\ I_1 J_{R1} \\ \vdots \\ I_p J_{Rp} \end{bmatrix} \qquad \bar{q}^e = \begin{bmatrix} f^e_1 \\ \vdots \\ f^e_p \\ \\ l^e_1 \\ \vdots \\ l^e_p \end{bmatrix} \qquad \bar{Q} = \begin{bmatrix} F_1 \\ \vdots \\ F_p \\ \\ L_1 \\ \vdots \\ L_p \end{bmatrix} . \tag{27}$$

Here $\overline{M}$ is a 6pxf-inertia matrix, $\overline{k}$ is a 6px1-vector of gyro and centrifugal forces, $\overline{q}^e$ is a 6px1-vector of the applied forces and Q is a 6pxq-distribution matrix of the constraint forces. The equation of motion follows from (26) according to d'Alembert's principle by pre-multiplication with the global Jacobian matrix of the system:

$$\overline{J} = \left[J_{T1}^T \; . \; . \; . \; J_{Tp}^T \quad J_{R1}^T \; . \; . \; . \; J_{Rp}^T\right]^T \; . \tag{28}$$

The result is a fx1-vector differential equation of second order,

$$M(y,t)\ddot{y}(t) + k(y,\dot{y},t) = q(y,\dot{y},t) \quad , \tag{29}$$

where the constraint forces are completely eliminated.
The fxf-inertia matrix

$$M(y,t) = \overline{J}^T\left[\text{diag} \;\; m_1 \; ... \; m_p \; I_1 \; ... \; I_p\right]\overline{J} \tag{30}$$

is symmetric and usually positive definit, the fx1-vector k describes the generalized gyro and centrifugal forces and the fx1-vector q is the vector of the generalized applied forces wellknown from Lagrange's equations.

All nonordinary systems are <u>general multibody systems</u>, particularly systems with PI-forces. Extending (26) by (24) the following equations are obtained:

$$M(y,t)\ddot{y} + k(y,\dot{y},t) = q(y,\dot{y},w,t) \quad , \tag{31}$$

$$\dot{w} = \dot{w}(y,\dot{y},w,t) \quad . \tag{32}$$

For many vehicle applications the equations of motion can be linearized. Then, it follows from (31), (32) for general multibody systems with time-invariant coefficient matrices:

$$M\ddot{y} + P\dot{y} + Qy + Rw \;=\; h(t) \quad , \tag{33}$$

$$\dot{w} + Ww + Yy + \dot{Y}\dot{y} = k(t) \quad . \tag{34}$$

There are obtained the fxf-matrices M, P, Q, the fxρ-matrix R, the ρxρ-matrix W, the ρxf-matrices Y, $\dot{Y}$, the fx1-vector h(t) and the ρx1-vector k(t). Equations (33), (34) are often written in state space representation,

$$\dot{x} = A\,x + b(t) \tag{35}$$

where the (2f+ρ)x1-state vector

$$x = \left[y^T \;\; \dot{y}^T \;\; w^T\right]^T \tag{36}$$

is introduced and the system matrix A and the excitation vector b(t) have the corresponding dimensions.

COMPUTERIZED DERIVATION OF EQUATIONS OF MOTION

The presented method uses the following input variables:

$$y,\;\; r_i(y,t),\;\; S_i(y,t),\;\; m_i,\;\; I_i,\;\; f_i^e,\;\; l_i^e \quad . \tag{37}$$

These variables have to be prepared by the engineer dealing with a specific application. Then, during the derivation of the equations of motion the following operations have to be performed:

Summation of vectors and matrices,
multiplication of vectors and matrices,
differentiation of vectors and matrices,
simplification of trigonometrical expressions,
linearization of expressions.

All these operations are often done by the engineer but they can also be done by the computer. However, a symbolical execution of these operations is very desirable to obtain symbolical equations saving computation time. Using the index coding tor symbolic manipulations a

program called NEWEUL is available, see Kreuzer[6].
For the memorization and execution of a symbolical expression given by summed terms an array of integers is used. The positive integers are defined as the indices of the array of variables while the negative integers represent the indices of the array of functions. The sign of a term is given by the sign of its numerical factor and all elements of a term (numerical factor, variables and functions) are automatically multiplied. Vectors and matrices can be formed using symbolical expressions as matrix elements. The programming of vector and matrix operations yields to the complete routine for the equations of motion of a multibody system where simplifications due to trigonometric and nonlinear expressions are automatically executed. The result may be obtained as a printed listing or as package of punch cards for the numerical solution of the equations.

A realistic example for the computerized derivation of equations of motion is presented in the author's lecture "Complex Nonlinear Vehicles".

REFERENCES

1. Schiehlen, W.O. and Kreuzer, E.J., Computerized Derivation of Equations of Motion, in: *Dynamics of Multibody Systems*, Ed. K. Magnus, Springer-Verlag, Berlin-Heidelberg-New York, 1978, 290-305.

2. Schiehlen, W., Dynamical Analysis of Suspension Systems, in: *Dynamics of Vehicles on Road and Railway Tracks*, Ed. by A. Slibar and H. Springer, Swets & Zeitlinger, Amsterdam, 1978, 40-48.

3. Magnus, K., *Gyrodynamics*, CISM Courses and Lectures No. 53, Springer-Verlag, Wien-New York, 1974.

4. Luftfahrtnorm LN 9300, Bezeichnungen in der Flugmechanik, Beuth-Vertrieb, Köln, 1959.

5. Schiehlen, W., Zustandsgleichungen elastischer Satelliten. *Z. angew. Math. Phys.* 23, 575-586.

6. Kreuzer, E., *Symbolische Berechnung der Bewegungsgleichungen von Mehrkörpersystemen*, VDI-Verlag, Düsseldorf, 1979.

MATHEMATICAL METHODS IN VEHICLE DYNAMICS

Peter C. Müller
Department of Safety Control Engineering
University of Wuppertal
POB 100127, D-5600 Wuppertal 1, FRG

1 INTRODUCTION

The dynamic analysis of deterministic and random vehicle vibrations and the consequences especially to passenger comfort requires an integrated study of three subproblems:

(i) modeling and characterization of guideway roughness,

(ii) prediction of vehicle motion for traversal of a given guideway,

(iii) prediction or characterization of passenger response to vibration exposure.

Here we assume the common causality of the three subproblems: guideway roughness causes vehicle motion and vehicle motion causes passenger response. Therefore, the complete analysis consists of a stepwise characterization of each subproblem by a suitable mathematical model where the input to problems (ii) and (iii) are given by the results of problems (i) and (ii), respectively, and a subsequent manipulation of these mathematical methods to obtain information about dynamical behaviour of vehicles and about ride quality.

The modeling of problems (i) and (ii) is presented in other parts of this book by Popp and Schiehlen. Therefore we assume, that the vehicle dynamics are described by a n-dimensional first-order vector differential equation

$$\dot{\underline{x}}(t) = \underline{A}\,\underline{x}(t) + \underline{f}(\underline{x}(t)) + \underline{B}\,\underline{\xi}(t) \tag{1.1}$$

where $\underline{x}$ denotes a vector of vehicle state variables, the $n \times n$ - matrix $\underline{A}$ characterizes the linear dynamical behaviour while the vector $\underline{f}(\underline{x})$ has regard to nonlinearities in the system; the $n \times r$ - input-matrix $\underline{B}$ describes how the guideway excitations $\underline{\xi}(t)$ influence the dynamical behaviour of the vehicle. The excitation process $\underline{\xi}(t)$ of a multi-axle vehicle consists of a composition of multiple scalar excitation processes $\zeta_i(t)$ representing the guideway roughness at a single contact between vehicle and guideway. Denoting the distance between the front axle and the axle i by l_i and the vehicle's velocity by v than the processes $\zeta_i(t)$ are usually generated by one time function $\zeta(t)$ with different time delays

$$t_i = \frac{l_i}{v}, \quad i = 1, \ldots, r, \tag{1.2}$$

$$\zeta_i(t) = \zeta(t - t_i), \quad 0 = t_1 < t_2 < \ldots < t_r, \tag{1.3}$$

$$\underline{B}\,\underline{\xi}(t) = \sum_{i=1}^{r} \underline{b}_i\,\zeta_i(t). \tag{1.4}$$

In earlier times this excitation function $\zeta(t)$ was assumed sinusoidally,

$$\zeta(t) = e \sin \omega t, \tag{1.5}$$

while recently $\zeta(t)$ is modeled by a stochastic process, e.g. by a stationary Gaussian coloured noise process which can be obtained from a Gaussian white noise process w(t) by means of a shape filter

$$\begin{aligned} \zeta(t) &= \underline{h}^T \underline{v}(t), \\ \underline{v}(t) &= \underline{F}\,\underline{v}(t) + \underline{g}\,w(t), \\ \mathrm{Re}\lambda(\underline{F}) &< 0, \quad w(t) \sim N(0,q_w). \end{aligned} \tag{1.6}$$

For details compare the article of Popp in this book.

Equations (1.1 - 1.6) represent an abstract description of the problems (i) and (ii). Additionally we also need a mathematical model characterizing the passenger response to vehicle motions. This problem will be discussed in more detail in section 5 where it is shown that the passenger response may be characterized by a perception variable $\bar{w}_k(t)$ which is connected to a vehicle vibration variable

$$w_k(t) = \underline{c}_k^T \underline{x}(t) + \underline{d}_k^T \underline{\xi}(t) \tag{1.7}$$

(usually $w_k(t)$ means the lateral or vertical acceleration measured on the vehicle floorboard or on the passenger/seat interface) by a perception shape filter

$$\begin{aligned} \bar{w}_k(t) &= \alpha \underline{h}_k^T \underline{v}_k , \\ \dot{\underline{v}}_k(t) &= \underline{F}_k \underline{v}_k(t) + \underline{g}_k w_k(t) , \quad \mathrm{Re}\,\lambda(\underline{F}_k) < 0 \end{aligned} \tag{1.8}$$

The complete set of equations (1.1 - 1.8) describes mathematically the problems (i), (ii) and (iii) in the time domain. By these equations the analysis of vehicle vibrations and of ride quality can be performed in the time domain by suitable mathematical methods which will be discussed later on. Another type of mathematical description of the problems stems from classical measurement techniques and uses frequency domain quantities such as frequency response or power spectral densities (PSD). For example, the stochastic excitation process $\zeta(t)$ may be characterised by a measured PSD

$$S_\zeta(\omega) , \tag{1.9}$$

or the passenger frequency response $\bar{w}_k(\omega)$ is related to the vehicle response $w_k(\omega)$ by a frequency relation

$$\bar{w}_k(\omega) = \alpha\, f_k(\omega)\, w_k(\omega) \tag{1.10}$$

in the case of sinusoidal excitation, and by a PSD-relation

$$S_{\bar{w}k}(\omega) = \alpha^2 \, | f_k(\omega) |^2 \, S_{wk}(\omega) \tag{1.11}$$

in the case of Gaussian stochastic excitation. By this frequency domain characterization of the problems (i) and (iii) we globally have a mixed problem formulation because problem (ii) is still described by (1.1) in the time domain. Therefore, we have to decide if the global problem is solved in the time domain introducing shape filters (1.6, 1.8) from (1.9, 1.11) or if the global problem is manipulated in frequency domain introducing a frequency response description of (1.1). Also this frequency domain approach will be discussed in the following sections.

After formulating the basic equations characterizing vehicle vibrations and passenger response we can now look for the mathematical methods to handle these equations for obtaining results such as vibration amplitudes of the vehicle, accelerations of the passenger seat or of the vehicle body, dynamic wheel loads etc. This will be done in the next sections. In section 2 simulation techniques are shortly mentioned, section 3 deals with the calculation of sinusoidal and stochastic responses in the linear case (i.e. $\underline{f}(\underline{x})$ in (1.1) is assumed to be neglected); time domain as well as frequency domain techniques are presented. In section 4 the nonlinear problem is solved by harmonical and statistical linearization methods. At least in section 5 the evaluation of human exposure to whole-body vibrations is discussed on the basic of the standards ISO 2631 and VDI 2057 leading to the specification of passenger response by (1.7 - 1.8). A collection of references for further study closes the text.

2 SIMULATION

The gouverning equations of vehicle dynamics are ordinary differential equations. Therefore, simulations of vehicle dynamics means the integration of the equations (1.1) where the input $\underline{\xi}(t)$ of the guideway roughness has to be given by suitable measurement data. The output of the simulation process may be any variable interesting for vehicle vibration such as (1.7) or a perception variable for passenger comfort like (1.8). In this later case simultaneously with (1.1) the equations (1.8) have to be integrated. The interesting quantities of the simulation are deterministic or stochastic time functions dependent on the character of $\underline{\xi}(t)$. But usually not only the time history is of interest but also some special data such as frequency spectrum, amplitudes, mean values, variances etc. Therefore, besides of integration techniques some signal processing techniques are required to obtain suitable results by simulation.

The integration techniques depend on the kind of computer. By the aid of analogue computer the integration is performed by hardware while using a digital computer the integration routines follow from numerical mathematics as a software package. Hybrid simulation techniques and digital simulation by parallel- and multi-processor systems try to combine the advantages of analogue and of digital computer systems: the fast integration time of analogue computer and the flexibility and accuracy of the digital computer. We would not like to discuss all those simulation techniques in more detail; therefore, it is refered to the literature. By Schmidt [1] an introduction is given to this field; by Korn et al. [2,3] analogue/hybrid and digital continuous-system simulations are considered. Numerical integration methods for digital simulations are described by Lapidus and Seinfeld[4] or by Grigorieff[5]. A comparison of different integration routines applied to vehicle dynamics is presented by Federl[6].

Also the wide field of signal processing techniques including the preparation of measurements for digital signal analysis, the discrete Fourier transform (DFT) and the fast Fourier transform (FFT) for calculation of frequencies, Fourier amplitudes and power spectral densities, the procedures determining mean values and variances etc. are not dealt with in this text. Related textbooks are references 7-12.

3 LINEAR SYSTEM ANALYSIS

In many applications of vehicle dynamics the effects of nonlinearities are neglected. For example, designing an automobile suspension system often a linear dynamic model is used neglecting the effects of progressive springs or nonlinear damping; only in a more detailed study nonlinearities are considered. Also the problem of designing a control system for a magnetically levitated vehicle is usually solved by a linear approach. Therefore, in a first essential step mathematical methods are discussed for linear models of vehicle dynamics. Instead of (1.1) a linear description

$$\dot{\underline{x}}(t) = \underline{A}\,\underline{x}(t) + \sum_{i=1}^{r} \underline{b}_i\, \zeta(t-t_i) \tag{3.1}$$

is assumed and the excitation is modeled with respect to (1.2 - 1.4).

3.1 Stability

It is well known[13] how the stability behaviour of (3.1) can be examined. If the inputs $\zeta(t-t_i)$ are bounded then the system response will be also bounded if and only if the matrix $\underline{A}$ is asymptotically stable, i.e. all eigenvalues λ_i, $i = 1,\dots, n$, of $\underline{A}$ have negative real parts. This stability test can be performed without explicit calculation of the eigenvalues by criteria of Hurwitz or Routh. But if the number n is larger than 5 or 6, then it is more convenient to calculate numerically the eigenvalues on a digital computer by a stable eigenvalue subroutine[14,15].

3.2 Sinusoidal Excitation (Frequency Response Analysis)

Although today the excitation process $\zeta(t)$ is usually assumed to be stochastically according to (1.6) we have a short look to the deterministic case of a sinusoidal excitation (1.5). The the inhomogeneous part of (3.1) is rewritten[13,16] as

$$\sum_{i=1}^{r} \underline{b}_i\, \zeta(t-t_i) = \sum_{i=1}^{r} \underline{b}_i\, e \sin\omega(t-t_i) = \underline{b}_c \cos\omega t + \underline{b}_s \sin\omega t \tag{3.2}$$

$$= \underline{b}\, e^{i\omega t} + \bar{\underline{b}}\, e^{-i\omega t} \tag{3.3}$$

where

$$\underline{b}_c = -e \sum_{i=1}^{r} \underline{b}_i \sin\omega t_i\ , \quad \underline{b}_s = e \sum_{i=1}^{r} \underline{b}_i \cos\omega t_i\ ,$$
$$\underline{b} = \frac{1}{2}(\underline{b}_c - i\,\underline{b}_s). \tag{3.4}$$

The real trigonometric description (3.2) is represented in (3.3) in a complex notation for a convenient computation later on. In an asymptotically stable system (3.1) the steady-state solution of (3.1) is characterised in a similar manner by

$$\underline{x}(t) = \underline{g}_c \cos\omega t + \underline{g}_s \sin\omega t = \underline{g}\, e^{i\omega t} + \bar{\underline{g}}\, e^{-i\omega t}\ , \tag{3.5}$$

$$\underline{g} = \frac{1}{2}(\underline{g}_c - i\,\underline{g}_s)\ ; \tag{3.6}$$

where the complex frequency response vector $\underline{g} = \underline{g}(\omega)$ is derived from

$$\underline{g}(\omega) = \underline{F}(\omega)\,\underline{b}, \tag{3.7}$$

where

$$\underline{F}(\omega) = (i\omega\underline{I} - \underline{A})^{-1} \tag{3.8}$$

is the n × n - frequency response matrix of the system. Each state variable $x_i(t)$, $i = 1, \ldots, n$, of the harmonic response (3.5) shows an amplitude a_i and a phase angle ψ_i:

$$x_i(t) = a_i \cos(\omega t - \psi_i), \tag{3.9}$$

$$a_i = \sqrt{g_{ci}^2 + g_{si}^2}, \quad \tan\psi_i = \frac{g_{si}}{g_{ci}}.$$

Obviously, the frequency response is completely characterized by the complex vector $\underline{g}(\omega)$ (3.7).

The steady-state response of a vehicle vibration variable (1.7) or of a perception variable (1.8) is determined in the same way:

$$\begin{aligned} w_k(t) &= w_c \cos\omega t + w_s \sin\omega t = a_{wk}\cos(\omega t - \psi_{wk}), \\ \overline{w}_k(t) &= \overline{w}_c \cos\omega t + \overline{w}_s \sin\omega t = a_{\overline{wk}}\cos(\omega t - \psi_{\overline{wk}}), \end{aligned} \tag{3.10}$$

where

$$w_c = \underline{c}_k^T \underline{g}_c - e \sum_{i=1}^{r} d_{ki} \sin\omega t_i, \quad w_s = \underline{c}_k^T \underline{g}_s + e \sum_{i=1}^{r} d_{ki} \cos\omega t_i, \tag{3.11}$$

$$(\overline{w}_c - i\,\overline{w}_s) = \alpha\, \underline{h}_k^T (i\omega\underline{I} - \underline{F}_k)^{-1}\, \underline{g}_k\, (w_c - i\,w_s) \tag{3.12}$$

(d_{ki} is defined by $\underline{d}_k^T \underline{\xi}(t) = \sum_{i=1}^{r} d_{ki}\zeta(t - t_i)$). Comparing (3.12) with (1.10) the complex frequency response function relating an objective variable to a subjective perception variable is given by

$$f_k(\omega) = \underline{h}_k^T (i\omega\underline{I} - \underline{F}_k)^{-1}\, \underline{g}_k\,. \tag{3.13}$$

The main numerical problem of this classical vibration analysis is the computation of the frequency response vector (3.7) If the system matrix $\underline{A}$ depend itself by the excitation frequency ω (that may arise in railway problems considering lateral motions) then $\underline{g}(\omega)$ is determined by solving the algebraic complex linear equation

$$(i\omega \underline{I} - \underline{A})\ \underline{g}(\omega) = \underline{b} \qquad (3.14)$$

for every frequency ω of interest. Suitable computer programs are available[17]. If however the matrix $\underline{A}$ is independent on ω (that is the usual case) a more efficient algorithm is applicable:

$$\underline{g}(\omega) = \sum_{j=1}^{n} \frac{1}{i\omega - \lambda_j}\ \underline{x}_j\ (\ \underline{y}_j^T\ \underline{b}) \ . \qquad (3.15)$$

Here, λ_j denotes the eigenvalues of $\underline{A}$ (we assume that the eigenvalues are different) and $\underline{x}_j$, $\underline{y}_j$ represent the corresponding right and left eigenvectors. Although in a first step the complete eigenvalue/eigenvector-problem has to be solved [14,15], the total computation time is reduced to about 10% of the time algorithm (3.14) if the frequency response is required in a large frequency region.

3.3 Stochastic Excitation (PSD Analysis, Covariance Analysis)

A more realistic excitation of vehicles is represented by a Gaussian stochastic process than by a harmonic time function. In a part of this book by Popp stochastic models for the guideway roughness are discussed. In the following we assume that the excitation ζ(t) at a single contact of the vehicle with its environment (guideway, but also stochastic influences of cross-wind or other stochastic effects) is characterized either by a (measured) power spectral density (PSD) $S_\zeta(\omega)$ (1.9) or by a shape filter (1.6).

It should be mentioned that the introduction of a shape filter is only easily done in relatively simple cases if the dimension of the filter can be chosen very low. In general, e.g. if a power spectral density matrix $\underline{S}(\omega)$ of a multi-dimensional coloured noise process is given by measurements, then the determination of an appropriate shape filter may be difficult. In this case a first agorithm for a spectral factorization of $\underline{S}(\omega) = \underline{H}(\omega)\ \underline{H}^T(-\omega)$ and a second algorithm of realizing the frequency response matrix H(ω) by a shape-filter-matrix-triple are required. Systematic solutions of these problems are described in [18,19] and more recently by Goßmann[20]. Therefore, even for more complicated stochastic processes both representations, PSD as well as shape filter, are assumed to be available.

In the following we shall describe two methods to analyse the stochastic response of vehicle systems to stochastic excitations. These two methods are the power spectral density analysis and the covariance analysis. The first method is performed in the frequency domain while the second method is a time domain approach. The objective of both methods consists in the determination of the mean values and the variances of interesting vibration and perception variables.

Usually the mathematical model of the vehicle dynamics is represented such that $\underline{x}(t) \equiv \underline{0}$ is an equilibrium position of the system without excitation. As the stochastic excitation is assumed to have vanishing mean value,

$$E\{\underline{\xi}(t)\} = \underline{0}\ , \tag{3.16}$$

it follows immediately

$$E\{\underline{x}(t)\} = \underline{0}\ . \tag{3.17}$$

Therefore, in the following we are essentially interested in calculating the variances

$$\sigma_{wk}^2 = E\{w_k^2(t)\}\ ,\quad \sigma_{\overline{w}k}^2 = E\{\overline{w}_k^2(t)\} \tag{3.18}$$

of the vibration and the perception variables (1.7) and (1.8).

3.3.1 Power Spectral Density Analysis

The basis of the PSD analysis are the fundamental relations

$$\underline{S}_x(\omega) = \underline{F}(\omega)\ \underline{B}\ \underline{S}_\xi(\omega)\ \underline{B}^T\underline{F}^T(-\omega) \tag{3.19}$$

and

$$\underline{P}_x := E\{\underline{x}(t)\ \underline{x}^T(t)\} = \int_{-\infty}^{\infty} \underline{S}_x(\omega)\ d\omega \tag{3.20}$$

The PSD matrix $\underline{S}_x(\omega)$ of the stationary vibration response $\underline{x}(t)$ is obtained from the PSD matrix $\underline{S}_\xi(\omega)$ of the stationary stochastic excitation $\underline{\xi}(t)$ by multiplying this matrix from left by the frequency response matrix and from right by the conjugate transposed frequency response matrix. While the PSD gives information about the density of the frequency spectrum in the stationary stochastic signal, in technical applications the variances, or more general the covariance matrix $\underline{P}_x$, is essential. The variances allow statements about the probability of the vibration amplitudes to remain under certain bounds. For example, from the theory of Gaussian stochastic processes well known confidence intervals are

$$\begin{aligned} \Pr[|w_k(t)| < \sigma_{wk}] &= 0{,}6827 \triangleq 68{,}27\,\%\,, \\ \Pr[|w_k(t)| < 2\sigma_{wk}] &= 0{,}9545 \triangleq 95{,}45\,\%\,, \\ \Pr[|w_k(t)| < 3\sigma_{wk}] &= 0{,}9973 \triangleq 99{,}73\,\%\,. \end{aligned} \tag{3.21}$$

The relation (3.20) is a special case of the inverse Fourier transform of the PSD to obtain correlation functions (for details compare chapter 9 in Müller-Schiehlen[13]).

Applying (3.19) and (3.20) to the analysis of vehicle dynamics the variances (3.18) have to be determined. The PSD matrix $\underline{S}_\xi$ of the excitation process follows from (1.9) by

$$\underline{S}_\xi(\omega) = [S_{ij}(\omega)]\,, \quad S_{ij}(\omega) = e^{-i\omega(t_i - t_j)}\, S_\zeta(\omega) \tag{3.22}$$

because of the time-delayed inputs. From (3.19) and (1.7) the PSD of $w_k(t)$ is represented by

$$S_{wk}(\omega) = [\underline{c}_k^T\, \underline{F}(\omega)\, \underline{B} + \underline{d}_k^T]\underline{S}_\xi(\omega)[\,\underline{B}^T\, \underline{F}^T(-\omega)\underline{c}_k + \underline{d}_k] \tag{3.23}$$

implying

$$\sigma_{wk}^2 = \int_{-\infty}^{\infty} S_{wk}(\omega)\, d\omega\,. \tag{3.24}$$

Corresponding to (1.11) and (3.13) it is also obtained

$$S_{wk}(\omega) = \alpha^2 |\underline{h}_k^T(i\omega\underline{I} - \underline{F}_k)^{-1}\underline{g}_k(\omega)|^2\, S_{wk}(\omega) = \alpha^2 |f_k(\omega)|^2\, S_{wk}(\omega)\,, \tag{3.25}$$

$$\sigma_{wk}^2 = \int_{-\infty}^{\infty} S_{wk}(\omega)\, d\omega\,. \tag{3.26}$$

The evaluation of (3.24) and (3.26) requires therefore the integration of the PSD over a theoretically infinite interval. Because of the decreasing of the PSD with increasing frequency and because of the property of an even function we evaluate the approximations

$$\sigma_{wk}^2 \approx 2\int_0^{\omega_1} S_{wk}(\omega)\, d\omega\,, \quad \sigma_{\overline{wk}}^2 \approx 2\int_0^{\omega_1} S_{\overline{wk}}(\omega)\, d\omega \tag{3.27}$$

where ω_1 will be a suitable upper bound. Nevertheless, the integration (e.g. by Simpson's rule) needs many points of support, and that implies the determination of the PSD's (3,23) and (3.25) for many points ω_i. Here we can apply again the numerical methods (3.14) or (3.15). However, the numerical evaluation of the power spectral density analysis is laborious.

3.3.2 Covariance Analysis

In contrast to the PSD analysis the time domain covariance analysis is more suitable to determine the variances. This method is a result of the good experience and the good numerical know-how of the Kalman-Bucy filter theory in stochastic dynamical systems. The main advantage consists in the direct computation of the covariance matrix $\underline{P}_x$ by the algebraic Liapunov matrix equation without calculating PSD's. E.g. for a problem with one contact point, $\underline{B}\,\underline{\xi}(t) = \underline{b}_1\zeta(t)$, the procedure of covariance analysis is performed using the shape filter (1.6) of the vehicle excitation, the linear equation (3.1) of vehicle dynamics, and the shape filter (1.8) of human perception resulting in a Liapunov matrix equation

$$\begin{bmatrix} \underline{F} & \underline{0} & \underline{0} \\ \underline{b}_1\,\underline{h}^T & \underline{A} & \underline{0} \\ \underline{g}_k\,d_{k1}\,\underline{h}^T & \underline{g}_k\,\underline{c}_k^T & \underline{F}_k \end{bmatrix} \underline{P} + \underline{P} \begin{bmatrix} \underline{F} & \underline{0} & \underline{0} \\ \underline{b}_1\,\underline{h}^T & \underline{A} & \underline{0} \\ \underline{g}_k\,d_{k1}\,\underline{h}^T & \underline{g}_k\,\underline{c}_k^T & \underline{F}_k \end{bmatrix}^T +$$

$$+ \begin{bmatrix} q_w\,\underline{g}\,\underline{g}^T & \underline{0} & \underline{0} \\ \underline{0} & \underline{0} & \underline{0} \\ \underline{0} & \underline{0} & \underline{0} \end{bmatrix} = \underline{0}\ . \qquad (3.28)$$

The matrix $\underline{P}$ includes all covariances of the shape filters as well as of the system response:

$$\underline{P} = \begin{bmatrix} \underline{P}_v & \underline{P}_{xv}^T & \underline{P}_{kv}^T \\ \underline{P}_{xv} & \underline{P}_x & \underline{P}_{kx}^T \\ \underline{P}_{kv} & \underline{P}_{kx} & \underline{P}_k \end{bmatrix}\ . \qquad (3.29)$$

Hence, the variances (3.18) of the vibration variable $w_k(t)$ and of the perception variable $\overline{w}_k(t)$ follows as

$$\sigma^2_{wk} = \underline{c}_k^T \underline{P}_x \underline{c}_k + 2d_{k1} \underline{c}_k^T \underline{P}_{xv} \underline{h} + d_{k1}^2 \underline{h}^T \underline{P}_v \underline{h} , \tag{3.30}$$

$$\sigma^2_{\overline{wk}} = \alpha^2 \underline{h}_k^T \underline{P}_k \underline{h}_k . \tag{3.31}$$

The covariance analysis was suggested in a series of papers[13,16,21,22]. The key of this method is the Liapunov matrix equation (3.28) and its numerical solution. That is effectively implemented by algorithms of Smith[23], Bartels-Stewart[24] or others[25]. The algorithms are much faster than the computation by the PSD analysis[20].

The covariance analysis was generalized to vehicle problems with successive contact points[26]. Equation (3.28) has to be modified by some additional terms. E.g. for a problem with two contact points, $\underline{B}\xi(t) = \underline{b}_1 \zeta(t) + \underline{b}_2 \zeta(t-t_2)$, the covariance matrix $\underline{P}$ of the global system response is obtained by the following Liapunov matrix equation

$$\begin{bmatrix} \underline{F} & \underline{O} & \underline{O} & \underline{O} \\ \underline{O} & \underline{F} & \underline{O} & \underline{O} \\ \underline{b}_1 \underline{h}^T & \underline{b}_2 \underline{h}^T & \underline{A} & \underline{O} \\ \underline{g}_k d_{k1} \underline{h}^T & \underline{g}_k d_{k2} \underline{h}^T & \underline{g}_k \underline{c}_k^T & \underline{F}_k \end{bmatrix} \begin{bmatrix} \underline{P}_v & \underline{P}_{v21}^T & \underline{P}_{xv1}^T & \underline{P}_{kv1}^T \\ \underline{P}_{v21} & \underline{P}_v & \underline{P}_{xv2}^T & \underline{P}_{kv2}^T \\ \underline{P}_{xv1} & \underline{P}_{xv2} & \underline{P}_x & \underline{P}_{kx}^T \\ \underline{P}_{kv1} & \underline{P}_{kv2} & \underline{P}_{kx} & \underline{P}_k \end{bmatrix} +$$

$$\begin{bmatrix} \underline{P}_v & \underline{P}_{v21}^T & \underline{P}_{xv1}^T & \underline{P}_{kv1}^T \\ \underline{P}_{v21} & \underline{P}_v & \underline{P}_{xv2}^T & \underline{P}_{kv2}^T \\ \underline{P}_{xv1} & \underline{P}_{xv2} & \underline{P}_x & \underline{P}_{kx}^T \\ \underline{P}_{kv1} & \underline{P}_{kv2} & \underline{P}_{kx} & \underline{P}_k \end{bmatrix} \begin{bmatrix} \underline{F} & \underline{O} & \underline{O} & \underline{O} \\ \underline{O} & \underline{F} & \underline{O} & \underline{O} \\ \underline{b}_1 \underline{h}^T & \underline{b}_2 \underline{h}^T & \underline{A} & \underline{O} \\ \underline{g}_k d_{k1} \underline{h}^T & \underline{g}_k d_{k2} \underline{h}^T & \underline{g}_k \underline{c}_k^T & \underline{F}_k \end{bmatrix}^T +$$

$$+ q_w \begin{bmatrix} \underline{gg}^T & e^{\underline{F}t_2} \underline{gg}^T & \underline{O} & \underline{O} \\ \underline{gg}^T e^{\underline{F}^T t_2} & \underline{gg}^T & \underline{gg}^T \underline{S}^T & \underline{O} \\ \underline{O} & \underline{Sgg}^T & \underline{O} & \underline{O} \\ \underline{O} & \underline{O} & \underline{O} & \underline{O} \end{bmatrix} = \underline{O} \tag{3.32}$$

where $\underline{S}$ results from

$$\underline{A}\,\underline{S} - \underline{S}\,\underline{F} = e^{\underline{A}t_2}\,\underline{b}_1\underline{h}^T - \underline{b}_1\underline{h}^T e^{\underline{F}t_2} \quad . \tag{3.33}$$

The interesting variances satisfy the relations

$$\sigma^2_{wk} = \underline{c}_k^T \underline{P}_x \underline{c}_k + 2d_{k1}\underline{c}_k^T \underline{P}_{xv1}\underline{h} + 2d_{k2}\underline{c}_k^T \underline{P}_{xv2}\underline{h} + \underline{h}^T[(d^2_{k1} + d^2_{k2})\,\underline{P}_v + 2d_{k1}d_{k2}\underline{P}_{v21}]\underline{h}\ , \tag{3.34}$$

$$\sigma^2_{\overline{wk}} = \alpha^2\,\underline{h}_k^T\,\underline{p}_k\,\underline{h}_k \tag{3.35}$$

While (3.35) is formally identical with (3.31) the computation of σ^2_{wk} additionally includes the effects of the second contact point.

Although the writing of above equations seems to be extensive, the solution is evaluated by the above-metioned algorithms quite fast. Since the variances are the essential results, the PSD method goes a detour while the covariance analysis directly determines the desired values by well-established, numerically subroutines of linear algebra.

The time-domain analysis of vehicle systems with successive contact points was firstly discussed[27] in the case of white noise excitation. A coloured noise example was published considering a four-degree-of-free-dom-model of the carbody and axles vibrations of an automobile[28]. There it was shown that the computation including the time-delay between front and rear excitation is necessary to get correct results; neglecting the time-delay leads to erreneous results especially near the seat position of the car.

It should be mentioned that an essential advantage of the covariance analysis, compared with PSD analysis, consists in its generalization to instationary processes. But this field will not be presented here.

4 NONLINEAR SYSTEM ANALYSIS

Exact solutions of nonlinear differential equations are quite rare. In general, there does not exist systematic methods to analyse exactly the behaviour of nonlinear dynamical systems. Therefore, approximation techniques are usually applied to obtain approximate solutions describing the unknown exact solution more or less correct. Most effective approximation methods are certain linearization techniques. Taylor series expansion is applicable if the nonlinearities are small; then usually instead of the nonlinear system a linearized system is discussed using the methods of section 3. More interesting are the techniques of quasilinearization: harmonic linearization in the case of periodic solutions, and statistical linearization in the case of stochastic excitations. These two methods will be discussed because of their importance in vehicle dynamics. E.g. rail vehicle hunting was investigated[29-33] by the method of harmonic linearization. The method of statistical linearization was successfully applied to rail vehicle analysis[34] and to the analysis of automobiles[35]. Therefore, for the analysis of nonlinear vehicle dynamics both quasilinearization techniques have to be introduced.

4.1 Harmonic Linearization

With respect to (1.1), (1.4), (1.5) and (3.2) we assume that the vehicle dynamics are described by the nonlinear vector differential equation

$$\dot{\underline{x}}(t) = \underline{A}\,\underline{x}(t) + \underline{f}(\underline{x}(t)) + \underline{b}_c\cos\omega t + \underline{b}_s\sin\omega t \;. \tag{4.1}$$

If the harmonic excitation drives the system we will look for a periodic solution $\underline{x}(t) = \underline{x}(t+T)$ with the same period $T = \frac{2\pi}{\omega}$ as the excitation period. In the case of a vanishing excitation ($\underline{b}_c = \underline{0}$, $\underline{b}_s = \underline{0}$) we will also look for a periodic solution $\underline{x}(t) = \underline{x}(t+T)$ ("limit cycle") where now the period $T = \frac{2\pi}{\omega}$ is unknown and has to be determined. Both problems arise in vehicle dynamics: the first one is typical for an automobile excitation by a harmonically waved road while the second one characterizes rail vehicle hunting.

The unknown periodic solution $\underline{x}(t)$ is approximated by a harmonic vector function with a shift term:

$$\underline{x}(t) \approx \underline{x}_a(t) = \frac{1}{\sqrt{2}}\underline{x}_o + \underline{x}_c\cos\omega t + \underline{x}_s\sin\omega t \;. \tag{4.2}$$

The constant vectors $\underline{x}_o$, $\underline{x}_c$ and $\underline{x}_s$, and the frequency ω if necessary, are unknown and have to be determined. It is remarked that for odd nonlinearities,

$$\underline{f}(\underline{x}) = -\underline{f}(-\underline{x}), \tag{4.3}$$

the shift term vanishes, $\underline{x}_o = \underline{0}$. The approximation technique consists in approximating the nonlinearity by a linear vector function with respect to (4.2),

$$\underline{f}(\underline{x}(t)) \approx \underline{F}_h \underline{x}(t) \;:\; \underline{e}(t) = \underline{f}(\underline{x}_a(t)) - \underline{F}_h \underline{x}_a(t), \tag{4.4}$$

such that the mean squared approximation error $\underline{e}(t)$ is minimized:

$$\int_0^T \underline{e}^T(t)\,\underline{e}(t)dt \rightarrow \underset{\underline{F}_h}{\text{minimum}}. \tag{4.5}$$

From (4.5) the gain matrix $\underline{F}_h$ is determined dependent on the unknown vectors $\underline{x}_o$, $\underline{x}_c$, $\underline{x}_s$ and the frequency ω:

$$\underline{F}_h \int_0^T \underline{x}_a(t)\,\underline{x}_a^T(t) = \int_0^T \underline{f}(\underline{x}_a(t))\,\underline{x}_a^T(t)dt. \tag{4.6}$$

Usually the solution $\underline{F}_h$ of (4.6) is non-unique; that allows to choose a proper solution (i.e. a solution which allows easy computation):

$$\underline{F}_h = \underline{F}_h(\underline{x}_o, \underline{x}_c, \underline{x}_s;\ \omega) . \tag{4.7}$$

By (4.4) we obtain an equivalent linear system

$$\dot{\underline{x}}(t) = [\underline{A} + \underline{F}_h]\,\underline{x}(t) + \underline{b}_c\cos\omega t + \underline{b}_s\sin\omega t \tag{4.8}$$

instead of (4.1). This is the key for the approximation. In the next step it is required that the approximation (4.2) satisfies (4.8):

$$-\omega\underline{x}_c\sin\omega t + \omega\underline{x}_s\cos\omega t = \tag{4.9}$$

$$= [\underline{A} + \underline{F}_h]\left(\frac{1}{\sqrt{2}}\underline{x}_o + \underline{x}_c\cos\omega t + \underline{x}_s\sin\omega t\right) + \underline{b}_c\cos\omega t + \underline{b}_s\sin\omega t .$$

Comparing the coefficients of the constant term, the sine- and cosine-functions nonlinear equations for the unknowns $\underline{x}_o$, $\underline{x}_c$, $\underline{x}_s$ are derived:

$$[\underline{A} + \underline{F}_h(\underline{x}_o, \underline{x}_c, \underline{x}_s;\ \omega)]\,\underline{x}_o = \underline{0}, \tag{4.10a}$$

$$\omega\underline{x}_s - [\underline{A} + \underline{F}_h(\underline{x}_o, \underline{x}_c, \underline{x}_s;\ \omega)]\,\underline{x}_c = \underline{b}_c, \tag{4.10b}$$

$$-\omega\underline{x}_c - [\underline{A} + \underline{F}_h(\underline{x}_o, \underline{x}_c, \underline{x}_s;\ \omega)]\,\underline{x}_s = \underline{b}_s . \tag{4.10c}$$

The equations (4.10) determine the approximate solution $\underline{x}_a(t)$. But the evaluation of this vector function requires the numerical solution of

(4.10). That can be done by a Newton-Raphson- or by a Fletcher-Powell-algorithm[36,37]. Two different cases have to be considered as mentioned above. If (4.1) represents an inhomogeneous system, i.e. the vehicle dynamics are driven by a harmonic excitation, then (4.10) is inhomogeneous, too, and the frequency ω is known. Here, (4.10a-c) consists of three nonlinear vector equations for the three unknown vectors $\underline{x}_o$, $\underline{x}_c$, $\underline{x}_s$. In the second case of a limit cycle, i.e. there exists a periodic solution without a harmonic excitation, the equations (4.10a-c) are homogeneous. To obtain a nontrivial solution the frequency has to be determined such that

$$\det[i\omega\underline{I} - (\underline{A} + \underline{F}_h(\underline{x}_o, \underline{x}_c, \underline{x}_s; \omega))] = 0. \tag{4.11}$$

It should be mentioned that from a theoretical point of view the accuracy of the approximation generally cannot be estimated. In very rare cases the method may fail[38]. Nevertheless, in practice the harmonic linearization technique is successfully applied.

4.2 Statistical Linearization

The method of statistical linearization is applied to the nonlinear system

$$\dot{\underline{x}}(t) = \underline{A}\,\underline{x}(t) + \underline{f}(\underline{x}(t)) + \underline{B}\,\underline{\xi}(t) \tag{4.12}$$

where $\underline{\xi}(t)$ is a stationary Gaussian stochastic process. The method is developed in a similar manner as the method of harmonic linearization. That means that the nonlinearity is approximated by a linear vector function with respect to an approximate solution

$$\underline{x}(t) \approx \underline{x}_a(t) = \underline{x}_o + \underline{x}_{st}(t) \tag{4.13}$$

where $\underline{x}_o$ is a constant vector (mean value of $\underline{x}_a$) and $\underline{x}_{st}(t)$ denotes a stationary Gaussian stochastic process with vanishing mean value and covariance matrix

$$\underline{P}_x = E\,\{\underline{x}_{st}(t)\,\underline{x}_{st}^T(t)\}\ : \tag{4.14}$$

$$\underline{f}(\underline{x}) \approx \underline{f}_o + \underline{F}_{st}\underline{x}_{st}\ :\ \underline{e}_{st}(t) = \underline{f}(\underline{x}_o + \underline{x}_{st}(t)) - \underline{f}_o - \underline{F}_{st}\underline{x}_{st}(t). \tag{4.15}$$

The mean vector $\underline{f}_o$ and the equivalent gain matrix $\underline{F}_{st}$ are determined such that the expectation of the squared error is minimized:

$$E\{\underline{e}_{st}^T(t)\,\underline{e}_{st}(t)\} \rightarrow \underset{\underline{f}_o,\,\underline{F}_{st}}{\text{minimum}} . \qquad (4.16)$$

That leads to the following results:

$$\underline{f}_o = \underline{f}_o(\underline{x}_o, \underline{P}_x) = E\{\underline{f}(\underline{x}_o + \underline{x}_{st}(t))\}, \qquad (4.17)$$

$$\underline{F}_{st} = \underline{F}_{st}(\underline{x}_o, \underline{P}_x) = E\{\underline{f}(\underline{x}_o + \underline{x}_{st}(t))\,\underline{x}_{st}^T(t)\}\,\underline{P}_x^{-1} . \qquad (4.18)$$

It has to be mentioned that the expectation values has to be evaluated with respect to the process $\underline{x}_{st}(t)$, i.e.

$$E\{[\,\bullet\,]\} = \int_{-\infty}^{\infty}\cdots\int_{-\infty}^{\infty}[\,\bullet\,]\,p(\underline{x})\,dx_1\ldots dx_n , \qquad (4.19)$$

$$p(\underline{x}) = \frac{1}{\sqrt{(2\pi)^n \det \underline{P}_x}}\exp\left(-\frac{1}{2}\underline{x}^T\underline{P}_x^{-1}\underline{x}\right) , \qquad (4.20)$$

where $\underline{P}_x$ denotes the unknown covariance matrix (4.14) of the solution process. Furthermore, the analogy to the harmonic linearization method will be more clear rewriting (4.18) as

$$\underline{F}_{st}\,E\{\underline{x}_{st}(t)\,\underline{x}_{st}^T(t)\} = E\{\underline{f}(\underline{x}_o + \underline{x}_{st}(t))\,\underline{x}_{st}^T(t)\} \qquad (4.21)$$

and comparing with (4.6). The averaging is performed in (4.6) by an integral over one period while in (4.21) the expectation operator is used. In both cases it is required that the correlation between nonlinearity and solution process is equal to the correlation between the equivalent linear description and solution process. At last we mention that for an odd nonlinear function (4.3) and for a vanishing mean value of the excitation the constant vectors vanish, too:

$$\underline{x}_o = \underline{0},\ \underline{f}_o = \underline{0} \text{ if } \underline{f}(\underline{x}) = -\underline{f}(-\underline{x}) \text{ and } E\{\underline{\xi}(t)\} = \underline{0} . \qquad (4.22)$$

A detailed discussion of the evaluation of the equivalent gain matrix (4.18) was represented by Müller-Popp-Schiehlen[35].

The equivalent linearized system is obtained by two equations: the one is a relation between mean values (shift vectors), the other is a linear differential equation for $\underline{x}_{st}(t)$:

$$\underline{A}\,\underline{x}_o + \underline{f}_o + \underline{B}\,E\{\underline{\xi}(t)\} = \underline{0} , \qquad (4.23)$$

$$\dot{\underline{x}}_{st}(t) = [\underline{A} + \underline{F}_{st}]\,\underline{x}_{st}(t) + \underline{B}[\underline{\xi}(t) - E\{\underline{\xi}(t)\}]. \qquad (4.24)$$

The calculation of $\underline{x}_o$ and $\underline{P}_x$ will be performed by the power spectral density analysis or by the covariance analysis as shown below.

4.3. Stability

The investigation of the stability behaviour of a nonlinear system is much more difficult than in the linear case. A general theory was developed by Liapunov, but its application is very cumbersome or unsatisfying. Therefore, the stability analysis is performed by above approximation techniques. Although the investigation of stability by approximate methods may fail in certain cases[38], reasonable results are usually obtained.

The equilibrium position $\underline{x}(t) \equiv \underline{0}$ of the system (4.1) without excitation ($\underline{f}(\underline{0}) = \underline{0}$) will be asymptotically stable, if the eigenvalues of the matrix $\underline{A} + \underline{F}_h$ have negative real parts also in the case $\underline{x}_o \to 0$, $\underline{x}_c \to \underline{0}$, $\underline{x}_s \to \underline{0}$. The solution of the system (4.1) with excitation will be bounded if the eigenvalues of $\underline{A} + \underline{F}_h$ have negative real parts in the case of solutions $\underline{x}_o$, $\underline{x}_c$, $\underline{x}_s$ of (4.10).

More complicated is the stability test for a limit cycle. Here we again discuss the eigenvalues of $\underline{A} + \underline{F}_h$, but we have to remember that the limit cycle solutions $\underline{x}_o$, $\underline{x}_c$, $\underline{x}_s$ leads to two eigenvalues $\lambda_{1,2} = \pm i\omega$ of $\underline{A} + \underline{F}_h$. Therefore we consider the eigenvalues of $\underline{A} + \underline{F}_h$ in a small neighbourhood of the limit cycle solution: $\mu\underline{x}_o$, $\mu\underline{x}_c$, $\mu\underline{x}_s$ with μ in a small region about $\mu = 1 : 1-\varepsilon < \mu < 1+\varepsilon$, $\varepsilon \to 0$. Then the limit cycle is called orbitally asymptotically stable if

$$\begin{aligned}
&\text{for } \mu = 1 &&: \lambda_{1,2} = \pm i\omega\,, && \operatorname{Re}\lambda_i < 0,\ i = 3, \ldots, n,\\
&\text{for } 1-\varepsilon < \mu < 1 &&: \operatorname{Re}\lambda_{1,2} > 0, && \operatorname{Re}\lambda_i < 0,\ i = 3, \ldots, n,\\
&\text{for } 1 < \mu < 1+\varepsilon &&: \operatorname{Re}\lambda_{1,2} < 0, && \operatorname{Re}\lambda_i < 0,\ i = 3, \ldots, n.
\end{aligned} \tag{4.25}$$

The real parts of the critical eigenvalues have to decrease if μ increase while the remaining eigenvalues always have negative real parts.

In the stochastic case stability behaviour is checked by the eigenvalues of the matrix $\underline{A} + \underline{F}_{st}$. The system (4.24) shows a bounded covariance matrix if all eigenvalues of $\underline{A} + \underline{F}_{st}$ have negative real parts.

4.4 Sinusoidal Excitation

The problem of sinusoidal excitation of a nonlinear system (4.1) was completely solved by the harmonic linearization technique of section 4.1. To get a better agreement with the notion of the linear problem in section 3.2 we consider in more detail the case (4.3) of an odd nonlinearity. Then the constant shift vector $\underline{x}_o$ vanishes. The remaining equations (4.10b-c) can be combined by multiplying equation (4.10c) by the negative imaginary unit (-i) and adding equation (4.10c):

$$[i\omega \underline{I} - (\underline{A} + \underline{F}_h(\underline{x}_c, \underline{x}_s; \omega))](\underline{x}_c - i\underline{x}_s) = \underline{b}_c - i\underline{b}_s . \tag{4.26}$$

This equation is a generalization of (3.14), i.e.

$$\underline{b} = \frac{1}{2}(\underline{b}_c - i\underline{b}_s) , \quad \underline{g} = \frac{1}{2}(\underline{x}_c - i\underline{x}_s) \tag{4.27a}$$

and

$$\underline{F}(\omega) = [i\omega \underline{I} - (\underline{A} + \underline{F}_h(\underline{x}_c, \underline{x}_s; \omega))]^{-1} . \tag{4.27b}$$

The frequency response matrix (4.27b) depends in the linear case on the complex vector $\underline{g}(\omega)$ while in the linear case it does not. In control theory this behaviour is well-known by the describing function which correspond to (4.27b).

4.5 Stochastic Excitation

Based on the statistical linearization technique we apply the PSD analysis or the covariance analysis to the equivalent linear system (4.23), (4.24). To show the analogy to the linear problem, the simplified problem of an odd nonlinearity (4.3) and a vanishing mean value of the excitation, $E\{\underline{\xi}(t)\} = \underline{0}$, will be considered. Then the mean vector $\underline{x}_o$ vanishes, $\underline{x}_o = \underline{0}$, and only equation (4.24) gouverns the problem:

$$\dot{\underline{x}}_{st}(t) = [\underline{A} + \underline{F}_{st}(\underline{P}_x)]\, \underline{x}_{st}(t) + \underline{B}\underline{\xi}(t). \tag{4.28}$$

Comparing with the linear problem, the same algorithms can be applied if the system matrix $\underline{A}$ of (3.1) is replaced by the system matrix $\underline{A} + \underline{F}_{st}$ of (4.28). But the one difference is that the results depend on the unknown covariance matrix $\underline{P}_x$. That has to be calculated by either PSD or covariance analysis.

4.5.1 Power Spectral Density Analysis

The frequency response matrix of (4.28) is written as

$$\underline{F}(\omega) = \underline{F}(\omega;\ \underline{P}_x) = [i\omega\underline{I} - (\underline{A} + \underline{F}_{st}(\underline{P}_x))]^{-1} \quad . \tag{4.29}$$

Therefore, the PSD matrix (3.19) is now obtained by

$$\underline{S}_x(\omega) = \underline{S}_x(\omega;\ \underline{P}_x) = \underline{F}(\omega;\ \underline{P}_x)\underline{B}\underline{S}_\xi(\omega)\underline{B}^T\underline{F}^T(-\omega;\ \underline{P}_x) \ , \tag{4.30}$$

i.e. $\underline{S}_x(\omega;\ \underline{P}_x)$ depends on $\underline{P}_x$, too. Using (3.20) a nonlinear equation determines the covariance matrix:

$$\underline{P}_x = \int_{-\infty}^{\infty} \underline{S}_x(\omega;\ \underline{P}_x)d\omega \quad . \tag{4.31}$$

As the quantities are numerically evaluated, equation (4.31) has to be solved numerically, too. E.g. a simple algorithm is given by

$$\underline{P}_x^{(i+1)} = \int_{-\infty}^{\infty} \underline{S}_x(\omega;\ \underline{P}_x^{(i)})\ d\omega \ ,$$
$$\lim_{i\to\infty} \underline{P}_x^{(i)} = \underline{P}_x . \tag{4.32}$$

Realizing this procedure the computation time will increase very much. Therefore, the PSD analysis is not well suited in the nonlinear case. Only in the special case, that the PSD matrix only depends on one variance, say σ_{wk}^2, the amount of computer time can be reduced solving

$$\sigma_{wk}^{2\,(i+1)} = \int_{-\infty}^{\infty} S_{wk}(\omega;\ \sigma_{wk}^{2(i)})\ d\omega \ ,$$
$$\lim_{i\to\infty} \sigma_{wk}^{2(i)} = \sigma_{wk}^2 \tag{4.33}$$

where S_{wk} is defined such as (3.23) with $\underline{F}(\omega) = \underline{F}(\omega;\ \sigma_{wk}^2)$. In multidimensional problems with many degrees of freedom this simplification is very rare.

If $\underline{P}_x$ (or σ_{wk}^2) are known by (4.32) (or (4.33)) then the other interesting variances (3.18) are computed by (3.27) like in the linear case.

4.5.2 Covariance Analysis

The covariance analysis of section 3.3.2 can be directly applied replacing again $\underline{A}$ by $\underline{A} + \underline{F}_{st}(\underline{P}_x)$. For example, the Liapunov matrix equation

(3.28) reads now

$$\begin{bmatrix} \underline{F} & \underline{O} & \underline{O} \\ \underline{b}_1\underline{h}^T & \underline{A}+\underline{F}_{st}(\underline{P}_x) & \underline{O} \\ \underline{g}_k d_{k1}\underline{h}^T & \underline{g}_k\underline{c}_k^T & \underline{F}_k \end{bmatrix} \begin{bmatrix} \underline{P}_v & \underline{P}_{xv}^T & \underline{P}_{kv}^T \\ \underline{P}_{xv} & \underline{P}_x & \underline{P}_{kx}^T \\ \underline{P}_{kv} & \underline{P}_{kx} & \underline{P}_k \end{bmatrix} + \qquad (4.34)$$

$$+ \begin{bmatrix} \underline{P}_v & \underline{P}_{xv}^T & \underline{P}_{kv}^T \\ \underline{P}_{xv} & \underline{P}_x & \underline{P}_{kx}^T \\ \underline{P}_{kv} & \underline{P}_{kx} & \underline{P}_k \end{bmatrix} \begin{bmatrix} \underline{F} & \underline{O} & \underline{O} \\ \underline{b}_1\underline{h}^T & \underline{A}+\underline{F}_{st}(\underline{P}_x) & \underline{O} \\ \underline{g}_k d_{k1}\underline{h}^T & \underline{g}_k\underline{c}_k^T & \underline{F}_k \end{bmatrix}^T + \begin{bmatrix} q_w\underline{g}\,\underline{g}^T & \underline{O} & \underline{O} \\ \underline{O} & \underline{O} & \underline{O} \\ \underline{O} & \underline{O} & \underline{O} \end{bmatrix} = \underline{O}\,.$$

Because of the nonlinearity the relation (4.34) is a nonlinear matrix equation which has to be iteratively solved. In contrast to the PSD algorithm (4.32) this iteration algorithm is directly related to the unknown covariance matrix. The convergence of a suitable iteration procedure was proven[39]. Applications[35] have also shown that the iteration cycles can be interrupted very soon. Therefore, in problems of nonlinear vehicle dynamics the covariance analysis is superior to the power spectral density method.

This text is performed without examples and applications. For these it is refered to other parts of this book and to above mentioned literature.

5 PASSENGER RESPONSE TO VEHICLE MOTIONS

An extensive amount of research dealing with human sensitivity to vibration has been reported in the literature over the past view decades. The subjective response of an individual to an imposed vibration depends not only on the physiological and biomechanical response of his body but also on a number of psychological and environmental factors. It has revealed that human reaction to vibration is not only a function of the amplitudes, accelerations, and frequencies to the body but also of the direction and character of motion. Also the time during which the human body is exposed to vibration is important.

Therefore, the evaluation of vehicle ride quality is extremely complex. The problems and results in this field were recently summarized by a survey paper of Smith[40] as follows (there are also a lot of further references): "Human are relatively insensitive to small changes in vibration level, passenger perception is thus difficult to quantify. This does not reduce the need to know and understand the relationships between passenger perception and vibration levels, however, because vehicle design, construction and manufacturing costs are often influenced by allowable vibration levels. If permissible vibration levels were too high, ride quality would be unacceptable to many passengers. Human sensitivity to vibration is probably nonlinear: below certain levels, sensitivity is low, but at higher levels small changes will be noticeable and annoying. Much of the early work associated with ride quality focused in the definition of a perception, or discomfort, boundary. Single axis, sinusoidal tests required test subjects to indicate perception level as a function of excitation frequency. The significant variation in many of these studies is presumably attributable to the difficulty of controlling the many possible variables that affect perception, e.g. posture, psychological mood, expectation. After careful review of harmonic motion data available prior to 1975 an international standard for vibration exposure, including vehicle comfort, was adopted by the International Standards Organization[41].

The application of sinusoidal sensitivity boundaries to the determination of ride quality as represented by a broadband vibration spectrum has frequently utilized. Comparison of broadband and sinusoidal data on the basis of rms levels is possible only if the bandwidth of the filter used to process the broadband data is specified. The ISO standard thus specifies a one-third octave bandwidth, stating that the rms acceleration level within that bandwidth must be below the prescribed boundary specified for the center frequency of the band. This method relates broadband random and sinusoidal data, but the extent to which this relationship is valid for predicting passenger perception has not been established.

It is desirable for the vehicle designer that any ride quality criteria be defined such that a single number ride index is available to serve as an objective function in design tradeoffs. Such criteria have been proposed. E.g. the power absorbed by the passenger from the vehicle is used as a measure of the ride. In application of this technique, the square of the magnitude of the contact impedance effectiviely becomes a frequency weighting function for the calculation of a weighted mean square measure of the vibration. Additionally, the ISO standard suggests that an alternate method to independent consideration of individual frequency bands be the use of a frequency weighted index using ISO boundaries as weighting curves.

Various investigators have taken field data relating some proposed ride quality measures to subjective passenger response. These studies indicate agreement as well as some discrepancy with the ISO standard. Thus, although many criteria have been proposed for relating passenger acceptance and vibration characteristics, no general equation has been defined, if indeed such an equation can be defined. Criteria that now exist should therefore be used with care and judgment, especially when they are extropolated to conditions significantly different from those for which they were derived."

The difficulties defining single-valued measures of ride comfort are also expressed by the recent planning stage of a VDI-standard[42]. Only further experiments relating field data and quality measures will give more insight in the problem of rating passenger response. E.g. Scheibe[43] suggests to have much more regard to the effects of interruptions of vibration exposure and of impulses within random vibrations; also it is proposed to extend the most sensitive frequency region from 4-8 Hz, ISO and VDI standard, to 4-12 Hz.

What is actually the recommendation of ISO or VDI? Shortly it is summarized that ISO defines and gives numerical values for limits of exposure to vibrations transmitted from solid surfaces to the human body in the frequency range 1 to 80 Hz. These limits cover human sensitivity to vertical, lateral, fore and aft vibrations of a periodic, nonperiodic or random nature. The exposure times are ranging from 1 minute to 24 hours. Periodic excitation is evaluated by the rms acceleration amplitude, while broadband excitation by the rms acceleration levels measured through one third octave band filter. In the ISO guide, the human passengers are more sensitive to vibration frequencies in the 4-8 Hz range for vertical motion and 1-2 Hz range for fore and aft motion, and the human tolerance of vibration decreases in a characteristic way with increasing exposure time.

In the following a procedure is presented how a measure of human response can be computed which corresponds to the ISO and VDI standards[41,42] and which is within the scope of the mathematical methods described above. The explanation follows reference[28].

The objective measurable mechanical vibrations acting from the vehicle on the human body can be evaluated by scalar vibration variables (1.7). These mechanical vibrations are subjectively perceptible by man characterizing passenger response. As above mentioned numerous physiological investigations have shown that the human perception of vehicle vibration is approximately proportional to the acceleration and depends on the dynamics of human organs which may be modeled by low order systems. Furthermore, it is assumed[41,42] that the passenger response can be characterized by scalar perception variables $\bar{w}_k(t)$ depending on the position of passenger and on the objective vibration variables $w_k(t)$. Therefore, each scalar perception variable can be represented by a shape filter (1.8) where a mechanical variable (1.7) is the input. The shape filter may be given in the time domain using differential equations such as (1.8) or in the frequency domain using rational functions such as (3.13). The general form of the frequency response of a single-input/ single-output shape filter is represented by

$$f_k(\omega) = \underline{h}_k^T(i\omega\underline{I} - \underline{F}_k)^{-1}\underline{g}_k \tag{5.1}$$

$$= \frac{b_o + b_1(i\omega) + b_2(i\omega)^2 + \ldots + b_{r-1}(i\omega)^{r-1} + b_r(i\omega)^r}{a_o + a_1(i\omega) + a_2(i\omega)^2 + \ldots + a_{s-1}(i\omega)^{s-1} + a_s(i\omega)^s}, \; r<s.$$

Determining the shape filter, differential equations (1.8) or frequency response (5.1), we have to fix the position of the passenger. Here, the ride quality of the vehicle with respect to the longitudinal position of passenger is considered. Then the interesting objective mechanical vibration variable is given by the vertical accelerations a(t) at the seat

($w_k(t) = a(t)$). The corresponding perception variable is denoted by $\bar{a}(t)$ ($\bar{w}_k(t) = \bar{a}(t)$). According to standards[41,42] the frequency response of the shape filter is shown in Fig.1. However, this given frequency response can only be realized by a high order shape filter. For law order shape filters permissible deviations are also given in the standards as shown in Fig. 1.

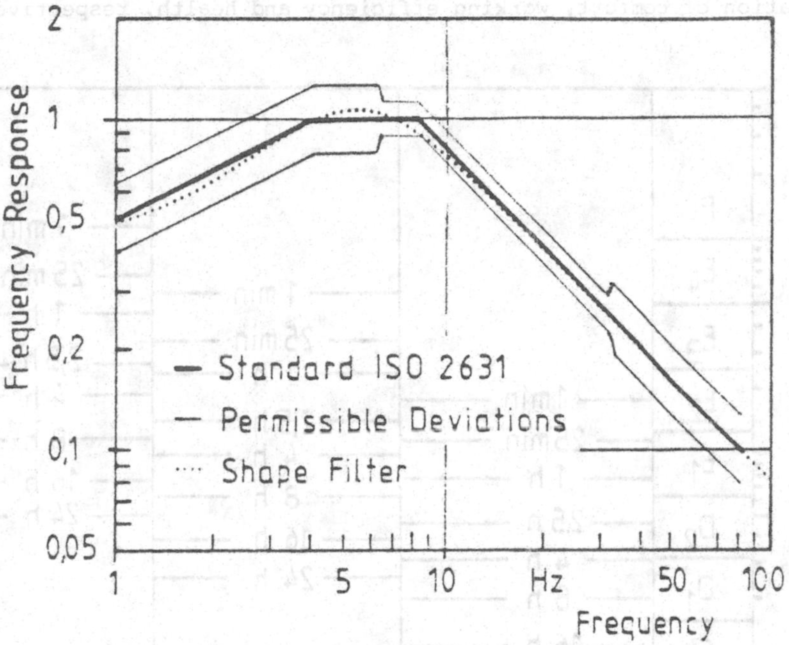

Fig. 1 Frequency response of a shape filter for passenger response to longitudinal vibrations.

For a second order shape filter the coefficients in (5.1) read as

$$b_o = 500\ s^{-2},\ b_1 = 50\ s^{-1},\ a_o = 1200\ s^{-2},\ a_1 = 50\ s^{-1} \tag{5.2}$$

and the normalizing constant in (1.10) is given by

$$\alpha = 20\ s^2 m^{-1} . \tag{5.3}$$

Fig. 1 shows that the second order shape filter fits very well. In the time domain the corresponding shape filter (1.8) is represented by

$$\underline{F}_k = \begin{bmatrix} 0 & 1 \\ -a_o & -a_1 \end{bmatrix}, \quad \underline{g}_k = \begin{bmatrix} 0 \\ 1 \end{bmatrix}, \quad \underline{h}_k = \begin{bmatrix} b_o \\ b_1 \end{bmatrix} . \tag{5.4}$$

By VDI standard[42] the standard deviation of the perseption variable is used to describe the perception

$$K = \sigma_{\bar{a}} \tag{5.5}$$

where $\sigma_{\bar{a}} = \sigma_{\overline{wk}}$ is evaluated by (3.26) or (3.35). Then [41], the tolerable exposure time can be found from Fig. 2 for the three main human criteria: preservation of comfort, working efficiency and health, respectively.

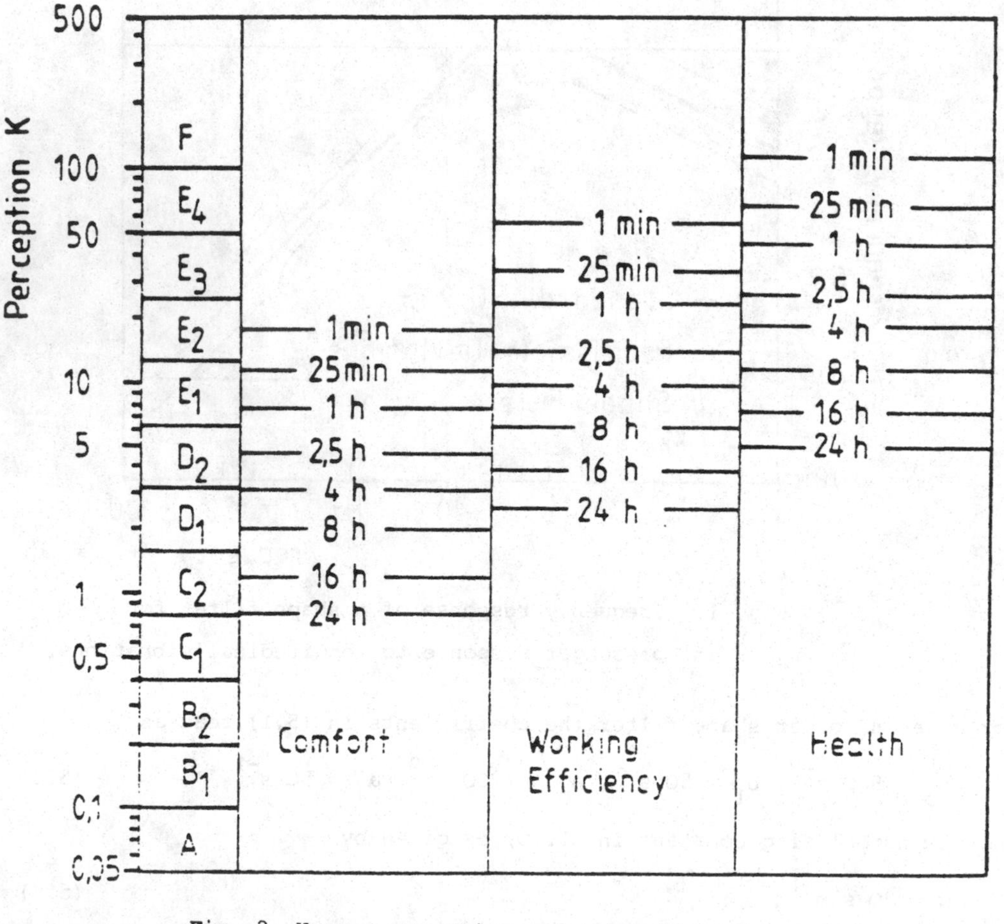

Fig. 2 Human perception and tolerable exposure time to vibration with respect to preservation of comfort, working efficiency and health.

Although the prediction or characterization of passenger response to a given vehicle vibration is controversial in literature because physical, psychological and physiological factors are involved, the present state of engineering allows a classification of ride quality as shown above. Perception shape filter or frequency response are introduced into the integrated investigation of guideway roughness, vehicle dynamics and passenger response. The uniform mathematical methods allow an effective analysis of the dynamics of high-speed vehicles and their ride qualities to passengers.

REFERENCES

1 Schmidt, G., Simulationstechnik, Oldenbourg, München-Wien, 1980.

2 Korn, G. A., Korn, T. M., Electronic analog and hybrid computers, Mc Graw-Hill, New York, 1972.

3 Korn, G. A., Wait, J. V., Digital continuous-system simulation, Prentice-Hall, Englewood Cliffs, 1978.

4 Lapidus, L., Seinfeld, J. H., Numerical solution of ordinary differential equations, Academy Press, New York-London, 1971.

5 Grigorieff, R. D., Numerik gewöhnlicher Differentialgleichungen, Vol. 1 and 2, Teubner, Stuttgart, 1972 and 1977.

6 Federl, U., Numerische Integration im Zeitbereich, in CCG Lehrgang O R 2.2, Simulationsmodell für spurgebundene Fahrzeuge, DFVLR, Oberpfaffenhofen, 1977.

7 Bendat, J. S., Piersol, A. G., Random data: Analysis and measurement procedures, Wiley, New York, 1971.

8 Beauchamp, K. G., Signal processing, George Allan & Unwin, London, 1973.

9 Brigham, E. O., The fast Fourier transform, Prentice-Hall, Englewood Cliffs, 1974.

10 Rabiner, L. R., Gold, B., Theory and application of digital signal processing, Prentice-Hall, Englewood Cliffs, 1975.

11 Stearns, S. D., Digital signal analysis, Haydon, Rochelle Park, 1975.

12 Yuen, C. K., Fraser, D., Digital spectral analysis, CSIRO-Pitman, Melbourne-London, 1979.

13 Müller, P. C., Schiehlen, W. O., Lineare Schwingungen, Akademische Verlagsgesellschaft, Wiesbaden, 1976.

14 Smith, B. T., Boyle, J. M., Garbow, B. S., Ikebe, Y., Klema, V. C., Moler, C. B., Matrix eigensystem routines - EISPACK guide, Lect. Notes in Computer Science, Vol. 6, Springer, Berlin, Heidelberg, New York, 1974.

15 Garbow, B. S., Boyle, J. M., Dongarra, J. J., Moler, C. B., Matrix eigensystem routines - EISPACK guide extension, Lect. Notes in Computer Science, Vol. 51, Springer, Berlin, Heidelberg, New York, 1977.

16 Müller, P. C., Schiehlen, W. O., Forced linear vibrations, CISM Courses and Lectures No. 172, Springer, Wien, New York, 1977.

17 Dongarra, J. J., Moler, C. B., Bunch, J. R., Stewart, G. W., LINPACK user's guide, Society of Industrial and Applied Mathematics (SIAM), Philadelphia, 1979.

18 Anderson, B. D. O., The inverse problem of stationary covariance generation, J. Stat. Physics 1, 133, 1969.

19 Schwarz, H., Mehrfachregelungen, Vol. 1, chapter IV.8 and Vol. 2, chapter VIII.5, Springer, Berlin, Heidelberg, New York, 1967 and 1971.

20 Goßmann, E., Kovarianzanalyse mechanischer Zufallsschwingungen bei Darstellung der mehrfachkorrelierten Erregungen durch stochastische Differentialgleichungen, Mitteilungen aus dem Institut für Mechanik, Ruhr-Universität Bochum, Nr. 24, February 1981.

21 Schiehlen, W., Kovarianzanalyse von Zufallsschwingungen, in VDI-Bericht 221, 105, VDI, Düsseldorf, 1974.

22 Müller, P. C., Lineare stationäre Zufallsschwingungen bei farbigem Rauschen, Z. Angew. Math. Mech. 58, T 164, 1978.

23 Smith, R. A., Matrix equation XA + BX = C, SIAM J. Appl. Math. 16, 198, 1968.

24 Bartels, R. H., Stewart, G. W., Solution of the matrix equation AX + BX = C, Comm. ACM 15, 820, 1972.

25 Koupan, A., Müller, P. C., Zur numerischen Lösung der Ljapunovschen Matrizengleichung $A^T P + PA = -Q$, Regelungstechnik 24, 167, 1976.

26 Müller, P. C., Popp, K., Kovarianzanalyse von linearen Zufallsschwingungen mit zeitlich verschobenen Erregerprozessen, Z. Angew. Math. Mech. 59, T 144, 1979.

27 Hedrick, J. K., Billington, G. F., Dreesbach, D. A., Analysis, design and optimization of high speed vehicle suspensions using state variable technique, J. Dyn. Syst. Meas. Control, Transact. ASME, Ser. G. 96, 193, 1974.

28 Müller, P. C., Popp, K., Schiehlen, W. O., Covariance analysis of nonlinear stochastic guideway - vehicle - systems, in Proc. 6th IAVSD Symp. Dynamics of Vehicles on Roads and Tracks, Berlin, Sept. 3-7, 1979, Willumeit, H.-P., Ed., Swets & Zeitlinger, Amsterdam - Lisse, 1980.

29 Cooperrider, N. K., Hedrick, J. K., Law, E. H., Malstrom, C. W., The application of quasilinearization techniques to the prediction of nonlinear railway vehicle response, Vehicle Systems Dynamics 4, no. 2-3, 1975.

30 Hull, R., Cooperrider, N. K., Influence of nonlinear wheel-rail contact geometry on stability of rail vehicles, Transact. ASME, J. Engng. Industry 99, no. 1, 1977.

31 Hannebrink, D. N., Lee, H. S. H., Weinstock, H., Hedrick, J. K., Influence of axle load, track gauge and wheel profile on rail vehicle hunting, Transact. ASME, J. Engng. Industry 99, no. 1, 1977.

32 Hauschild, W., The application of quasilinearization to the limit cycle behaviour of the nonlinear wheel-rail system, In Proc. 6th

IAVSD Symp. Dynamics of Vehicles on Roads and Tracks, Berlin, Sept. 3-7, 1979, Willumeit, H.-P., Ed., Swets & Zeitlinger, Amsterdam - Lisse, 1980.

33 Hauschild, W., Grenzzykelberechnung am nichtlinearen Rad-Schiene-System mit Hilfe der Quasilinearisierung, Dissertation D 83, Technische Universität Berlin, 1981.

34 Hedrick, J. K., Cooperrider, N. K., Law, E. H., The application of quasi-linearization techniques to rail vehicle analysis, Final Report for U.S. Department of Transportation, DOT - TSC - 902, 1978.

35 Müller, P. C., Popp, K., Schiehlen, W. O., Berechnungsverfahren für stochastische Fahrzeugschwingungen, Ing.-Arch. 49, 235, 1980.

36 Ortega, J. M., Rheinboldt, W. C., Iterative solution of nonlinear equations in several variables, Academic Press, New York, San Francisco, London, 1970.

37 Schwetlick, H, Numerische Lösung nichtlinearer Gleichungen, Oldenbourg, München-Wien, 1979.

38 Müller, P. C., Approximate methods for investigating the stability of limit cycles, General lecture, Euromech Colloquium 141 "Stationary Motions of Nonlinear Mechanical Systems", Twente University of Technology, Enschede, June 23-25, 1981.

39 Müller, P. C., Kovarianzanalyse von stationären nichtlinearen Zufallsschwingungen, Z. Angew. Math. Mech. 59, T 142, 1979.

40 Smith, C. C., Literature review - automobile ride quality, The Shock and Vibration Digest 12, no. 4, 15, 1980.

41 International Standard ISO 2631, Guide for the evaluation of human exposure to whole-body vibrations, International Organization for Standardization, 1974.

42 VDI - Richtlinie VDI 2057 (Entwurf), Beurteilung der Einwirkung mechanischer Schwingungen auf den Menschen, Blatt 1 (1975), Blatt 2 (1976), Blatt 3 (1977), Verein Deutscher Ingenieure, Düsseldorf.

43 Scheibe, W., Beurteilung von Belastung, Aktivität und Beanspruchung des Menschen bei kontinuierlicher und unterbrochener Exposition mit vertikalen Fahrzeugschwingungen in Simulations- und Feldexperimenten, Fortschr.-Ber. VDI - Z. Reihe 11, Nr. 31, Verein Deutscher Ingenieure, Düsseldorf, 1979.

HORIZONTAL MOTION OF AUTOMOBILES

Theoretical and Practical Investigations

Dr. Peter Lugner
Daimler-Benz AG
Postfach 202
Stuttgart 60, FRG

1 INTRODUCTION

When driving an automobile today the reliable function of the total system under various environmental conditions has become a matter of course. Experimentation and theoretical studies since 1886 have led to the automobile of today. The "functioning" of the dynamic system motor vehicle, - whether it is comfortable, easy to handle, racy, etc., can be explained by the interaction of the sub-systems. For example the engine and the drive train provide the energy for forward motion while the suspension systems are responsible for comfort and wheel guidance. The goal of this paper is to explain the dynamics of a vehicle as a connection of subsystems based on theoretical and experimental studies. The emphasis is placed on the motion of a passenger car on an even, dry roadway within the normal driving range.

1.1 The motor vehicle as a kinetic system and its driving range

Considered as an individual means of transportation the car offers the driver the possibility to influence the driving speed and the driving direction within the area of a specially prepared roadway, but not limited to a track. Ignoring the aerodynamic forces, the forces for changing the direction and speed are transferred solely by the adhesion of the contact surfaces of the tires.

The average driver can handle today's cars without any particular problems on dry roads - naturally he must be able to concentrate his attention primarily on the traffic The lateral accelerations encountered in a curve are in the range of approx. 0.3 g and therefore considerably below the permissible maximum values for dry roadways. The driver only becomes aware of vehicle dynamics limits in regard to the top speed and in the acceleration and climbing ability of the car.

However, the vehicle should also remain controllable for the driver even in extreme situations. Here the maximum possible adhesion between the tires and the roadway is of decisive importance for the physical limits. On a dry roadway decelerations of up to approximately 1 g can be achieved and in steady state turning lateral accelerations of approximately 0,8 g are possible. However, if the vehicle is braked in a curve, it is only possible to control a resulting acceleration of less than 0,8 g. A further limitation of the driving range results from the interaction of the motion resistance and the drive or may be also from the stability characteristics of the vehicle.

Figure 1.1 shows schematically such a driving range for a passenger car. The limits are not only affected by the driving conditions but also by vehicle parameters such as brake balance, the load, center of gravity above the roadway, type of drive.

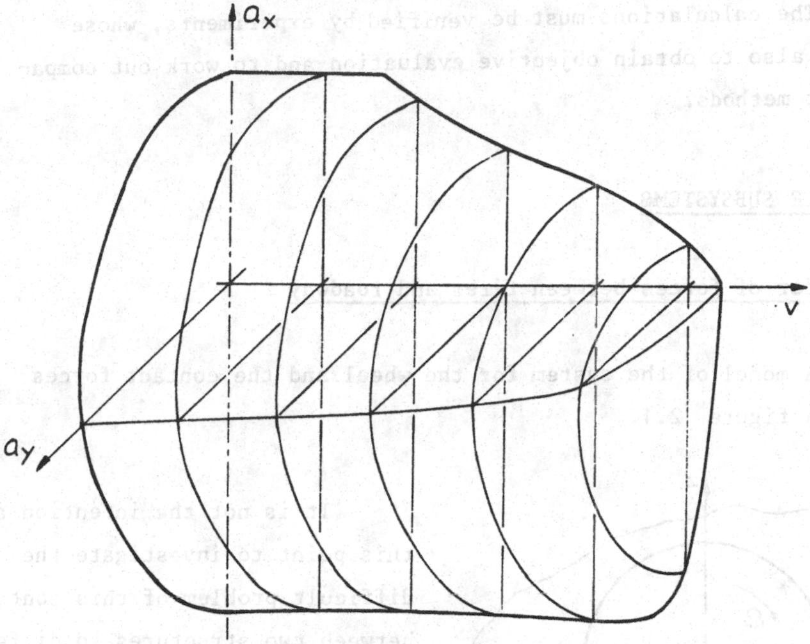

Fig. 1.1 Driving range schema of a passenger car, horizontal dry roadway

Expansion of this driving range can be achieved through special vehicle designs. E. g. more powerful engines, reduction of the motion resistance through lower weight, lower aerodynamic resistance as well as greater acceleration values through the use of special tires lead to sportscars. But simultaneously we have the problem to improve the vehicle handling and operation.

Calculations based on theoretical models of the vehicle or of substructures offer the possibility to show the basic relationships for this task and also to come to quantitative estimates. The vehicle and substructure models to be used depend upon the assignment and the requirements for the results. A universal vehicle model can neither be represented nor is it necessary.

The calculations must be verified by experiments, whose purpose is also to obtain objective evaluation and to work out comparative test methods.

2 VEHICLE SUBSYSTEMS

2.1 Transfer of forces between tires and roadway

A model of the system for the wheel and the contact forces is shown in figure 2.1.

It is not the intention at this point to investigate the difficult problem of this contact between two structures so different as an elastic tire with its complex geometry and material characteristics and the solid roadway, but only to illustrate the most important relationship for the vehicle dynamics. Transient phenomena, the reaction of the tire to rapid changes in kinetic values, will only be taken into consideration in a simplified way in respect to the lateral tire force and also only when an essential effect can be expected.

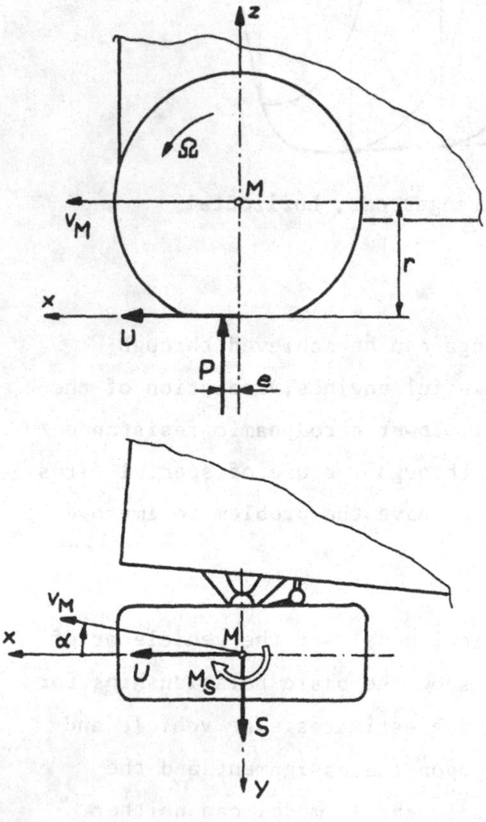

Fig. 2.1 Model of the vehicle wheel

2.1.1 Longitudinal force

For the longitudinal force U the longitudinal slip, determined by the relative speed between the wheel and the roadway, is decisive.

$$s_{LT} = \frac{r\Omega - v_M}{r\Omega} \quad \text{for driving, } r\Omega > v_M$$

$$s_{LB} = \frac{v_M - r\Omega}{v_M} \quad \text{for braking, } v_M > r\Omega \qquad (2.1)$$

In a slip curve typical for a passenger car tire - figure 2.2 - the longitudinal force coefficient

$$\mu_L(s_L) = \frac{U}{P} \qquad (2.2)$$

achieves its maximum at approximately 15 % longitudinal slip and decreases at increasing slip to the sliding coefficient $\mu_{L,g}$.

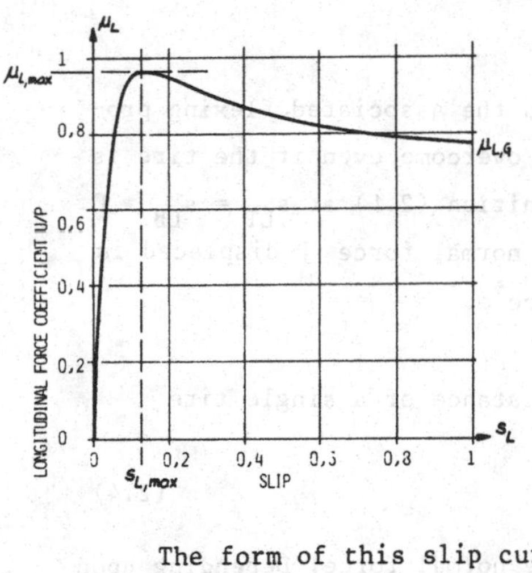

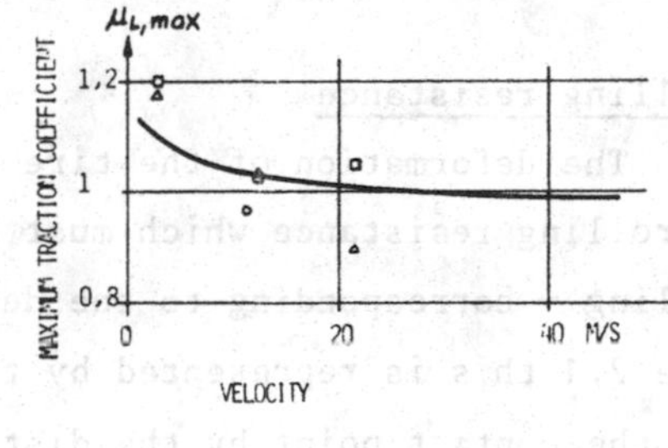

Fig. 2.2 Longitudinal friction coefficient U/P, dry surface

The form of this slip curve, which increases very fast to the maximum $\mu_{L,\,max}$ and shows a decrease afterwards, remains qualitatively unchanged for other road surfaces, tire parameters and driving speeds.

An essential reduction of μ_L can result from great side - slip angles of the tire - see fig. 2.4.

In normal driving the slip values remain small, e. g. less than 3 % on dry roadways. At extreme acceleration or braking torques the entire slip curve is run through up to full spinning or lock up of the wheels, usually connected with loss of steering ability or instability of the vehicle motion. The maximum traction coefficient is one of the limitations to the driving range.

A possible analytical approximation in the range up to $\mu_{L,\, max.}$ using the equation (2.3) is within the line magnitude of the curve in figure 2.2.

$$\frac{|U|}{P} = \mu_{L,\, max} \sum_{i=0}^{6} a_i \left(\frac{s_L - s_{L,max}}{s_{L,max}}\right)^i, \quad s_L \leqslant s_{L,\, max}$$

$$U < 0 \quad v_M > r\Omega, \qquad U > 0 \quad r\Omega > v_M \tag{2.3}$$

2.1.2 Rolling resistance

The deformation of the tire and the associated flexing produces a rolling resistance which must be overcome even if the tire is only rolling - corresponding to the definition (2.1) at $s_{LT} = s_{LB} = 0$. In figure 2.1 this is represented by the normal force P displaced in front of the contact point by the distance e.

The coefficient of rolling resistance of a single tire

$$f_R = \frac{e}{r} \tag{2.4}$$

can be considered as independent from the normal force. Depending upon the type of tire it remains nearly constant or shows only a small increase with speed up to a speed of approximately 150 km/h and then increases progressively - see part II, chapter 3.1.

2.1.3 Lateral force, aligning torque

Lateral forces must be exerted by the tires and wheels to change the vehicle direction of motion. This lateral guidance function is accomplished mainly by the wheel running at a slip angle, see figure 2.1. The lateral sliding velocity $v_M \cdot \tan\alpha$ leads to a force system which can be represented by the lateral force S and the aligning torque M_S at the tire contact point.

With the introduction of the pneumatic trail or tire offset n_S the aligning torque can also be described using the lateral force.

$$M_S = n_S \cdot S \tag{2.5}$$

For longitudinal force U = 0 the representation of the lateral force using the slip angle - figure 2.3 - represents an

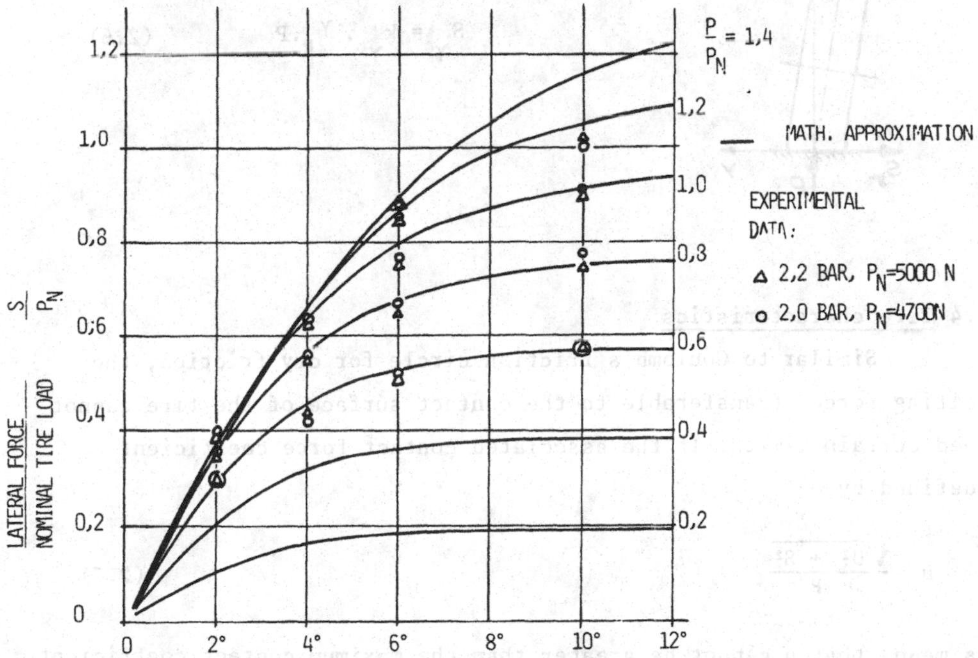

Fig. 2.3 Lateral force as a function of slip angle

analogy to the longitudinal force slip curve, figure 2.2. Also here the maximum coefficient of lateral force $\mu_{S,\,max.} = S_{max.}/P$ is reached at a relatively small slip angle. Above all the type of roadway surface and its condition, the type of tire and the tread depth again affect the magnitude of the lateral force, while the basic form of the curves remains unchanged. At increasing speed the value $\mu_{S,\,max.}$ decreases only slightly on dry pavement, less than 8 % from 40 km/h to 150 km/h.

An inclination of the wheel plane in relation to the plane of contact, as illustrated in the sketch below, also leads to a lateral force.The following can be used as a good approximation for the lateral force at the generally small camber angly γ_P at a slip angle of $\alpha = 0$:

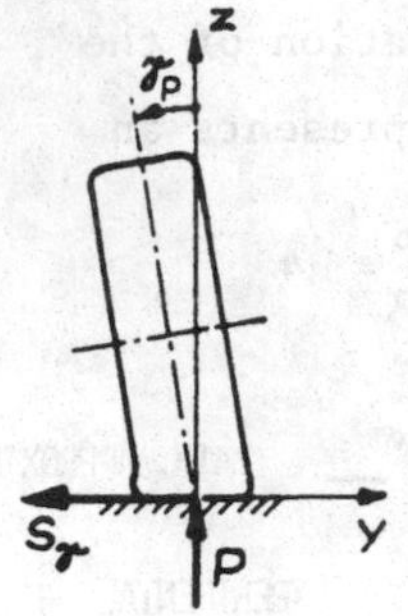

$$S_\gamma = k_\gamma \cdot \gamma_P \cdot P \qquad (2.6)$$

2.1.4 Tire characteristics

Similar to Coulomb's friction circle for dry friction, the resulting forces transferable to the contact surface of the tire cannot exceed certain limits. If the associated contact force coefficient is defined by

$$\mu = \frac{\sqrt{U^2 + S^2}}{P} \qquad (2.7)$$

this means that μ cannot be greater than the maximum contact coefficient μ_{max} (e. g. for the lateral force $S = 0$ μ_{max} is then equal to $\mu_{L,\,max}$). For locked wheels or skidding the resultant vector of the contact forces is in the opposite direction to the relative speed. The value of the resultant determined by the sliding coefficient μ_g is therefore smaller than their value at μ_{max}. Figure 2.4 represents such characteristics of a radial tire for a constant normal force. Possible limit curves for μ_{max}, μ_g where entered in the graph. These tire characteristics show primarily the force relationships to be expected in the regular driving range disregarding the curve branches which run inward beginning at their maximum U/P. In this situation the wheel cannot transfer any higher longitudinal force and will quickly lock up or spin at higher braking or driving torques.

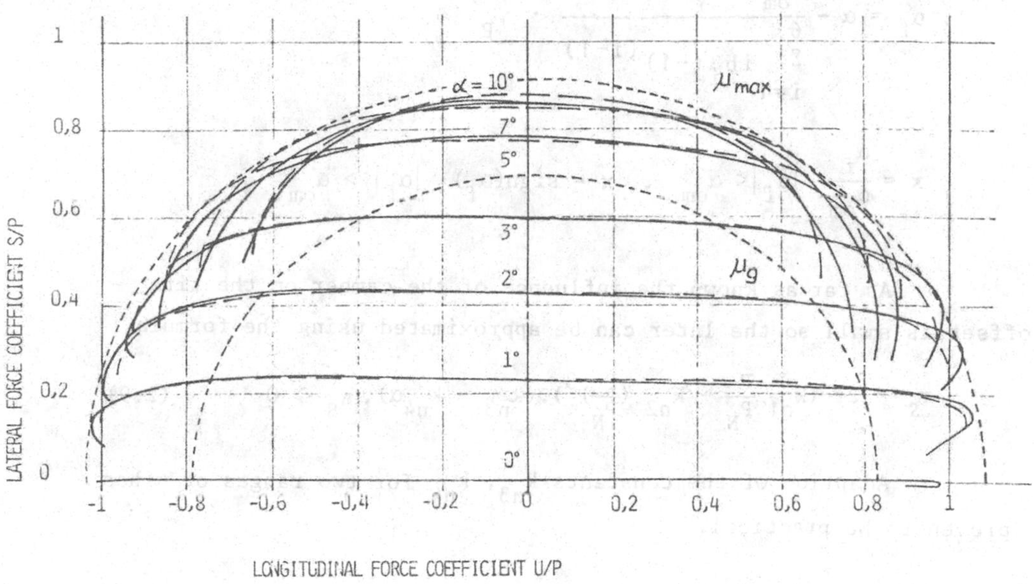

Fig. 2.4 Tire forces at constant vertical load

A possible mathematical approximation of such a tire characteristic is

$$S = f_S(\alpha_f,\ P,\ U)$$

$$f_S = \mu_{S,\ max}.P.\ (\sum_{i=0}^{6} b_i.(|x| - 1)^i).\mathrm{sign}(x).f_u\ (\alpha_f,\ P,\ U)$$

$$f_u = \frac{k_1 + \sqrt{1 - \left|\frac{U - Uo}{P.\mu_{L,\ max}.(1 - k_3|x|)}\right|^{\kappa 1}}}{k_1 + 1}\ (1 - k_2\frac{U - Uo}{P\mu_{L,\ max}})$$

$$\alpha_{om} = k_s\ (1 + k_d\ (\frac{P}{P_N})^{q_2}) \qquad (2.8)$$

$$\alpha_f = \alpha - \frac{\alpha_{om}\ .\ k_\gamma}{\sum_{i=1}^{6} ib_i(-1)^{(i-1)}}\ .\ \gamma_P$$

$$x = \frac{\alpha_f}{\alpha_{om}} \quad |\alpha_f| < \alpha_{om} \quad , \quad x = \mathrm{sign}(\alpha_f) \quad |\alpha_f| > \alpha_{om}$$

As far as known the influence of the camber on the tire offset is small so the later can be approximated using the formula

$$n_S = 2r\ (k_{n1}.\frac{P}{P_N} + k_{n2}.(\frac{P}{P_N})^3).(k_{n3} - k_{n4}\alpha),\ n_S \geqslant 0 \qquad (2.9)$$

Adaption of the constants k_{n3}, k_{n4} for two ranges of α has proven to be practical.

Extension of the driving range by increasing $\mu_{max.}$ leads to adaptation of the tires to specific applications: snow tires, tires with spikes, racing tires.

2.2 Aerodynamic forces

The aerodynamic problems have increased considerably in importance due to the higher travelling speeds as well as the increase in fuel price.

The air flowing around the vehicle produces forces and moments which can be illustrated as shown in figure 2.5. At driving speed v and wind velocity w_W the resultant air velocity vector and aerodynamic angle can be calculated as follows.

$$\vec{v}_{res} = -\vec{v} + \vec{w}_W$$

$$\tau = \arctan\left(-\frac{\vec{v}_{res}^{\,T}\cdot\vec{e}_y}{\vec{v}_{res}^{\,T}\cdot\vec{e}_x}\right) \qquad (2.10.)$$

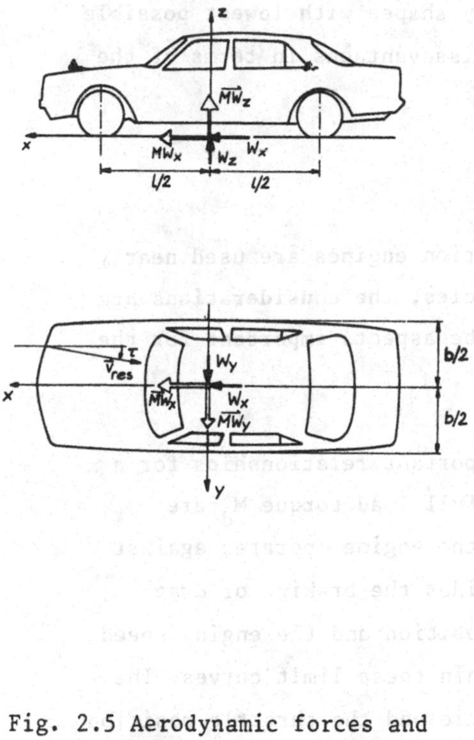

Fig. 2.5 Aerodynamic forces and moments

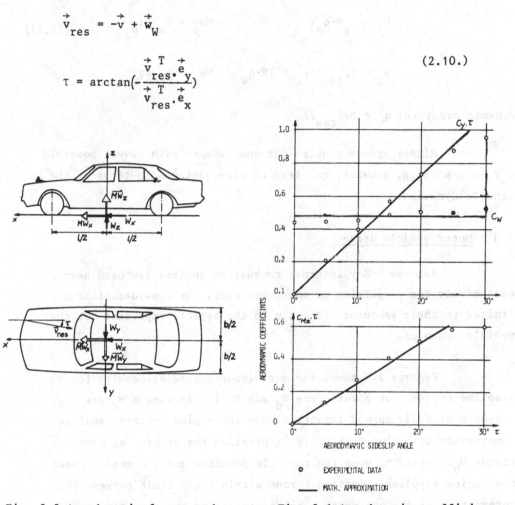

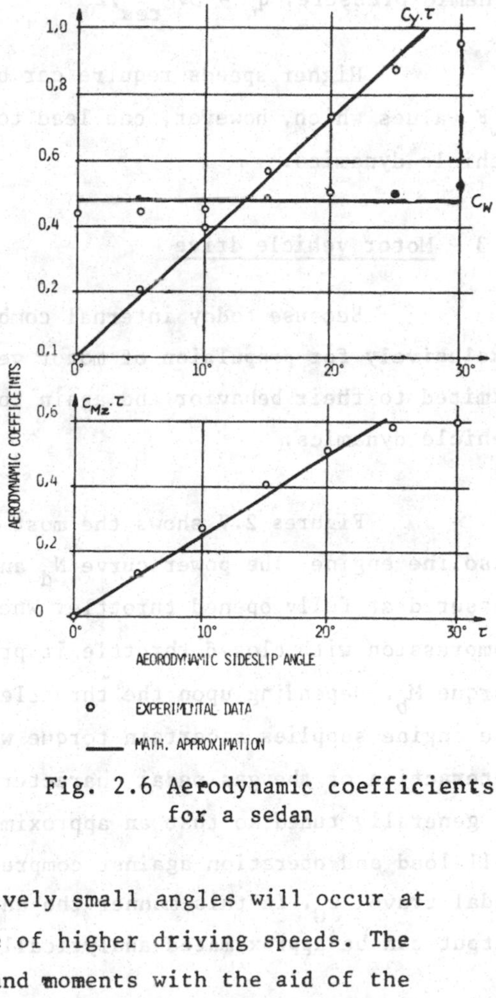

Fig. 2.6 Aerodynamic coefficients for a sedan

On the average, only relatively small angles will occur at the aerodynamically interesting range of higher driving speeds. The following description of the forces and moments with the aid of the frontal area F of the vehicle is applicable up to approx. $\tau = 20°$ - see example figure 2.6.

Aerodynamic drag: $W_L = -W_x = c_W \cdot F \cdot q_w \qquad MW_x = c_{Mx} F q_w \tau$

$$W_y = c_y F \cdot q_w \tau \qquad MW_y = c_{My} F q_w \tag{2.11}$$

$$W_z = (c_{z1} + c_{z2}\tau^2) F \cdot q_w \qquad MW_z = c_{Mz} F q_w \tau$$

dynamic pressure: $q_w = \rho v^2{}_{res}/2$

Higher speeds require car body shapes with lowest possible $c_w F$ values which, however, can lead to disadvantages in terms of the vehicle dynamics.

2.3 Motor vehicle drive

Because today internal combustion engines are used nearly exclusively for propulsion of motor vehicles, the considerations are limited to their behavior and again to the aspects important for the vehicle dynamics.

Figures 2.7 shows the most important relationships for a gasoline engine. The power curve N_d and full load torque M_d are measured at fully opened throttle. When the engine operates against compression with closed throttle it provides the braking or drag torque M_b. Depending upon the throttle position and the engine speed the engine supplies a certain torque within these limit curves. The interaction of the gas pedal characteristic and the throttle position is generally tuned so that an approximately linear graduation between full load and operation against compression results depending upon the pedal travel u_G. In this manner the engine torque M_M and the engine output can be approximated analytically as follows

$$M_M = \frac{u_G \cdot M_d}{u_{G,max}} + \left(1 - \frac{u_G}{u_{G,max}}\right) M_b \tag{2.12}$$

$$N_M = M_M\,(\omega_M,\, u_G)\,.\,\omega_M, \qquad \omega_M = \frac{n_M \cdot \pi}{30} \tag{2.13}$$

$$M_d = M_{d0} + \sum_{i=1}^{4} k_{Mi}\,\omega_M^i$$

$$M_b = M_{b0} - k_{M5} \cdot \omega_M \tag{2.14}$$

The approximation of the full load and the drag torque and the equation (2.12) are applicable from just above the idle speed ω_{MO} to the maximum $\omega_{M,\ max.}$ and full throttle position $u_G = u_{G,\ max.}$

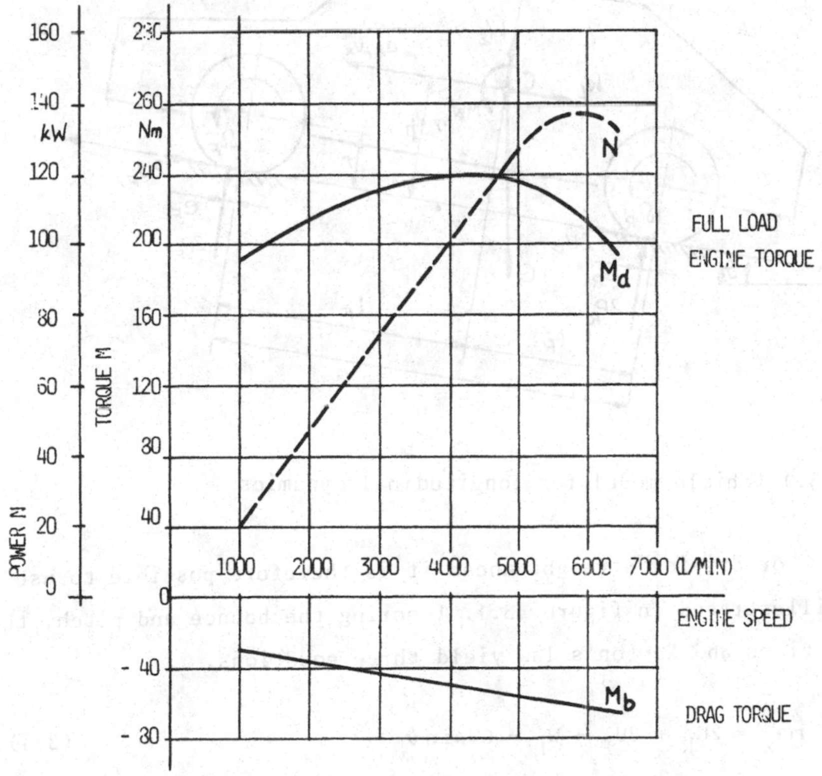

Fig. 2.7 Performance of a 2800 ccm gasoline injection engine

3 LONGITUDINAL DYNAMICS

3.1 Basic equations

The behavior of a vehicle when driving straight ahead or at very small lateral acceleration values is defined as the longitudinal dynamics. The calculation and valuation of acceleration, braking, climbing ability and top speed can be accomplished with a plane model. All factors which also cause considerable unsymmetry when driving straight ahead (side wind, extremely one sided load) belong to the lateral dynamics.

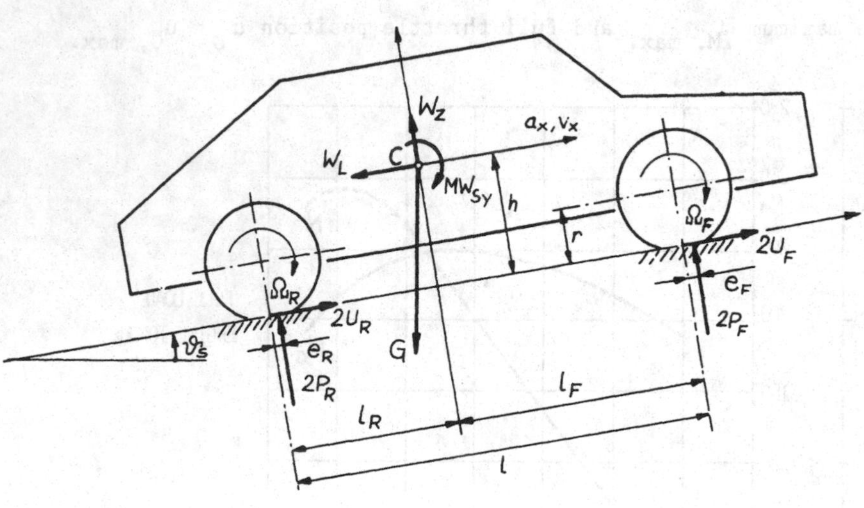

Fig. 3.1 Vehicle model for longitudinal dynamics

For driving straight ahead it is therefore possible to use a model as illustrated in figure 3.1. Ignoring the bounce and pitch, the Euler equations and Newton's law yield three equations.

$$ma_x = 2U_R + 2U_F - W_L - G.\sin\vartheta_s \tag{3.1}$$

$$0 = 2P_R + 2P_F + W_z - G.\cos\vartheta_s \tag{3.2}$$

$$2\Theta_F\dot{\Omega}_F + 2\Theta_R\dot{\Omega}_R = 2P_R.(1_R - e_R) - 2P_F.(1_F - e_F)$$

$$- (2U_F + 2U_R).h + MW_{yS} \tag{3.3}$$

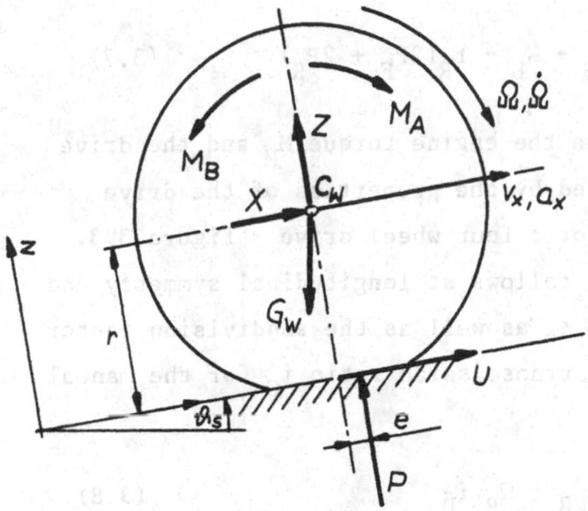

Fig. 3.2 Forces and moments at a single wheel

Three additional equations can be derived for each individual wheel. By figure 3.2 and using (2.4) you get:

$$m_W a_x = X + U - G_W.\sin\vartheta_s \tag{3.4}$$

$$0 = Z + P - G_W.\cos\vartheta_s \tag{3.5}$$

$$\Theta\dot{\Omega} = M_A - M_B - r.(U + f_R.P) \tag{3.6}$$

The wheels upon which a driving torque M_A actually is applied depends upon the type of drive. The braking moment M_B can be applied independently of M_A.

The normal forces on the wheels can be determined using equations (3.2) ÷ (3.3) and the forces X,Z in the wheel bearing determined

using (3.4) ÷ (3.5) once the motion of the vehicle is known.
From equations (3.1), (3.6) with equal rolling resistance coefficients for all wheels an equation for the longitudinal motion of the vehicle can be derived.

$$ma_x + 2\left(\frac{\Theta_F\dot{\Omega}_F}{(r^2)} + \frac{\Theta_R\dot{\Omega}_R}{(r^2)}\right) = 2\left(\frac{M_{AF} - M_{BF}}{r} + \frac{M_{AR} - M_{BR}}{r}\right)$$

$$- G.\sin\vartheta_S - W_L - f_R(2P_F + 2P_R) \qquad (3.7)$$

The relationship between the engine torque M_M and the drive moments at the wheels are determined by the properties of the drive train. The equations are derived for a four wheel drive - figure 3.3. The kinematic relationships are as follows at longitudinal symmetry and considering the differential ratio i_D as well as the subdivision factor ν of the distribution unit and the transmission ratio i_G for the manual transmission with clutch engaged:

$$\Omega_{KF} = \Omega_F.i_D \qquad \Omega_{KR} = \Omega_R.i_D \qquad (3.8)$$

$$\omega_M = i_G\cdot\frac{\nu\Omega_{KR} + \Omega_{KF}}{\nu + 1} \qquad (3.9)$$

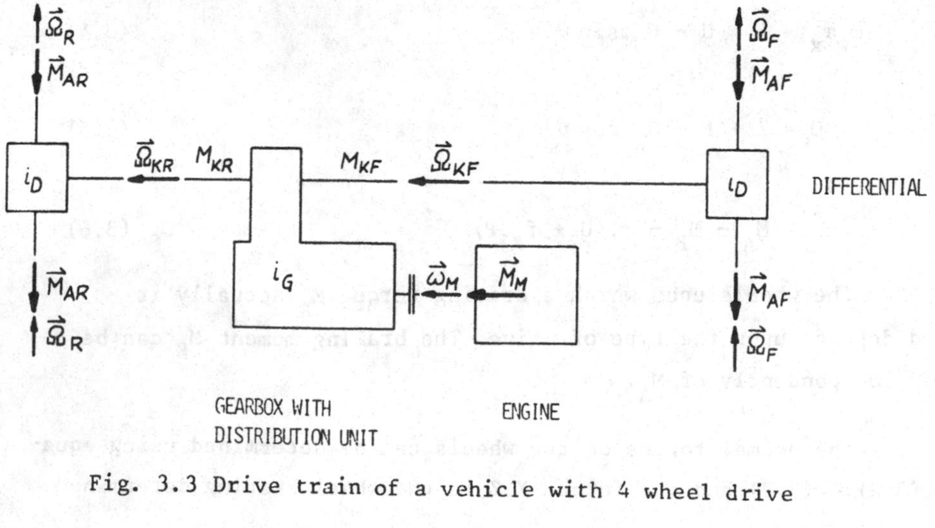

Fig. 3.3 Drive train of a vehicle with 4 wheel drive

The same values for the wheel torque on the left and right due to the symmetry and the axle differentials are entered in figure 3.3. The subdivision of the torque values resulting from the distribution unit is

$$M_{KR} = \nu M_{KF} \tag{3.10}$$

This unit offers the possibility of sudividing the torque dependent upon the load on the axles and therefore to utilize the adhesion between the wheels and the roadway as uniformly as possible. However, an additional distribution unit makes the transmission efficiency η of the drive train poorer. On the other hand a simple central distribution unit ($\nu = 1$) - such as used for passenger cars - can compensate the different angular velocities of the front and rear wheels without any significant sacrifices to the transmission efficiency. When the distribution unit is locked the subdivision of the torque between front and rear is determined by the conditions of contact forces between wheels and roadway.

The derivation of further relationships is restricted to a four wheel drive without locked distribution unit and $\nu = 1$. With equal drive moments M_A at all wheels follows

$$M_A = \frac{M_M i_G i_D \eta}{4} - \frac{1}{4}\left(\Theta_M (i_G i_D)^2 + \Theta_G i_D{}^2\right) \frac{\dot{\Omega}_F + \dot{\Omega}_R}{2} \tag{3.11}$$

The transmission losses in the drive train up to and including the wheel bearings are included in the efficiency η. Only the main parts of the rotating masses are taken into consideration.

Using (3.11) equation (3.7) can now be converted.

$$\frac{M_M i_G i_D \cdot \eta}{r} = ma_x + (\dot{\Omega}_F + \dot{\Omega}_R)\left(\frac{2\Theta}{r} + \frac{\Theta_G i_D^2 + \Theta_M i_D^2 i_G^2}{2r}\right)$$

$$+ \frac{2M_{BR}}{r} + \frac{2M_{BF}}{r} + G.\sin\vartheta_s + W_L + f_R(2P_F + 2P_R) \qquad (3.12)$$

The transmission of the force will result in a slip between the tire and the roadway - chapter 2.1.2. However, since the slip remains small in the normal driving range on dry surface and since the percentages of the rotating masses are small in comparison to tho total mass of the vehicle, the second term on the right side can be simplified by the rolling condition.

$$\Omega = \Omega_F \cong \Omega_R \cong \frac{v_x}{r} \quad , \quad \dot{\Omega} \cong \frac{a_x}{r} \qquad (3.13)$$

Using (3.13) the individual terms from (3.12) can be associated with clear concepts which apply to all drive variants:

Driving force $\qquad V = \frac{M_M i_G i_D \cdot \eta}{r} \quad , \quad M_M > 0$

Reduced mass $\qquad m_{red} = m + (4\Theta + \Theta_G i_D^2 + \Theta_M i_D^2 i_G^2)/r^2$

Climbing resistance $\qquad W_s = G.\sin\vartheta_s \qquad (3.14)$

Rolling resistance $\qquad W_R = f_R(2P_F + 2P_R)$

Braking force $\qquad B = (2M_{BF} + 2M_{BR})/r$

$$m_{red} \cdot a_x = V - W_s - W_L - W_R - B$$

When driving with dragged engine the drive force in (3.14) becomes a braking force. Since the friction losses in the drive train help when braking, this is expressed with

$$V_B = \frac{M_M i_G i_D}{r.\eta} \qquad M_M < 0 \tag{3.15}$$

The engine power is calculated from (3.14) and (3.9) for four wheel drive, $\nu = 1$

$$N_{M,4} = \frac{V.r}{\eta_4} \quad \frac{\Omega_F + \Omega_R}{2} \tag{3.16}$$

For rear wheel drive and front wheel drive (3.14) leads to

$$N_{M,R} = \frac{Vr}{\eta_R}\,\Omega_R \;,\; N_{M,F} = \frac{Vr}{\eta_F}.\,\Omega_F \tag{3.17}$$

If the losses through the wheel slip are considered, the actual angular velocities of the wheels must be inserted in (3.16) ÷ (3.17).

The following approximation applies to all types of drive in the normal driving range, neglecting the losses of the wheel slip:

$$N_M = Vv_x/\eta \tag{3.18}$$

3.2 Driving condition diagram

In equation (3.14) M_M represents the torque supplied by the engine as explained in section 2.3. On dry pavement the slip at the driving wheels is small so that the following equation is a good approximation of the angular velocity of the engine (see also 3.13).

$$\omega_M = i_G i_D \frac{v_x}{r} \tag{3.19}$$

Using (3.19) the driving force V and V_B can now be represented in terms of the vehicle velocity. In figure 3.4 these curves for a car with rear wheel drive - corresponding to figure 2.8 - are illustrated for a four speed transmission. The area below $v_{x,\,min.}$ must be bridged by engaging the clutch while the transmission is responsible for adaptation of the engine performance to the vehicle. The basic resistance is formed by the aerodynamic drag and the rolling resistance always present. The curves running parallel to this basic resistance curve correspond to a slope of ±10% or an acceleration of $a_x = \pm 0,1.g.(m/m_{red})$.

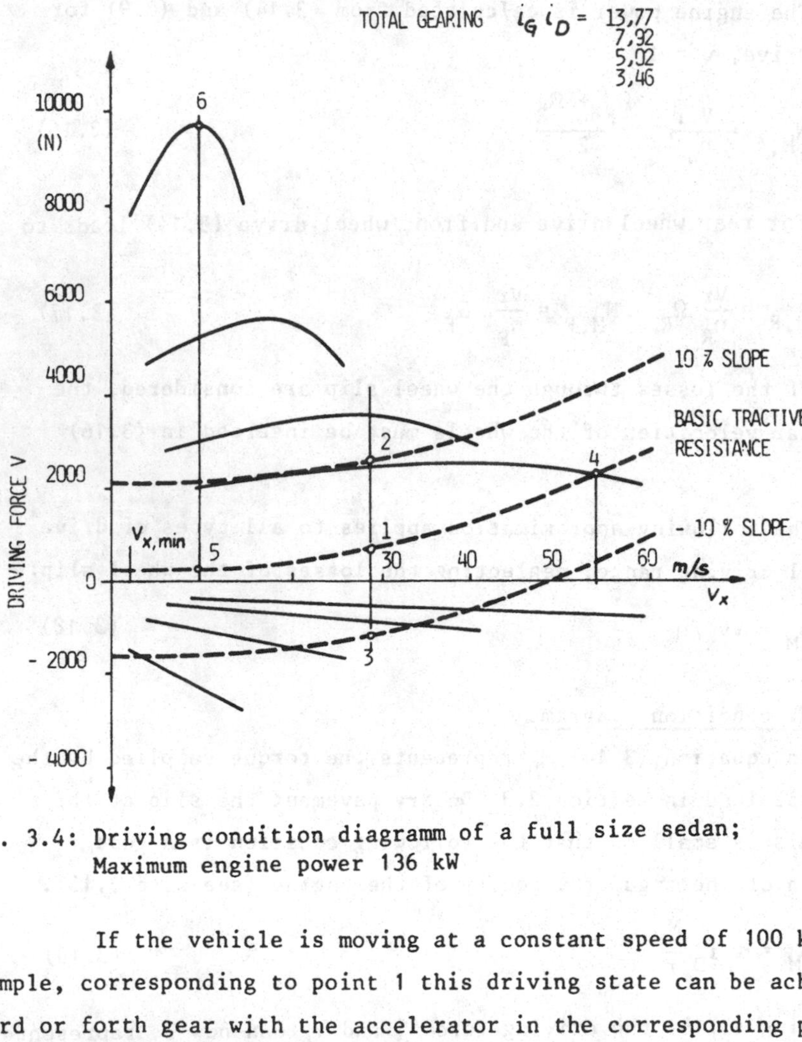

Fig. 3.4: Driving condition diagramm of a full size sedan; Maximum engine power 136 kW

If the vehicle is moving at a constant speed of 100 km/h, for example, corresponding to point 1 this driving state can be achieved in third or forth gear with the accelerator in the corresponding position in each case. An additional slope of +10 % (point 2) will require the third gear under all circumstances, while a slope of -10 % (point 3) cannot be handled using the engine drag alone, but requires additional braking. The maximum speed on a horizontal roadway is determined by point 4. The maximum climbing ability or the maximum acceleration results from the distance between the base resistance curve and the full load torque curve in first gear (points 5 and 6).

An increase in the maximum speed can be achieved through lower basic resistances and/or higher engine power, provided that the stability for straight line driving is secured - see chapter 4.1.2. A greater climbing ability can be achieved through the use of a different transmission ratio i_G (e.g. for commercial vehicles). The value for the maximum transferable drive torque is also limited by the maximum possible force coefficient between the drive wheels and the roadway, see chapter 3.3.

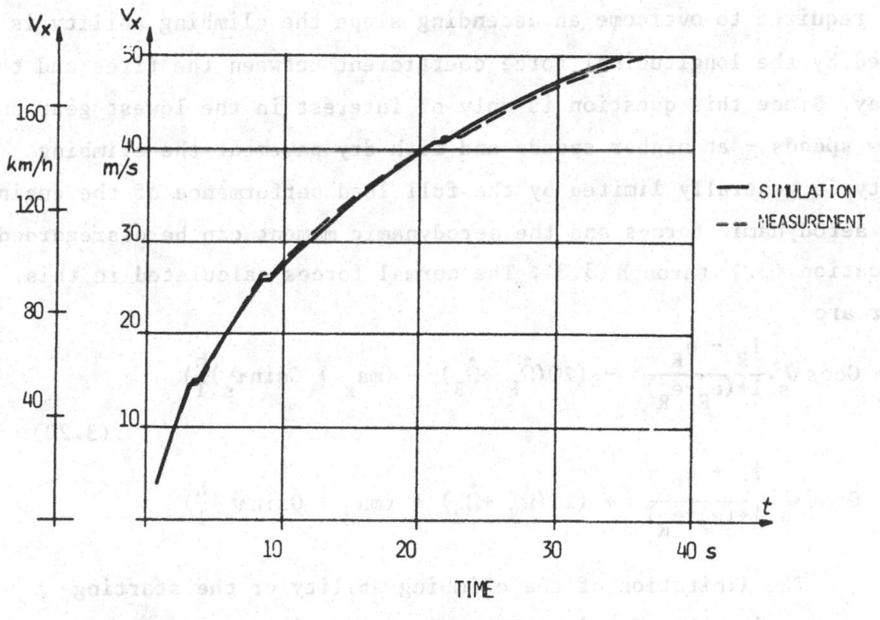

Fig. 3.5: Velocity history using maximum engine torque

The acceleration time up to 100 km/h, for example, which is often given in tests can be explained with illustration 3.4 and calculated using equations (3.12) through (3.14). With certain simplified assumptions regarding the actuation and function of the clutch figure 3.5 gives a speed curve according to figures 2.8 and 3.4 utilizing the maximum driving force.

For slippery pavement for high driving forces and therefore higher slip values in figure 3.4 the full load performance curves will shift to the left. More precise calculations would have to be made taking into consideration the slip curve for the actual tire road contact. The illustration given in figure 3.4 can be considered to be a good approximation.

3.3 Climbing ability, maximum starting acceleration

Assuming that the vehicle drive is able to deliver the driving force required to overcome an ascending slope the climbing ability is limited by the longitudinal force coefficient between the tires and the roadway. Since this question is only of interest in the lowest gear and at low speeds - at higher speeds and with dry pavement the climbing ability is generally limited by the full load performance of the engine - the aerodynamic forces and the aerodynamic moment can be disregarded in equation (3.1) through (3.3). The normal forces calculated in this manner are

$$2P_F = G\cos\vartheta_s \cdot \frac{l_R - e_R}{l+(e_F-e_R)} - (2\Theta(\dot{\Omega}_F + \dot{\Omega}_R) + (ma_x + G\sin\vartheta_s)\frac{h}{l})$$

$$2P_R = G\cos\vartheta_s \frac{l_F + e_F}{l+(e_F-e_R)} + (2\Theta(\dot{\Omega}_F + \dot{\Omega}_R) + (ma_x + G\sin\vartheta_s)\frac{h}{l}) \tag{3.20}$$

The limitation of the climbing ability or the starting acceleration is defined using the maximum traction coefficient for the driving wheels.

$$\frac{2U}{2P} < \mu_{L,max} \tag{3.21}$$

The longitudinal forces result from (3.1) and (3.6). As an example the minimum required force coefficient μ_{erf} for transferring the longitudinal forces for a vehicle with rear wheel drive is calculated with

$$\mu_{erf} = \frac{2U_R}{2P_R} = \frac{(ma_x + 2\Theta\dot{\Omega}_F/r) + G\sin\vartheta_s + 2P_F f_R}{2P_R} \quad (3.22)$$

whereby the expressions from (3.20) must be substitued for P_F and P_R.

Analog expressions result for front wheel drive and four wheel drive, whereby for the latter it must be checked which of the two driving axles will first reach the maximum traction coefficient.

In order to be able to rapidly illustrate the primary influences the small expressions are not taken into consideration in equation (3.22) and the analog relationships. Using the slope q and the "equivalent slope" q_E

$$q = \tan\vartheta_s\,, \qquad q_E = \frac{a_x}{g.\cos\vartheta_s} + q \quad (3.23),\ (3.24)$$

the primary relationships can be expressed as follows:

Rear wheel drive: $$\mu_{erf} = \frac{q_E}{(l_F/l)+q_E(h/l)} \quad (3.25)$$

Front wheel drive: $$\mu_{erf} = \frac{q_E}{(l_R/l)-q_E(h/l)} \quad (3.26)$$

Four wheel drive, $\nu = 1$:

$$\mu_{erf} = \frac{q_E/2}{(l_R/l)-q_E(h/l)} \qquad P_F < P_R \text{ or } \frac{l_R}{l} - q_E\frac{h}{l} < \frac{l_F}{l}+q_E\frac{h}{l} \quad (3.27\text{ a})$$

$$\mu_{erf} = \frac{q_E/2}{(l_F/l)+q_E(h/l)} \qquad P_F > P_R \quad (3.27\text{ b})$$

The ideal value results for a four wheel drive with blocked distribution unit:

$$\mu_{erf} = q_E \tag{3.27 c}$$

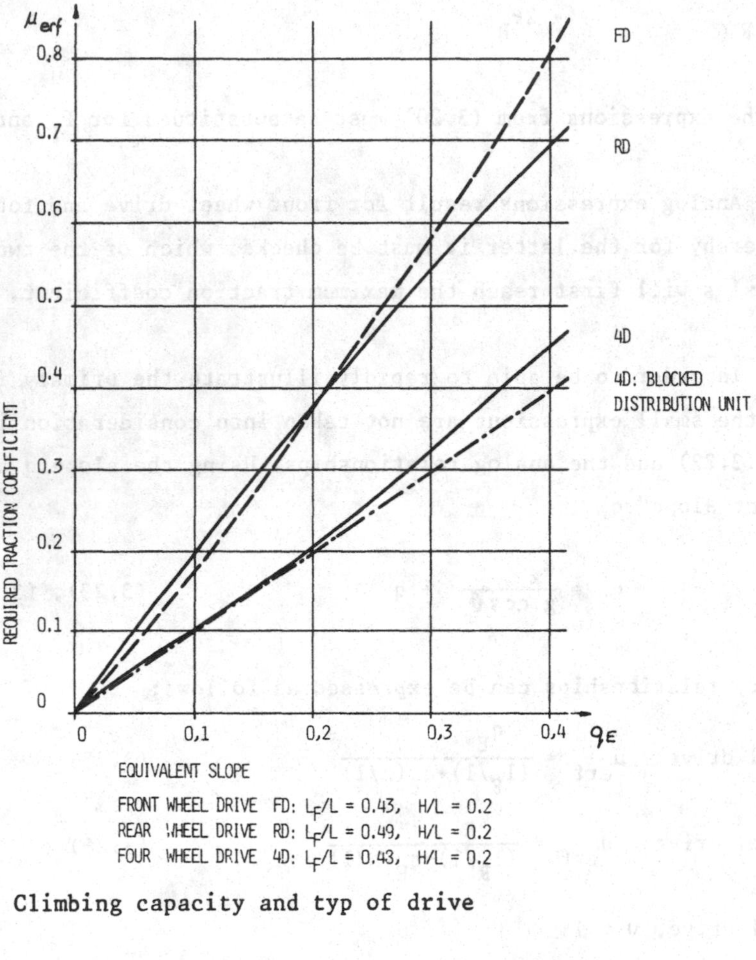

Fig. 3.6: Climbing capacity and typ of drive

The advantage of four wheel drive is obvious in figure 3.6. For the assumed data (vehicle loaded lightly) and on slippery pavement ($0.2 \leqq \mu_{L,max.} \leqq 0,4$) it can handle approximately double the equivalent slope as a single axle drive. A clear advantage for starting and accelerating in winter. Rear wheel drive has advantages over front wheel drive when the contact force of the drive axle is increased by loading. On dry pavement with the vehicle fully loaded rear wheel drive can even achieve the values of four wheel drive, because for the latter the front axle, which is also driven and unloaded additionally due to the equivalent slope, reaches the maximum traction coefficient first.

3.4 Braking

A further limitation of the driving range is the maximum longitudinal deceleration determined by the longitudinal force coefficient and/or the design of the vehicle brakes. In order to achieve the optimum braking for a vehicle the braking coefficients of all wheels would have to be the maximum value simultaneously.

$$\mu_{L,max} = \frac{2U_F}{2P_F} = \frac{2U_R}{2P_R} \tag{3.28}$$

If we confine the representation to the primary influential values (3.28) in combination with (3.1) through (3.3) and (3.6) results in

$$\frac{M_{BF}-M_{AF}}{M_{BR}-M_{AR}} = \frac{\frac{l_R}{l}\cos\vartheta_s + \frac{h}{l}a + (\frac{h}{l}W_L - MW_{yS} - \frac{l_R}{l}W_z)\frac{1}{mg}}{\frac{l_F}{l}\cos\vartheta_s - \frac{h}{l}a - (\frac{h}{l}W_L - MW_{yS} + \frac{l_F}{l}W_z)\frac{1}{mg}} \tag{3.29}$$

with the deceleration

$$a = -\frac{a_x}{g} = (\cos\vartheta_s - \frac{W_z}{mg})\mu_{L,max.} + \frac{W_L}{mg} + \sin\vartheta_s \tag{3.30}$$

These two equations determine the relationship of the braking moments - a special value can be calculated for each load distribution, driving speed, type of drive and road condition. An antiblock system (ABS) with its electronic control comes closest to providing the necessary M_{BF}, M_{BR} and therefore the maximum achieveable deceleration. However, most brake systems in passenger cars today are designed with a constant ratio M_{BF}/M_{RR} so that the maximum braking coefficient cannot be fully utilized! Highly simplified expressions can be derived from equations (3.1) to (3.3) and (3.6) for the basic design of such brake systems. The following equations apply for a horizontal roadway disregarding the aerodynamic resistance and the engine.

$$\frac{B_F}{G} = \frac{a\frac{h}{l}}{\frac{l_R}{l} + a\frac{h}{l}} \quad , \quad B_F = \frac{2M_{BF}}{r} \tag{3.31}$$

$$\frac{B_R}{G} = \frac{a\frac{h}{l}}{\frac{l_F}{l} - a\frac{h}{l}} \quad , \quad B_R = \frac{2M_{BR}}{r} \tag{3.32}$$

$$a = \frac{B_F + B_R}{G} \tag{3.33}$$

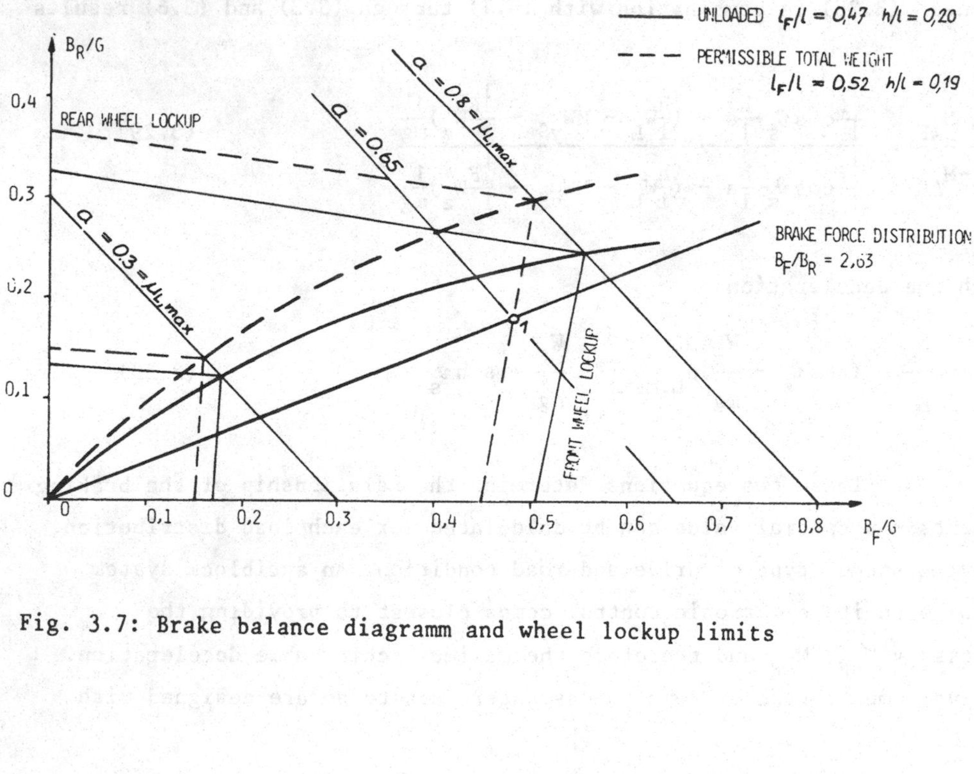

Fig. 3.7: Brake balance diagramm and wheel lockup limits

In the brake balance diagram, figure 3.7, the ideal layout $B_F/G(B_R/G, \mu_{L,max.})$ is represented by a parabola, whose position is determined by the vehicle center of gravity. The lines for the constant deceleration are calculated from (3.33). If only the front wheels lock for example

$$\frac{B_F}{2P_F} = \frac{B_F}{G(\frac{1_R}{1} + a\frac{h}{1})} = \mu_{L,max.} \tag{3.34}$$

a linear relationship again follows with (3.33) - see figure 3.7. The line for the brake layout B_F/B_R is determined so that for all values $\mu_{L,max.}$ the front wheels will lock before the rear wheels - the line must be positioned below the parabola for all road conditions, even for dry pavement. Point 1 shows that it is therefore only possible to achieve a deceleration of a'= 0.65 in contrast to the ideal value of a = 0.8 with a fully loaded vehicle.

4 LATERAL DYNAMICS

Models which allow the lateral forces and the yawing of the vehicle to be described, must be used for theoretical studies on cornering, transient steering manoeuvers as well as the directional stability when driving straight ahead. Inclusion of the body motion in interaction with the wheel suspension, the nonlinear behavior of the tires, the effects of the drive and steering system lead to very complex models with the help of which an attempt can be made to simulate the vehicle handling over a period of time - see chapter 4.2. However, a few of the basic considerations can be shown well using a linearized model.

4.1 Linearized vehicle model, constant driving speed

To some extent analytical expressions for describing the vehicle handling on a horizontal roadway can be derived from a model represented in figure 4.1, which is based primarily on the work of Riekert and Schunk |5|. One of the basic simplifications is to neglect changes in the wheel loads. The wheels on one axle are represented by one wheel in the middle of the axle, the axle loads correspond to their static values. All angles remain small and the lateral forces of the tires on one axle are represented by linearized expressions

$$S_F = k_{SF}\alpha_F \ , \ S_R = k_{SR}\alpha_R \tag{4.1}$$

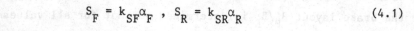

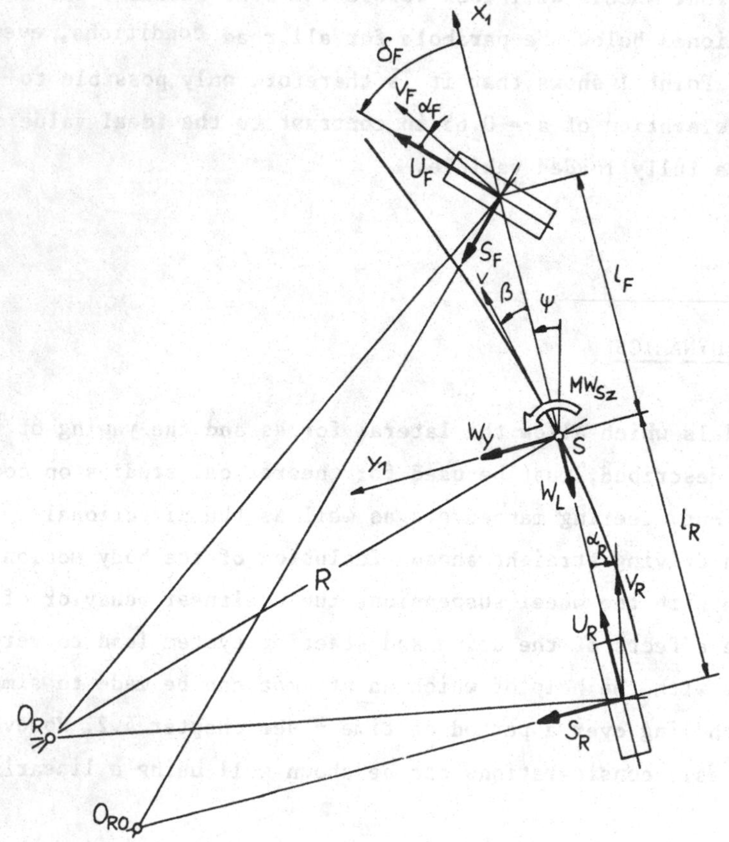

Fig. 4.1: 2 wheel - vehicle model

The slip angles of the wheels can be calculated using the yaw velocity of the vehicle.

$$\alpha_F = \delta_F - \beta - \frac{l_F \dot{\Psi}}{v}$$
$$\alpha_R = - \beta + \frac{l_R \dot{\Psi}}{v} \quad (4.2)$$

The steering system is simplified so that only the elasticities as well as the aligning torque resulting from the tire offset and the castor offset or wheel offset n_K are taken into consideration. The equation for the steering subsystem with steering ratio i_L and steering stiffness c_L give a relationship between the steering wheel angle δ_H and the steer angle δ_F of the wheel

$$c_L\left(\frac{\delta_H}{i_L} - \delta_F\right) = (n_K + n_S)\, S_F \quad (4.3)$$

The lateral forces at the axle can be represented with equations (4.2) and (4.3).

$$S_F = k_{SF}^x \left(\delta_F^x - \beta - \frac{l_F \dot{\Psi}}{v}\right)$$
$$S_R = k_{SR} \left(- \beta + \frac{l_R \dot{\Psi}}{v}\right) \quad (4.4)$$
$$k_{SF}^x = k_{SF}/\left(1 + \frac{(n_K + n_S)}{c_L} k_{SF}\right),\ \delta_F^x = \delta_H/i_L$$

The aerodynamic moment reduced to the vehicle center of gravity at the further assumed calm wind condition - see chapter 2.2, $\tau = \beta$, is:

$$W_y = -k_y \beta v^2$$
$$MW_{zS} = -W_y\left(\frac{l}{2} - l_F\right) + MW_z = -k_M \beta v^2 \quad (4.5)$$
$$k_y = c_y F \varrho /2,\ k_M = \left(c_{Mz} - c_y\left(\frac{l}{2} - l_F\right)\right) F \varrho /2$$

Newton's law and the Euler equations yield three equations for the plane motion of the vehicle on a horizontal roadway at constant speed when additional small terms are neglected.

$$m\beta(\dot{\Psi} + \dot{\beta})v = 2U_R + 2U_F - W_L - S_F\delta_F \tag{4.6}$$

$$m\,(\dot{\Psi} + \dot{\beta})v = S_F + S_R + W_y + 2U_F\delta_F \tag{4.7}$$

$$\Theta_z\ddot{\Psi} = (S_F + U_F\delta_F)l_F - S_R l_R + MW_{zS} \tag{4.8}$$

Equation (4.6) corresponds to the already known equation (3.1). The coupling of the two equations for the lateral dynamics (4.7), (4.8) to (4.6) exists only by terms containing small angles.

Equations (4.4), (4.5), (4.7) and (4.8) and further generally justifiable simplifications lead to two linearized differential equations.

$$\ddot{\beta}+2K_1\dot{\beta}+K_2\beta = \frac{k^x_{SF}}{mv}\cdot\dot{\delta}^x_F - \frac{k^x_{SF}(l_F mv^2 - k_{SR}l_R l)}{\Theta_z mv^2}\cdot\delta^x_F \tag{4.9}$$

$$\frac{d}{dt}\,(\ddot{\Psi}+2K_1\dot{\Psi}+K_2\Psi) = \frac{l_F k^x_{SF}}{\Theta_z}\dot{\delta}^x_F + \frac{k^x_{SF}\,k_{SR}l}{\Theta_z mv}\,\delta^x_F \tag{4.10}$$

$$K_1 = \frac{\Theta_z(k_{SR}+k^x_{SF})+m(k^x_{SF}l^2_F+k_{SR}l^2_R)}{2\Theta_z mv}$$

$$K_2 = \frac{l^2 k_{SR}k^x_{SF} + (k_{SR}l_R - k^x_{SF}l_F)mv^2 - mk_M v^4}{\Theta_z mv^2} \tag{4.11}$$

With the use of this highly simplified vehicle model, quantitive correspondence to measurements can only be expected in the range of small lateral acceleration (up to approx. 0,4 g).

4.1.1 Steady state turning

For steady state turning all of the time derivatives in equation (4.9) and (4.10) are equal to zero and the yaw velocity of the vehicle around the vertical axis as well as the lateral acceleration are calculated from the vehicle speed and the radius R.

$$\dot{\psi}_s = \frac{v}{R}\,, \quad a_{ys} \cong a_{qs} = v\dot{\psi}_s = \frac{v^2}{R} \tag{4.12}$$

The required steer angle δ^x_{Fs} and the associated side-slip angle β_s of the vehicle follows from (4.9) through (4.12) resulting in

$$\left(\frac{\delta_H}{\delta_{HO}}\right)_s = \left(\frac{\delta^x_F}{\delta^x_{FO}}\right)_s = 1 + \frac{k_{SR}l_R - k^x_{SF}l_F}{k_{SR}k^x_{SF}l^2} \cdot mv^2 - \frac{k_M}{k_{SR}k^x_{SF}l^2}\, mv^4 \tag{4.13}$$

$$\left(\frac{\beta}{\beta_0}\right)_s = 1 - \frac{l_F}{k_{SR}l_R l}\, mv^2\,, \quad \delta^x_{FO} = 1/R,\ \beta_0 = l_R/R \tag{4.14}$$

The change of the steer angle with the vehicle speed is primarily dependent upon the difference $k_{SR}l_R - k^x_{SR}l_F$ because k_{SR}, k^x_{SF} are always positive. On a vehicle with understeering properties a greater steering angle is required at increasing vehicle speed. The last term in (4.13), which counteracts the tendency to understeer due to the generally positive value of k_M has little influence on an average passenger car for smaller radii.

In figures 4.2, the values were selected for the understeering vehicle so that a comparison to the measured values and to the curves calculated using a non-linear model, figure 4.8 is possible. For an oversteering vehicle $(\delta_H/\delta_{HO})_s$ becomes smaller at increasing speed and lateral acceleration. For high oversteer and for large radii negative steering angles may even be required for steady state turning.

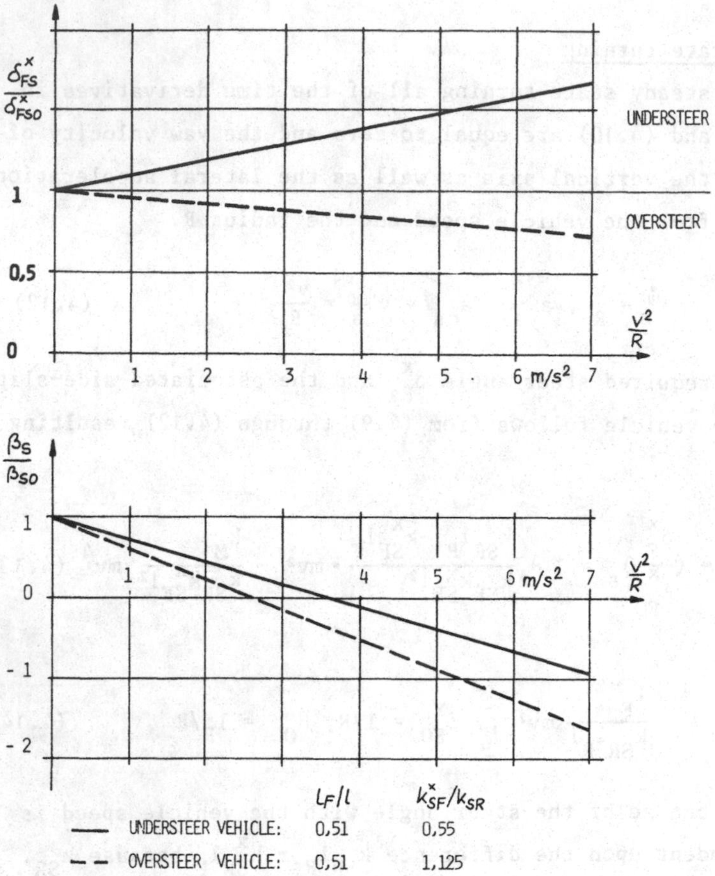

Fig. 4.2: Steer properties and side slip angle of a 2-wheel model
R = 42,5 m; steady state turning.

The side-slip angles decrease at increasing speed and become negative. The front of the vehicle points out of the curve at low speeds as illustrated in figures 4.1, while the vehicle turns into the curve increasingly at increasing speed. This tendency is independent of the steering behavior.

4.1.2 Stability characteristics

The stability of the vehicle handling, described with equation (4.9) and (4.10), is characterized by the eigen values of the homogenous differential equations.

$$\lambda_{1,2} = -K_1 \pm \sqrt{K_1^2 - K_2} \qquad (4.15)$$

Since K_1 is always positive according to (4.11), the behavior of the system is determined by the value and the sign of K_2.

$$K_2 > 0 \quad \text{stability}$$
$$(4.16)$$
$$K_2 < 0 \quad \text{instability}$$

A relationship between the stability and the steering properties can be derived using the results of the steady state turning (4.13):

$$K_2 = \frac{l^2 k_{SR} k^x_{SF}}{\Theta_z mv^2} \left(\frac{\delta^x_F}{\delta^x_{FO}}\right)_s \qquad (4.17)$$

Since the first factor is always positive, a required negative steer angle for steady state turning simultaneously means instability. The limit velocitiy results from $K_2 = 0$.

Basic predictions regarding the effects of the design upon the handling stability can best be shown using the terms in equation (4.13). A high front axle load ($l_F < l_R$) and a smaller cornering stiffness of the front axle ($k^x_{SF} < k_{SR}$) is better for a positive value of $k_{SR} l_R - k^x_{SF} l_F$. The value of k_M - (4.5) - also becomes smaller when the vehicle center of gravity is positioned near the front axle. However, since the center of gravity is primarily dependent upon the manner in which the vehicle is loaded a smaller front concerning stiffness is achieved through the axle lay-out (elasticities, roll steer, anti-roll bars). Tractive forces on the front wheels also mean a decrease in k^x_{SF}. The increase of k_{SR} through greater loading on the rear wheels is not sufficient enough to compensate the negative effects of the greater distance between the center of gravity and the front axle.

In figure 4.3, the effects of the axle cornering stiffness on the stability characteristics are illustrated based in each case upon the greatest real part Re (λ_1) of the eigen values (4.15). Since the axle stiffness can not be increased or decreased without restrictions $k^x_{SF} + k_{SR}$ was held constant. Using an example with the center of gravity in the middle $l_F/l = 0.5$ and a vehicle speed v = 30 m/s we see that the real part hardly changes up to a limit, here $k^x_{SF}/k_{SR} = 0.98$. In the system behavior two conjugated complex eigen values cause a damped vibration. The limit corresponds to the periodic critical case; beginning at $k^x_{SF}/k_{SR} = 1{,}32$ it becomes unstable. For $l_F/l = 0{,}4$ no instability occurs in the entire range, while an extremely heavily loaded vehicle becomes unstable even at $k^x_{SF}/k_{SR} < 1$.

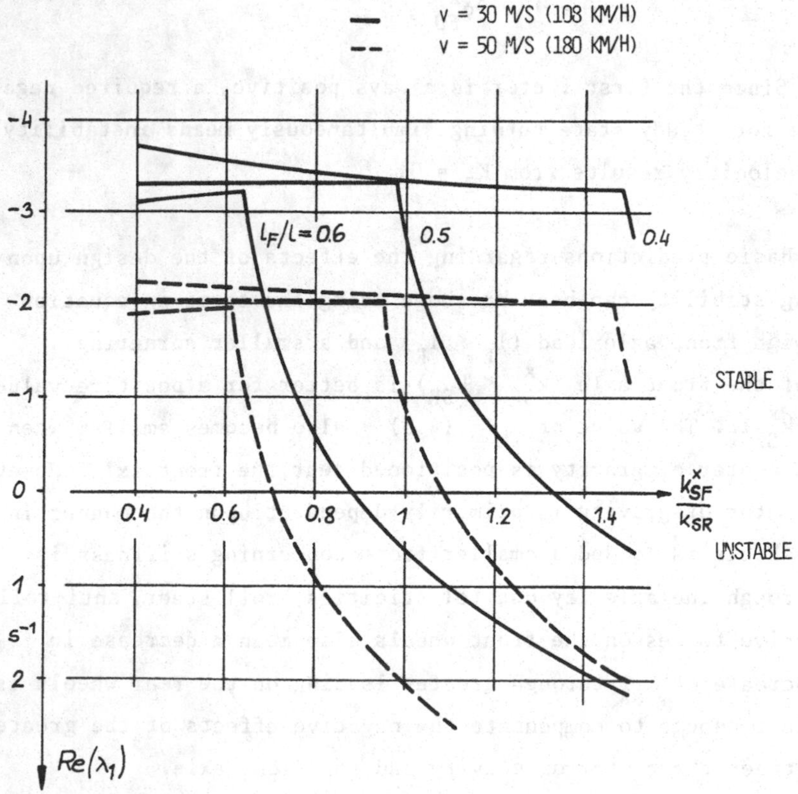

Fig. 4.3: Influence of cornering stiffness ratio on the stability behavior

At higher speeds, the absolute values $|Re(\lambda_1)|$ are lower and the limits for the aperiodicity have lower values. Disturbances die down more slowly and smaller values of k^x_{SF}/k_{SR} are required for stable driving. Design values of $k^x_{SF}/k_{SR} < 0.6$ must be realized to assure stability at top speed and with an unfavorable loading state.

Disregarding extreme driving conditions the driving stability of today's passenger car achieved through the axle layout does not result in any limitation of the driving range which generally is determined by the engine power.

4.1.3 Frequency response

A further possibility of characterizing the behavior of the linear system is the reaction of the vehicle to harmonic exitation

$$\delta^x_F = a_F e^{i\varepsilon t} \tag{4.18}$$

The system response in the form of the complex amplification function A can be calculated from (4.9) through (4.11) for the yaw velocity $\dot{\psi}$ and the lateral acceleration $a_y = v\,(\dot{\psi} + \dot{\beta})$ with (4.13) and (4.14)

$$A_{\dot{\psi}} = \left(\frac{\dot{\psi}}{\delta^x_F}\right)_s \cdot \frac{1 + T_z i\varepsilon}{1 + 2Di\varepsilon/\nu_n - \varepsilon^2/\nu_n^2}$$

$$A_{ay} = \left(\frac{a_y}{\delta^x_F}\right)_s \cdot \frac{1 + T_1 i\varepsilon - T_2\varepsilon^2}{1 + 2Di\varepsilon/\nu_n - \varepsilon^2/\nu_n^2}$$

$$\nu_n^2 = K_2\ , \quad D = K_1/\sqrt{K_2} \tag{4.19}$$

$$T_z = \frac{l_F m v}{k_{SR} l}\ , \quad T_1 = \frac{l_R}{v}$$

$$T_2 = \frac{\Theta_z}{k_{SR} l}$$

Figure 4.4 shows the DB values for the ratio of the amplification function of the lateral acceleration to its value at an excitation frequency of ε = 0, the value for steady state turning.

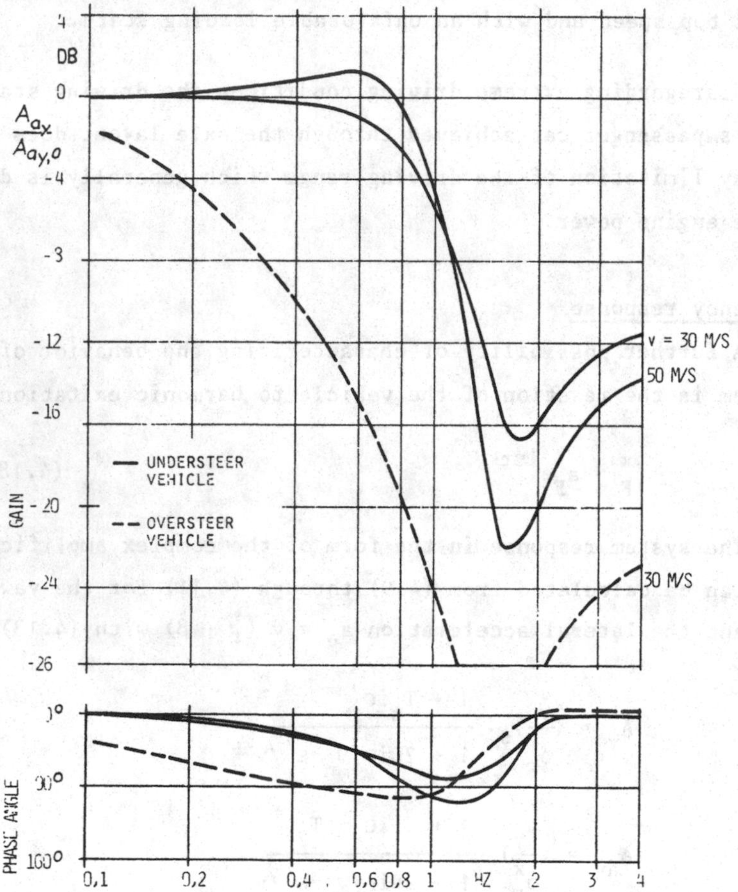

Fig. 4.4: Lateral acceleration frequency response of two vehicles with different steer properties

On an understeering vehicle - vehicle 1 from figure 4.2 - there are only slight changes of gain at phase angles below 40° up to approx. ε = 0.5 Hertz. For normal steering movements the vehicle reacts with a short delay nearly independently of the steering speed and the driving speed has only a small influence.

The oversteering vehicle - vehicle 2 in figure 4.2 - already shows great changes in the system behavior with relatively long delay times at 30 m/s up to 0.5 Hertz - an unfavorable type of behavior for the average driver. This vehicle becomes unstable after approx. 38 m/s. At steering frequencies of over 1.2 Hertz, both vehicles follow the straight line with very small deviations.

Measured frequency response curves, also for the yaw velocity, are shown in part II, chapter 3.3.2.

4.2 Non-linear vehicle models

In order to be able to describe the vehicle handling up to the limits of the driving range, the non-linear characteristics of the springs, dampers, tires, etc., and the geometric non-linearities of the steering or the wheel suspensions must be calculated and considered. Numerical integration of the system equations results in predictions about the vehicle handling within the time domain. Only in the case of steady state turning it is possible to determine the steady state values of the steer angle, side-slip angle, etc., through iteration without time simulation.

4.2.1 Steady state turning

Since no relative motions of the vehicle components to one another occur in this driving state (disregarding the rotation of the drive train), a corresponding "static" model can be used to derive the system equations. The model used is illustrated in figure 4.5. As a simplification in comparison to the model used in chapter 4.2.2, the body motion is described with the roll and pitch axis of the static initial configuration (horizontal roadway, speed v = 0) using the three variables ϑ, φ, h. The numbering of the four wheels designates their position on the vehicle and in the curve. The index "s", steady state turning, is used only in the comparative figures for chapter 4.1.1.

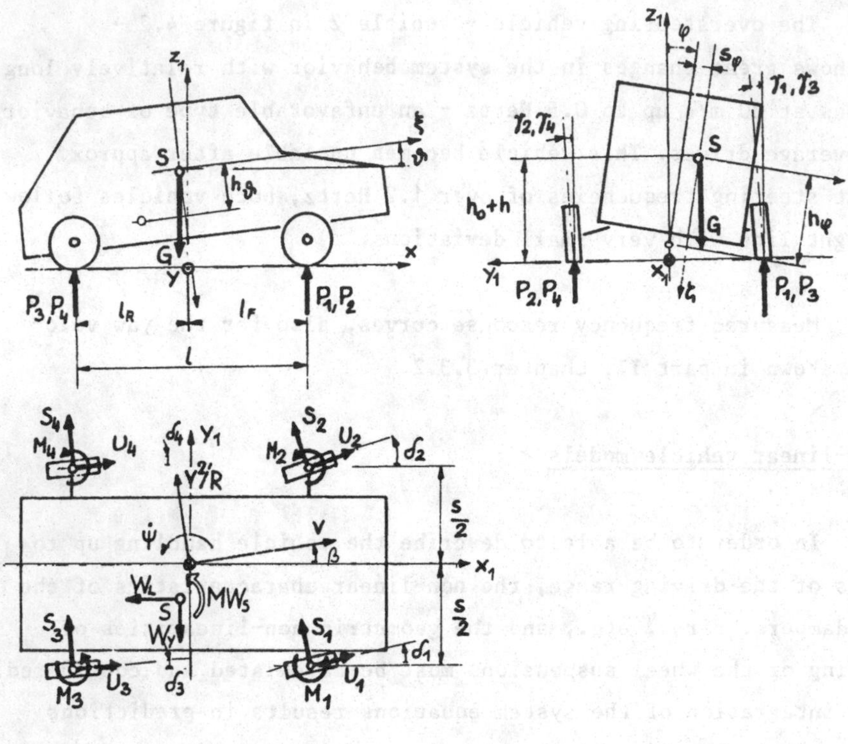

Fig. 4.5: 4 wheel vehicle model for steady stateturning

In laying down the equations of motion the gyroscopic moments of the wheels were neglected. The center of gravity was assumed at a fixed body point and the axes ξ, η, ζ are body fixed main axes of inertia.

With small pitch and roll angles the equations of motion are

$$- \frac{mv^2}{R} \sin\beta = \sum_1^4 A_i - W_L$$

$$- \frac{mv^2}{R} \cos\beta = \sum_1^4 B_i + W_y \qquad (4.20)$$

$$0 = mg - \sum_1^4 P_i$$

$$\left.\begin{aligned} 0 &= Mx - \vartheta My \\ 0 &= My + \varphi Mz \\ 0 &= Mz - \varphi My + \vartheta Mx \end{aligned}\right\} \quad (4.21)$$

$$\left.\begin{aligned} A_i &= U_i \cos\delta_i - S_i \sin\delta_i, \quad B_i = U_i \sin\delta_i + S_i \cos\delta_i \\ Mx &= (h_0+h)\sum_1^4 B_i + \sum_1^4 s_i P_i + (\varphi h_\varphi + s_\varphi)\sum_1^4 P_i \\ My &= (h_0+h)\sum_1^4 A_i + \sum_1^4 l_i P_i + \vartheta h_\vartheta \sum_1^4 P_i \\ Mz &= \sum_1^4 (s_i A_i - l_i B_i) + (\varphi h_\varphi + s_\varphi)\sum_1^4 A_i - \vartheta h_\vartheta \sum_1^4 B_i + MW_{zS} \end{aligned}\right\} \quad (4.22)$$

$$s_1 = s_3 = -s/2, \; s_2 = s_4 = s/2, \; l_1 = l_2 = l_F, \; l_3 = l_4 = -l_R \qquad (4.23)$$

The normal forces P_i, can be calculated using the spring deflection w_i and the non-linear spring characteristics - see figure 4.7 as an example. Any anti-roll bars are taken into consideration by the roll angle. Influences of the longitudinal and lateral forces resulting from the wheel suspension geometry are considered in a simplified manner using constant coefficients:

k_{Bi}, k_{Ai}:

$$\left.\begin{aligned} P_i &= c_i(w_i) + 2c_T k_i \varphi + k_{Bi} B_i + k_{Ai} A_i + (1 - j_i l_i) mg/2 \\ w_i &= -h - l_i \vartheta - s_i \varphi \\ k_1 &= k_3 = 1, \; k_2 = k_4 = -1, \; j_1 = j_2 = 1, \; j_3 = j_4 = -1 \end{aligned}\right. \qquad (4.24)$$

The effect of the masses of the wheel and the wheel suspension components are neglected within the substructure of the wheel suspension. The wheel is assumed to be laterally rigid, the vertical stiffness is taken into consideration in an approximate manner in the suspension spring characteristics.

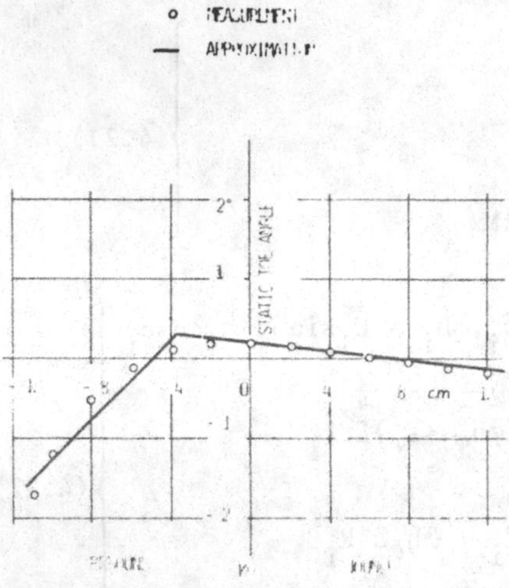

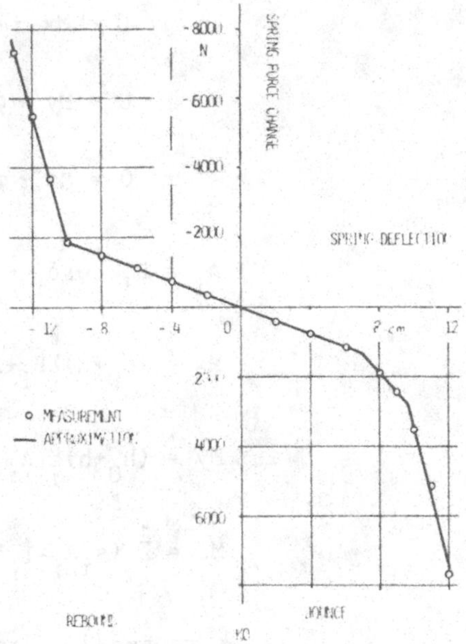

Fig. 4.6: Static toe-in angle of the frontwheel

Fig. 4.7: Suspension rate of the rear wheels, jounce and rebound mode

In order to make it possible to calculate limits for the driving range, it is necessary to use complex tire characteristics such as (2.8), (2.9).

$$S_i = f_{Si}(\alpha_i, P_i, U_i), \quad M_{Si} = n_{Si} S_i$$

$$n_{Si} = n_{Si}(\alpha_i, P_i) \qquad i = 1, ..4 \tag{4.25}$$

The slip angles α_i can be calculated by analogy to (4.2) as functions of the yaw velocity $\dot{\Psi}$, β and the steering angles δ_i

$$\alpha_i = \alpha_i(\delta_i, \dot{\Psi}, \beta) \qquad i = 1, ..4 \tag{4.26}$$

Determining the wheel camber γ_P in respect to the vertical direction of the road surface the body roll and the wheel camber γ in respect to the body have to be taken into account.

$$\gamma_{Pi} = \gamma_{Pi}(\varphi, \gamma_i) \qquad i = 1, ..4 \qquad (4.27)$$

The longitudinal forces can be reduced to the driving force V, see chapter 3.1. The following applies as the calculated example of a vehicle with rear wheel drive.

$$U_i = -f_R P_i \qquad i = 1,2$$

$$(4.28)$$

$$U_i = -f_R P_i + V/2 \qquad i = 3,4$$

The geometric steer angles δ_{Gi} of the front wheels as functions of the mean steer angle δ can be used in the program either by description of the steering linkage or via measured or given realtionships. For the calculated example with relatively small angles a parallel steering fits best.

$$\delta_{Gi} = \delta \qquad i = 1,2$$

$$(4.29)$$

$$\delta_{Gi} = 0 \qquad i = 3,4$$

For the steer angle δ_i of the wheel, the toe-in angle δ_{Ki} and its changes resulting from jounce and rebound of the suspension are taken into consideration via measured wheel elevation curves - figure 4.6 shows an example including polygonal approximation. Also the influences of elasticities must be considered.

$$\delta_{Ki} = \delta_{Ki}(w_i, S_i, U_i, M_{Si}), \quad \delta_{GEi} = \delta_{Gi} + \frac{M_{SPi}}{c_{SLi}}$$

$$(4.30)$$

$$\delta_i = \delta_{GEi} + \delta_{Ki}$$

$$i = 1, ..4$$

$$n_{Ki} = n_{Ki}(w_i, \delta_i) \qquad (4.31)$$

The steering moment M_{SP} in respect to the steering axis is determined using the longitudinal force, lateral force and normal force with wheel and tire offset, and suspension geometry including king pin inclination and caster.

$$M_{SPi} = M_{SPi}(\delta_i, n_{Ki}, S_i, P_i, U_i, M_{Si})\ i=1,2 \quad (4.32)$$

The steering wheel torque and the steering angle δ_H are determined taking into account the steering ration i_L and the elasticity of the steering column. For parallel steering, neglecting friction, follows

$$M_H = \frac{M_{SP1} + M_{SP2}}{i_L}, \quad \delta_H = i_L \delta + \frac{M_H}{c_{LC}} \quad (4.33)$$

For iterative evaluation the equations can be separated into two groups of three equations each, for which there is only a weak mutual coupling. A quick convergence with a relatively low amount of calculation time can be achieved through preiteration for the body variables and an iteration including the state variables β, δ, V.

A few results are illustrated in figures 4.8 through 4.10. The vehicle data listed in the appendix for a large passenger vehicle with rear wheel drive as well as the performance graphs or approximations shown in figures 2.4, 4.6 , and 4.7 serve as the basis.

In comparison the figure 4.8 shows the steering wheel angle ratio $(\delta_H/\delta_{HO})_s$ and side-slip angle ratio $(\beta/\beta_O)_s$ for a vehicle equivalent to the mentioned understeering 2-wheel model. In a range of up to approx. 4 m/s² lateral acceleration the measured values and both calculations correspond relatively well. The high degree of understeer of the vehicle and the limit values for the lateral acceleration caused by the limitation of contact force transfer, however, can only be calculated using the non-linear model. As a result of the highly different loads on the wheels and the additional tractive forces to be transferred, the maximum contact force coefficient of $\mu_{max} = 0{,}85$ can not be fully utilized.

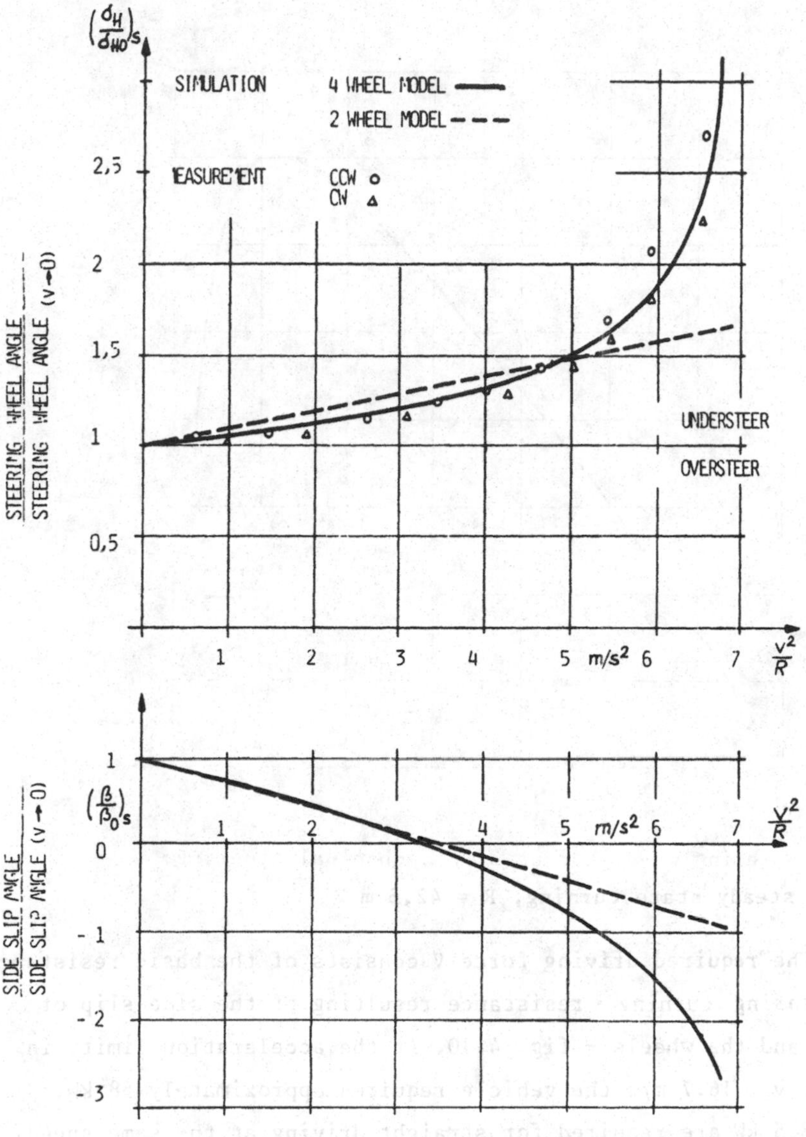

Fig. 4.8: Steer properties and side slip angle, steady turning; R = 42,5 m

The steering wheel torque M_H as well as the corresponding measured values are illustrated in figures 4.9. Following an approximately linear rise the torque decreases sharply mainly as a result of the rapid decrease of the tire offset.

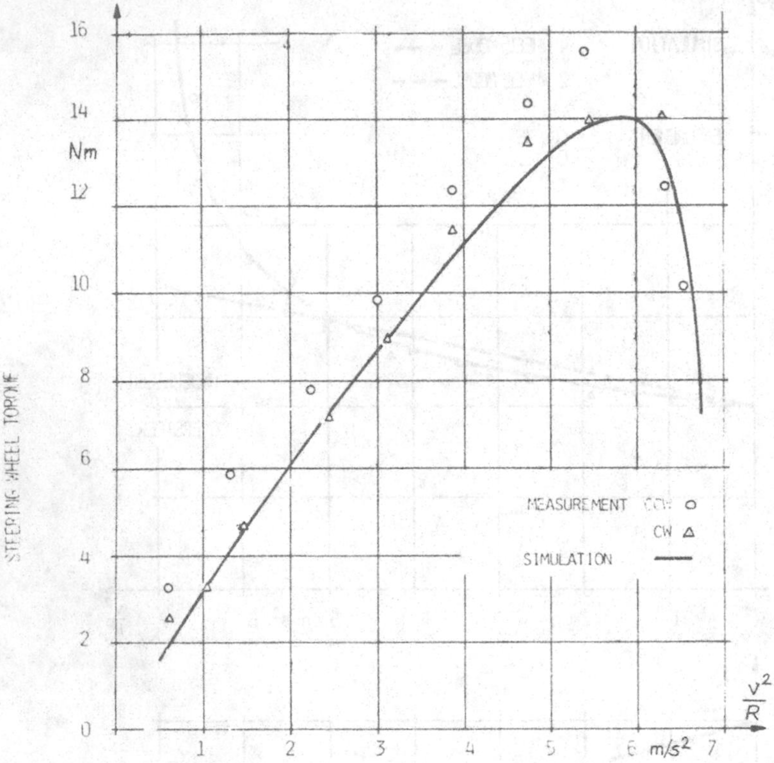

Fig. 4.9: Steering wheel torque of a mechanical steering; **steady state turning, R = 42,5 m**

The required driving force V consists of the basic resistance and an increasing turning resistance resulting of the side slip of the vehicle and the wheels - fig. 4.10. At the acceleration limit, in this case at v = 16.7 m/s the vehicle requires approximately 58 kW, while only 8.5 kW are required for straight driving at the same speed.

Therefore, for the motor vehicle on dry surface the maximum possible lateral acceleration in sharp curves is caused by the maximum contact force coefficient. As a result of the motion resistance at higher curve radii, a further limitation of the driving rang will result through the available engine power.

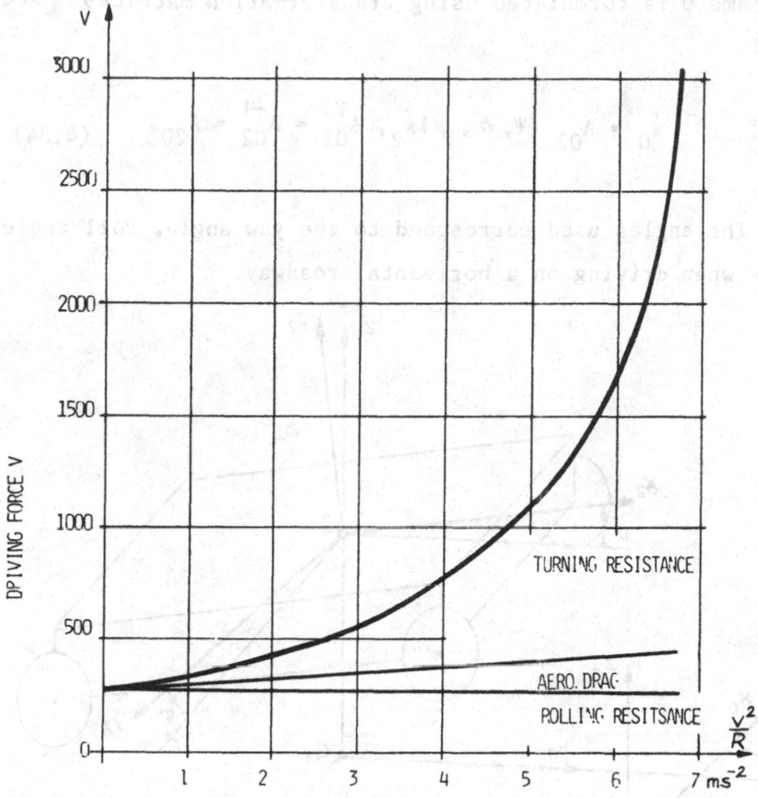

Fig. 4.10: Necessary driving force for steady state turning, R = 42,5 m

4.2.2 Simulation of general vehicle motions

The requirements for describing the vehicle reactions even for rapid steering movements and high accelerations lead to models such as described in |7, 9, 10|. The model considered should, on the one hand, allow studies using given driver reactions while also allowing real time simulation in an advanced driving simulator, with the least possible degree of adaptation.

The description of the motion of the vehicle and the vehicle body with the coordinate system 2 fixed to the body in relation to a inertial frame 0 is formulated using transformation matrices - see figure 4.11.

$$\vec{x}_0 = A_{02}\ (\Psi, \vartheta, \varphi)\vec{x}_2,\ A_{02}^T = A_{02}^{-1} = A_{20} \qquad (4.34)$$

The angles used correspond to the yaw angle, roll angle and pitch angle when driving on a horizontal roadway.

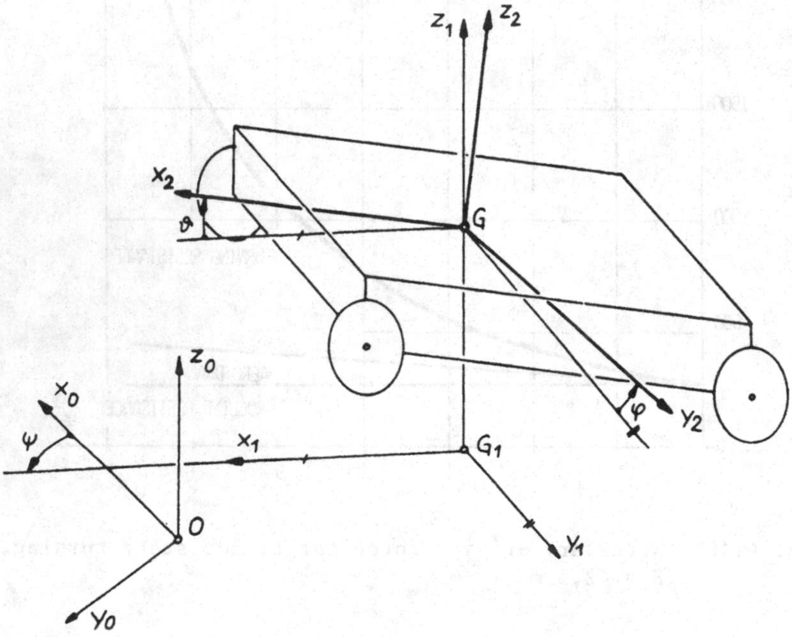

Fig. 4.11: Position of the vehicle (static configuration)

Point G is the center of gravity of the vehicle on an even, horizontal roadway at the actual loading state in the static configuration. The position vector $\vec{x}_G$ in the system 0 must be calculated using the system equations.

The momentary position of a wheel - the wheel numbering and therefore the subscripts be inserted correspond to those in chapter 4.2.1 - in relation to the body (figure 4.12) can be determined using its position M_0 in the design location. After deflection of the spring and rotation around the steering axis of the wheel the position of the center M_δ is:

$$\vec{x}_{M\delta G,2} = \vec{x}_{MOG,2} + \vec{z}_{F,2} + A_{23}(w)(I-D_{33}(\delta_{GE})\vec{n}_{\nu,3}) \quad (4.35)$$

The vector $\vec{z}_F$ and the position of the coordinate system 3 fixed to the steering axis in relation to coordinate system 2 can be described as a function of the spring deflection w and the elasticities of the suspension from the measured wheel elevation curves or their calculation (see |12| for example). In this manner for example figure 4.6 represents a rotation angle of the transformation matrix $A_{23}(w)$. In design location the axes of coordinate systems 2 and 3 are parallel. In the rotation matrix D_{33} the king pin inclination and the caster angle are taken into consideration, while the point of intersection of the steering axis with the plane x_3y_3 is defined by $\vec{n}_\nu$. The steer angle δ_{GE} corresponds to that from equation (4.30).

In respect to system 3 the wheel axis vector $\vec{y}_4$ and system 4 have camber and toe-in angles, described with matrix O_{34}. So for the vector $\vec{y}_4$ presented in system 2 follows

$$\vec{y}_{4,2} = A_{23}D_{33}O_{34}\vec{y}_4 \quad (4.36)$$

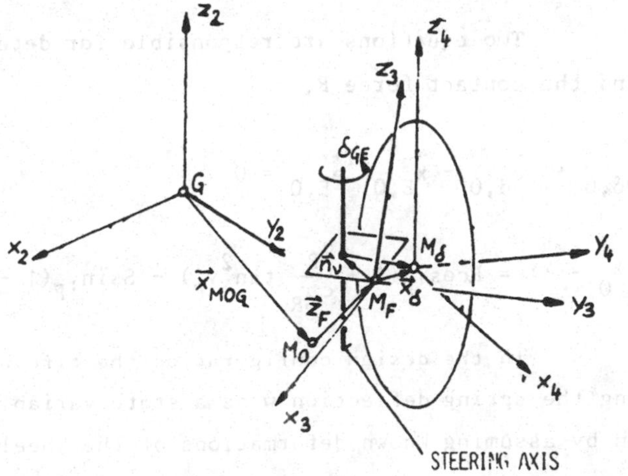

Fig. 4.12: Wheel geometry

The contact of the individual wheel with the road surface is calculated using the tire as an approximately linear spring with radial stiffness c_R and lateral stiffness c_{RS}. A given roadway contour $z_{R,0}(x_0,y_0)$ in system 0 will be replaced vertically under M_δ by its tangential plane. The position vector $\vec{x}_{E,0}$ and a normal unit vector $\vec{e}_E$ are used to define this plane. With the aid of figure 4.13 the position of the point A_M can now be explained. The unit vectors $\vec{e}_U$, $\vec{e}_d$ are determined from the vector products

$$\vec{e}_{U,0} = \frac{\vec{e}_{y4,0} \times \vec{e}_{E,0}}{|\vec{e}_{y4,0} \times \vec{e}_{E,0}|}, \quad \vec{e}_{d,0} = \vec{e}_{y4,0} \times \vec{e}_{U,0} \tag{4.37}$$

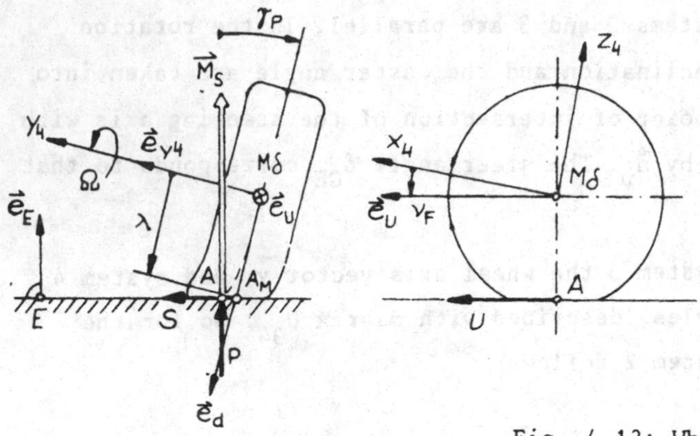

Fig. 4.13: Wheel roadway contact

Two equations are responsible for determination of the distance λ and the contact force P,

$$(\vec{x}_{M\delta,0} + \lambda\vec{e}_{d,0} - \vec{x}_{E,0})^T \cdot \vec{e}_{E,0} = 0 \tag{4.38}$$

$$c_R(\lambda_0 - \lambda) = P\cos\gamma_P\left(1 + \frac{c_R}{c_{SR}}\tan^2\gamma_P\right) - S\sin\gamma_P\left(1 - \frac{c_R}{c_{SR}}\right) - P_0 \tag{4.39}$$

In the design configuration the tire has the contact force P_0. Using the spring deflection w as a state variable and through simplification by assuming known deformations of the wheel suspension (e.g. from the last integration step in the time simulation) the equations can be decoupled and equation (4.39) is the equation for determination of the contact force.

For the position of the wheel contact point must be written

$$\vec{x}_{AM\delta,0} = \lambda\vec{e}_{d,0} + \frac{S}{c_{SR}}(\vec{e}_{E,0} \times \vec{e}_{U,0}) \tag{4.40}$$

The force vectors and the aligning torque M_S, described in system 0 follow using (4.37)

$$\vec{U}_0 = U\vec{e}_{U,0}, \quad \vec{P}_0 = P.\vec{e}_{E,0}$$
$$\vec{S}_0 = S(\vec{e}_{E,0} \times \vec{e}_{U,0}), \quad \vec{M}_{S,0} = -M_S.\vec{e}_{E,0} \tag{4.41}$$

Since steady state values for the lateral force can no longer be expected for rapid steering manoeuvres the tire behavior is described using a well known differential equation which takes the lateral deformation velocity of the tire into consideration.

$$\dot{S}\frac{l_E}{v_A} + S = f_S(\alpha, \gamma_p, U, P), \quad l_E = \frac{\partial fs}{\partial \alpha}/c_{SR} \quad , l_E \geq l_{E,min} \tag{4.42}$$

The function f_S presents the steady state tire characteristics corresponding to (2.8). The transition distance l_E of the tire, the path which the tire requires to adjust to the new conditions, is assumed as function of the normal force, but independent of other variables.

Equations (2.5) and (2.9) still apply for the aligning torque M_S. The longitudinal force U can be calculated using the slip curve, figure 2.2, whereby here the slip angle, that means the lateral slip velocity is taken into consideration. That approximation is a qualitative description of the general behaviour used only as an extension of the tire characteristics given by figure 2.5.

The slip angle and tire slip is determined using the velocity $v_{A,0}$ of the wheel contact point and the rolling radius r which shows little change with changing normal force P in a wide range.

$$\vec{v}_{A,0} = \dot{\vec{x}}_{G,0} + \frac{d}{dt}(A_{02}(\vec{x}_{MOG,2} + \vec{z}_{F,2} + \vec{x}_{\delta,2}) + \vec{x}_{AM\delta,0}) \qquad (4.43)$$

$$\left.\begin{aligned} &\alpha = -\arctan\left(\frac{\vec{v}^{\,T}_{A,0}\cdot\vec{e}_{S,0}}{\vec{v}^{\,T}_{A,0}\cdot\vec{e}_{U,0}}\right), \qquad \vec{e}_{S,0} = \vec{e}_{E,0} \times \vec{e}_{U,0} \\ &S_{LB} = \frac{(\vec{v}_{A,0} - \Omega r\vec{e}_{U,0})^T\cdot\vec{e}_{U,0}}{\vec{v}^{\,T}_{A,0}\cdot\vec{e}_{U,0}}, \quad S_{LT} = \frac{(\vec{v}_{A,0} - \Omega r\vec{e}_{U,0})^T\cdot\vec{e}_{U,0}}{r\Omega} \end{aligned}\right\} \qquad (4.44)$$

Since only the velocity components in the tangential plane without time derivatives of the lateral deformation are required for the slip angle and slip, the time derivative from $|\vec{x}_{AM\delta}|$ can be considered to be zero.

Individual obstacles (pot holes, surface waves) can be taken into consideration using this description of the movement of the individual wheel as long as their dimensions are sufficiently greater than the tire contact length. The description of the lateral forces used presently can, however, only practically consider camber angles γ_P up to a maximum of 10°.

For the total vehicle, six equations of motion have been derived using Newton's law and the Euler equations in relation to the reference point G.

$$m\ddot{\vec{x}}_{G,0} + A_{02}\sum_1^4 m_{TRi}\ddot{w}_i\vec{e}_{z,2i} = \sum_1^4(\vec{U}_{0i} + \vec{P}_{0i} + \vec{S}_{0i}) + \vec{W}_0 - mg\vec{e}_{z,0} \qquad (4.45)$$

$$\Theta_2\dot{\vec{\omega}}_{20,2} + \vec{\omega}_{20,2}\times(\Theta_2\vec{\omega}_{20,2}) + \sum_1^4(\vec{x}_{MOG,2i}\times(m_{TRi}\ddot{w}_i\vec{e}_{Z,2i})) +$$

$$\sum_1^4(\Theta_i\dot{\Omega}_i\vec{e}_{y4,2i} + \vec{\omega}_{20,2}\times(\Theta_i\Omega_i\vec{e}_{y4,2i})) =$$

$$A_{20}\sum_1^4(\vec{x}_{AG,0i}\times(\vec{U}_{0i} + \vec{P}_{0i} + \vec{S}_{0i}) + M_{S,0i}) + M\vec{W}_{G,2} \qquad (4.46)$$

The angular velocity $\omega_{20,2}(\Psi, \vartheta, \varphi)$ describes the rotation of the static configuration in relation to the fixed system 0, written down in system 2. The meanings of the other variables are listed in the nominations.

The equations of motion for the individual wheel are derived using the d'Alembert principle as shown in |8|. For each wheel we obtain two equations corresponding to steer and spring deflection which also contain second derivatives of the body degrees of freedom. Here only the principle relationships are presented in place of the long formulas

$$h_i(\ddot{\delta}_{Ei}, \ddot{\vec{x}}_G, \dot{\vec{\omega}}_{20}, \dot{\delta}_{Ei}, \delta_{Ei}, \ldots U_i, P_i, S_i, M_{Si}, \delta_{Gi}) = 0$$

$$\delta_{GEi} = \delta_{Gi} + \delta_{Ei} \qquad i = 1, \ldots 4 \tag{4.47}$$

$$g_i(\ddot{w}_i, \ddot{\vec{x}}_G, \dot{\vec{\omega}}_{20}, \ldots U_i, P_i, S_i, M_{Si}, F_i) = 0 \quad i = 1, \ldots 4 \tag{4.48}$$

Non-linear spring/damper forces and the effects of anti-roll bars are symbolized by the function $F_i(w_i, w_j, \dot{w}_i)$. The geometric steer angles δ_{Gi} (δ) are determined by the reference value $\delta(t)$ at the output of the steering gear box and the geometry of the steering linkage.

For the wheel spin velocity relationships similar to equation (3.6) can be derived, which are written down only for a rear wheel drive with the clutch engaged.

$$\left.\begin{aligned} &\dot{\Omega}_i\left(\Theta + \frac{\Theta_M i_G^2 i_D^2 \eta}{2}\right) = \frac{M_M i_G i_D \eta}{2} - (U_i + P_i f_R)\lambda_i - M_{Bi} + \\ &\frac{\Theta_M i_G^2 i_D^2 \eta}{4\Theta}\left((U_j + P_j f_R)\lambda_j + M_{Bj} - (U_i + P_i f_R)\lambda_i - M_{Bi}\right) \\ &M_M > 0; \quad i = 3,4; \quad j = 4,3 \end{aligned}\right\} \tag{4.49}$$

The longitudinal force U_i contains the state variable Ω_i via the wheel slip. The engine torque M_M is a function of $(\Omega_3 + \Omega_4)/2$ due to the differential analog to (3.9).

Using the derived equations, the vehicle motion can now be described via the six degrees of freedom of the body as well as per wheel jounce and rebound, steering and spin. Through the use of the differential equations (4.42) the system is expanded by four state variables, the lateral forces of the individual wheels. The input variables are the steering wheel motions, represented by the steer angle at the output of

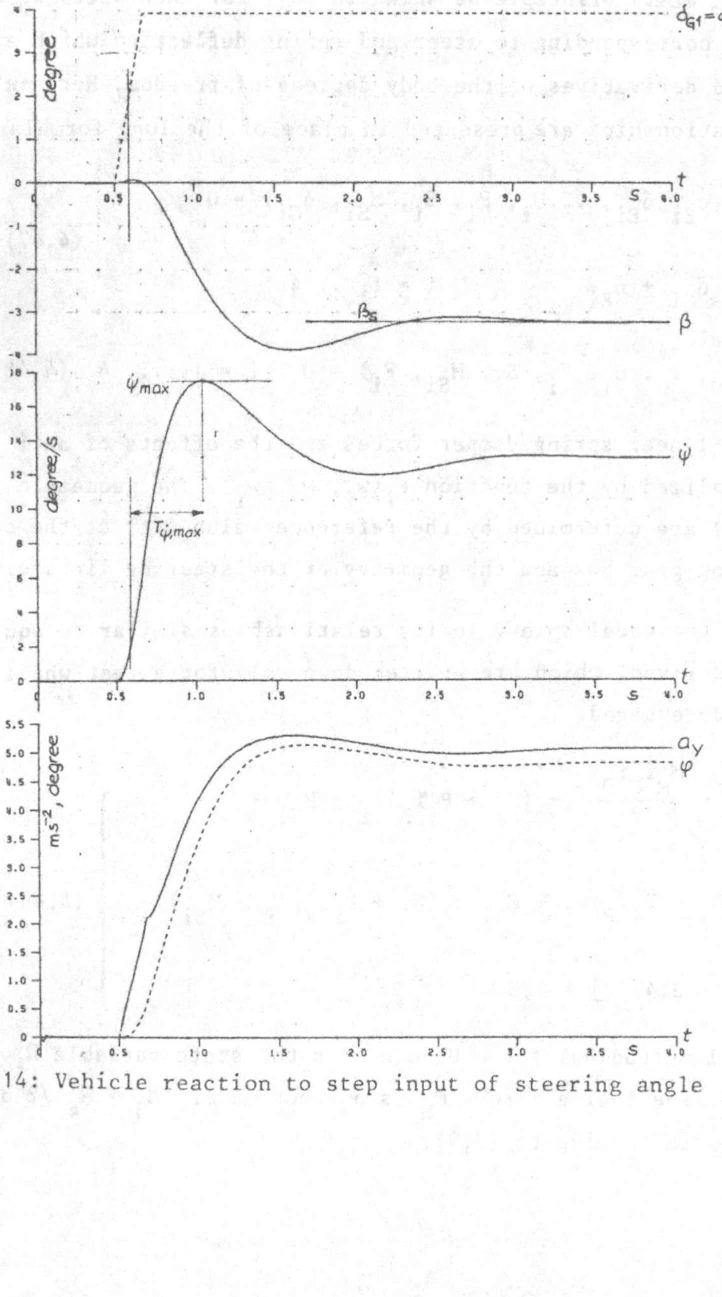

Fig. 4.14: Vehicle reaction to step input of steering angle

the steering gear box and the accelerator position, in the calculations represented by the driving moments at the wheels. For the numerical simulation integration increments of 1ms were used due to the high eigen dynamics of some parts of the system.

Figure 4.14 shows the vehicle response to a step input of the steering angle corresponding to chapter 3.3.1 of part II. The vehicle drives 0.5 s straight ahead until the driver forces the vehicle as fast as possible into a steady state turning. The change of the steering angle δ at the output of the steering gear box is adequate to a steering wheel speed of about 400°/s. The steady state value of δ produces a steady state turning with $a_y \cong 5\ m/s^2$. The drive moments at the rear wheels where chosen in such a way, that the velocity changes less than 1 % in the shown time period. After 3,5 sec the steady state driving conditions are already reached. Caused by the statement of the steering angles $\delta_{G1} = \delta_{G2} = \delta$ - the vehicle shows an immediate change of the lateral acceleration. The yaw velocity $\dot{\psi}$, but especially the roll angle φ and the side slip angle β increase with a distinct time lag. By the time-history of $\dot{\Psi}$ and δ_{G1} the vehicle factor TB can be calculated: $TB = T_{\dot{\Psi},max} \cdot \beta_s = 2,86$ degrees. The comparison of the vehicle factors TB of different cars can be used to get valuations for the vehicle handling - see part II.

To get objective informations in respect to the vehicle handling another essential test demands the braking of the vehicle starting with a steady state condition - see chapter 3.4, part II. With fixed steering wheel angle the brakes are applied as fast as possible up to a constant value of deceleration.

Because the description of the tire characteristics at great slip values s_L and simultaneously greater slip angles α is a more qualitative assumption, this test is only simulated for relatively small decelerations. In figure 4.15 the vehicle has a steady state condition of $v_x = 80$ km/h (22,2 m/s), $a_y = 5\ m/s^2$. The slow down starts at 0,4 s

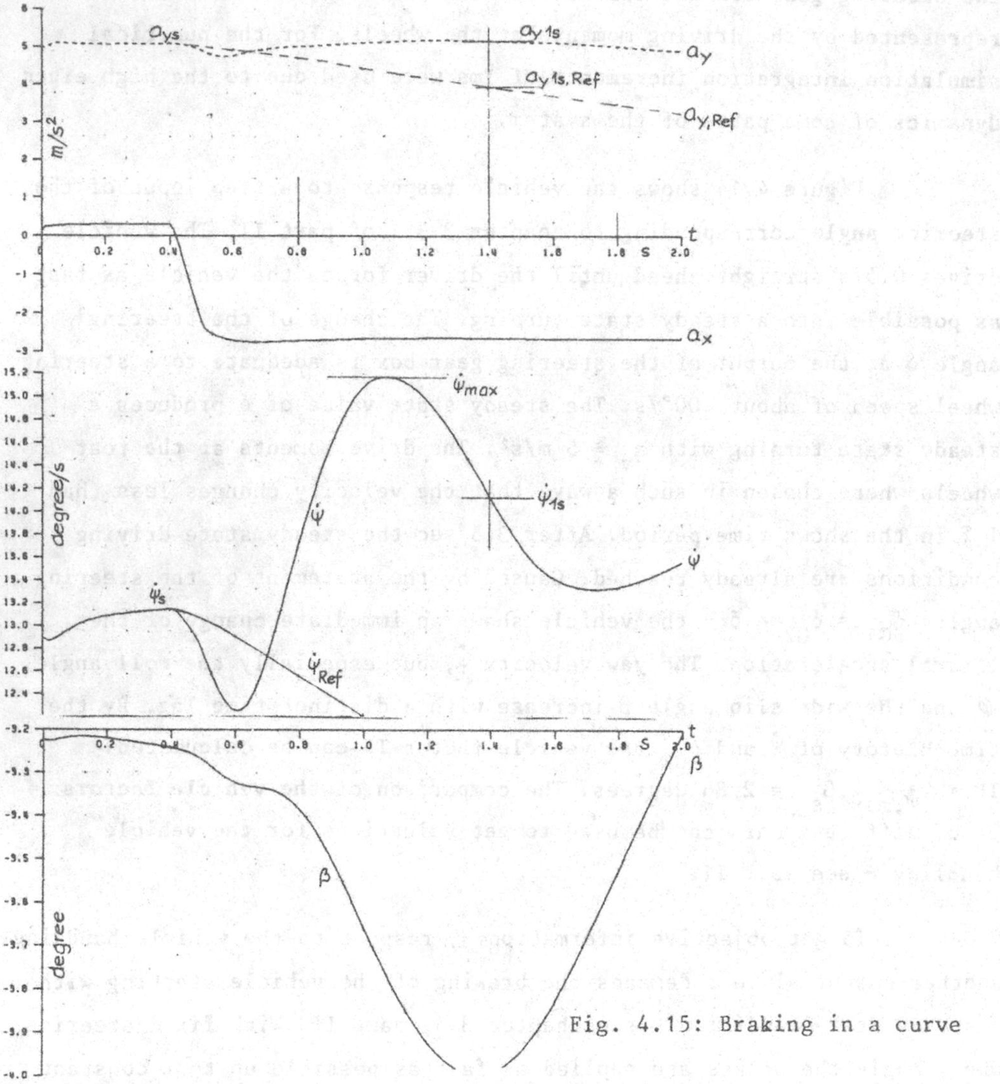

Fig. 4.15: Braking in a curve

and after 0,2s a deceleration $-a_x \cong 2,7$ m/s² is reached. The yaw velocity first decreases but afterwards increases to a maximum of $1,16\ \dot{\Psi}_s$. In correspondence to the reaction time of the driver after 1 s the yaw velocity is still $\dot{\Psi}_{1s} = 1,08\ \dot{\Psi}_s$ in comparison to its value $\dot{\Psi}_{1s,Ref} = 0,87\ \dot{\Psi}_s$ provided the vehicle does not leave the circular path. The corresponding values of the lateral acceleration are $a_{y1s} \cong a_{ys}$ and

$a_{y1s,Ref} = 0{,}76\ a_{ys}$. The absolute value of the vehicle side slip angle β changes only slightly. The vehicle turns towards the center of the curve and drives along with decreasing radius but without an uncontrollable break away.

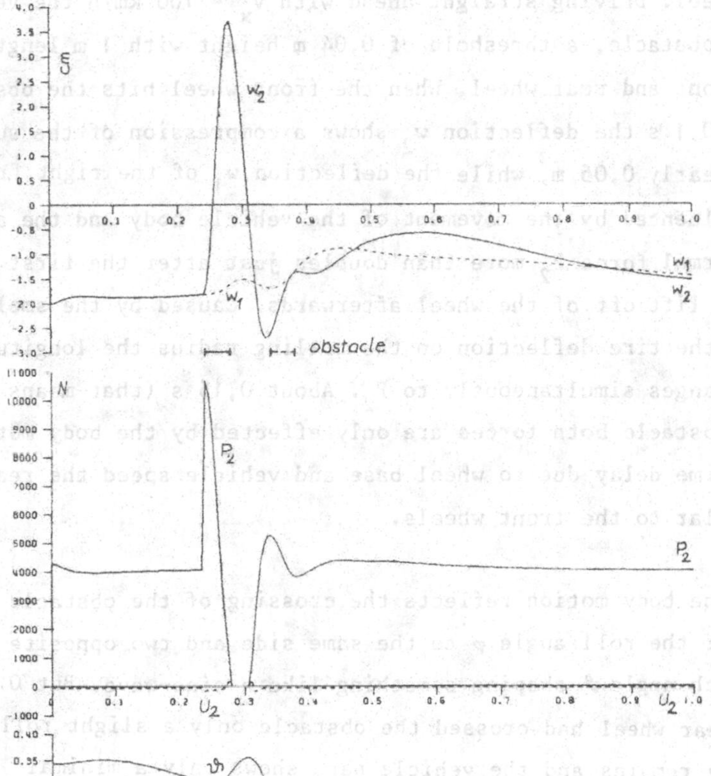

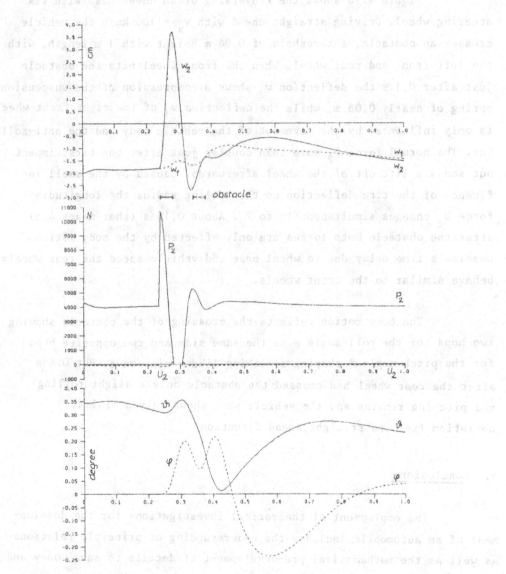

Fig. 4.16: Crossing of an obstacle, v = 100 km/h
Reaction of front wheels and car body

Higher decelerations will show the limits of the driving range due to the joint effects of lateral and longitudinal accelerations - see measurements in part II.

Figure 4.16 shows the traversing of an unevenness with fix steering wheel. Driving straight ahead with v_x = 100 km/h the vehicle crosses an obstacle, a threshold of 0,04 m height with 1 m length, with the left front and rear wheel. When the front wheel hits the obstacle just after 0,1 s the deflection w_2 shows a compression of the suspension spring of nearly 0,05 m, while the deflection w_1 of the right front wheel is only influenced by the movement of the vehicle body and the anti-roll bar. The normal force P_2 more than doubles just after the first impact but shows a lift off of the wheel afterwards. Caused by the small influence of the tire deflection on the rolling radius the longitudinal force U_2 changes simultaneously to P_2. About 0,15 s (that means 4 m) after the obstacle both forces are only effected by the body motion. Despite a time delay due to wheel base and vehicle speed the rear wheels behave similar to the front wheels.

The body motion reflects the crossing of the obstacle showing two hops for the roll angle φ to the same side and two opposite hops for the pitch angle ϑ shaping something like a sine-wave. But 0.4 s after the rear wheel had crossed the obstacle only a slight rolling and pitching remains and the vehicle path shows only a minimal deviation from the straight ahead direction.

5. Conclusions

The employment of theoretical investigations for the development of an automobile includes the understanding of principle relations as well as the mathematical pre-development of details to save money and time.

Already the most simple models make it possible to demonstrate and understand the effects of larger differences in the conception of a car. They also allow to interpret fundamentally the behavior of the vehicle analyzed by experiments. Taking into consideration how fast, without problems and using only few essential vehicle data the results can be obtained, the use of a 2-wheel-model is not only an important completion to the evaluations by more complex models but sometimes also the single opportunity to get informations and estimations in time.

More detailed investigations in respect to the vehicle behaviour at the limits of the driving range require complex 4-wheel vehicle models. Thereby not only the mathematical formulation of the vehicle model but also of the tire forces have to correspond with the problem. The introduced 4-wheel-model including the steering and bouncing of each single wheel makes it possible to describe the properties of different wheel suspensions and the motion of the vehicle in the time domain. For steady state turning a slightly simplified model is used. Hereby the iterative evaluation of the state variables saves expensive computer time.

Road tests, the measured reality, are the criteria of the usefulness of all calculations. On the other hand the theoretical background is essential for the analysation of the test results. The verification of effects of system modifications by tests gives a final prove of their feasibility.

The continuously improving possibilities of sensors and data processing equipments, statistical evaluation and the fast availability of results lead to a refinement and enlarge the field of vehicle dynamic tests.

Theorie and test together build the basement to improve the vehicle behaviour until the limits of the driving range and also to expand these limits.

NOTATION

a_q	m/s^2	centripedal acceleration
a_x	m/s^2	longitudinal vehicle acceleration
a_y	m/s^2	lateral acceleration
B	N	braking force
c(w)	N	spring force
c_L	Nm	steering system elasticity rate
c_{LC}	Nm	steering column elasticity rate
c_R	N/m	tire spring rate
c_{SL}	N	steering linkage elasticity rate
c_{SR}	N/m	lateral tire spring rate
c_T	Nm	spring rate of the anti roll bar
c_x, c_y, c_{z1}, c_{z2}	-	coefficients related to the aerodynamic
c_{Mx}, c_{My}, c_{Mz}	-	forces and moments
D	-	damping ratio
D_{33}		matrix describing the rotation in respect to the steering axis of the wheel
$\vec{e}$		unity vector
f_R	-	rolling resistance force coefficient
F	m^2	frontal area of the vehicle
F_i	N	spring and damper forces of wheel suspension
G = mg	N	vehicle weight
G_w	N	wheel weight
h, h+ h_o, h_φ, h_ϑ	m	position of CG above ground, components of CG position
i_D	-	transmission ratio of differential
i_G	-	ratio of manual transmission
i_L	-	steering ratio
I		unity matrix
k_S	N	cornering stiffness
k^x_{SF}	N	abbreviation, see (4.4)
k_y, k_M		abbreviations, see (4.5)
K_1, K_2, K^x_2		abbreviations, see (4.11)

l, l_F, l_R	m	wheel base, front and rear distance of CG or reference point from wheel axis
m, m_{red}	kg	vehicle mass, reduced vehicle mass
m_w, m_{TR}	kg	wheel mass, mass of wheel and part of wheel suspension
M_A, M_B	Nm	drive moment and braking moment at the wheel
M_M	Nm	engine torque
M_S	Nm	aligning torque of the tire
M_{SP}	Nm	steering moment in respect to steering axis
MW_{zS}	Nm	aerodynamic moment in respect to CG, z-axis
MW_x, MW_y, MW_z	Nm	aerodynamic moments
$\vec{n}_\nu$	m	position vector of steering axis
n_K	m	wheel offset (castor offset)
n_M	U/min	engine speed
n_S	m	tire offset (pneumatic trail)
N_M	W	engine power
P, P_N	N	normal tire force, nominal tire load
q, q_E	-	slope, street inclination; see (3.24)
q_w	N/m²	aerodynamic pressure
r	m	rolling radius of the wheel
R	m	radius at steady state turning
s	m	wheel track
s_L, s_{LT}, s_{LB}	-	slip, driving slip, braking slip
$s_{L,max}$	-	slip at $\mu_{L,max}$
s_φ	m	component of CG position
S, S_γ	N	lateral force, lateral force due to wheel camber
T_z	s	numerator time constant
$u_G, u_{G,max}$	m	accelerator position and its maximum value
U	N	longitudinal tire force
$v, \vec{v}$	m/s	velocity of CG, value and vector
v_x	m/s	longitudinal velocity of the vehicle
V, V_B	N	driving force and braking force of drive train

w	m	spring deflection
$\vec{w}_W$	m/s	air velocity vector
$W_L = -W_x$	N	aerodynamic drag
W_x, W_y, W_z	N	aerodynamic forces
$\vec{x}$	m	position vectors
X, Z	N	forces at the wheel bearing
$\vec{z}$	m	position vector of wheel center due to spring deflection
α, α_f		slip angle, fictive slip angle
β		side slip angle of the vehicle
γ, γ_P		wheel camber in respect to car body and in respect to the vertical direction of the road surface
δ		steering angle assigned to the steering gear exit
δ_E		elastic steer angle in respect to steering linkage
$\delta_F, \delta_F^x = \delta_H/i_L$		front wheel steering angle, 2 wheel model
δ_G		geometrical steer angle of the wheel
δ_H		steering wheel angle
δ_K		toe-in angle caused by suspension geometry and elasticity
Θ	kgm²	wheel moment of inertia in respect to spin axis
$\mathbb{\Theta}$	kgm²	inertia tensor of the vehicle in respect to point G
Θ_G	kgm²	inertia of transmission
Θ_M	kgm²	equivalent engine inertia
Θ_z	kgm²	moment of inertia of the vehicle in respect to the vertical axis, CG
η	-	efficiency of drive train including the wheel bearings
ϑ		pitch angle
ϑ_s		angle of road slope
λ	m	loaded wheel radius

μ	-	contact force coefficient
μ_{erf}	-	required traction coefficient
μ_L, $\mu_{L\,max}$, μ_{Lg}	-	longitudinal force coefficient and maximum value and sliding coefficient
$\mu_{S,max}$	-	maximum lateral force coefficient
ν	-	subdivision factor of distribution unit
ν_n	1/s	natural frequency without damping
τ		aerodynamic angle
φ		roll angle
$\dot{\psi}$	1/s	yaw velocity
ω_M	1/s	angular velocity of the engine
Ω	1/s	wheel spin velocity

Subscripts

capital letters	specification of special points, see diagrams, figures
numbers, i	number of coordinate system or number of wheel
O	starting configuration or values for $v \rightarrow 0$
F	front
R	rear
s	steady state condition

VEHICLE DATA

Chapter 4.1: Linearized 2-wheel-model

l_F = 1.44 m	oversteer	understeer
l_R = 1.36 m		
m = 1900 kg	k_{SF} = 90000N	60 000 N
Θ_z = 2900 kgm²	k_{SR} = 80000N	110 000 N
k_M = 1,16Ns²/m		
k_y = 2,9Ns²/m²		

Chapter 4.2.1: Steady state turning, 4-wheel-model

l_F = 1,44 m	m = 1900 kg
l_R = 1,36 m	k_M = 1,16 Ns²/m
h_o = 0,58 m	k_y = 2,9 Ns²/m²
s = 1,47 m	f_R = 0,015
s_φ = 0	c_{SL1} = c_{SL2} = 23 000 Nm
h_φ = 0,48 m	c_{LC} = 35 Nm
h_ϑ = 0,30 m	i_L = 22,36

Anti roll bars: c_{TF} = 7400 Nm, c_{TR} = 1500 Nm

Parallel steering : $\delta_{G1} = \delta_{G2} = \delta$

Position and inclination of the steering axis in design configuration: camber angle 12,2°, caster angle 9°, steering offset 0, wheel offset 0.027 m.

Wheel elevation curves, factors k_{Ai}, k_{Bi}, $n_{Ki}(w_i)$: approximated experimental data, see e. g. fig. 4.6.

Tire characteristics corresponding to fig. 2.3, radial tire, $\mu_{S,max.} = \mu_{L,max.}$ = 0,85.

Forces of suspension springs: approximated experimental data, see e.g. fig. 4.7.

Rear wheel drive.

Chapter 4.2.2: 4-wheel-model, only differences or additional datas to chapter 4.2.1 are listed below.

l_F	= 1,37 m	m	= 1760 kg
l_R	= 1,43 m	m_{TRi}	= 30 kg
h_o	= 0,56 m	Θ	= 0,9 kgm²
c_R	= 166000 N/m	Θ_M	= 0,1 kgm²
c_{SR}	= 120000 N/m		

$$\Theta = \begin{pmatrix} 620 & 0 & 0 \\ 0 & 2650 & 0 \\ 0 & 0 & 2820 \end{pmatrix} \text{kgm}^2$$

Radius of the unloaded tire r_o = 0,317 m

Aerodynamic forces and moments: approximated experimental data, see e.g. fig. 2.8.

Damper characteristics, friction, suspension elasticities: approximated experimental data.

Longitudinal force coefficient corresponding to fig. 2.2, as a function of slip angle α, $\mu_{L,max.}$ = 0,85.

Brake balance: M_{BF}/M_{BR} = 2.65

LITERATURE:

|1| H. Pfitzer: Erstellung eines real-time Rechenprogrammes zur Simulation der Längsdynamik eines PKWs
Diplomarbeit, Universität Stuttgart 1981

|2| M. Mitschke: Dynamik der Kraftfahrzeuge
Springer Verlag, Berlin, Heidelberg, New York 1972

|3| K. Desoyer: Fahrzeugdynamik
Skriptum zur Vorlesung 309.155, 1. Teil
Techn. Universität Wien 1980

|4| K. Desoyer; P. Lugner: Bewertung der kritischen Fahrzustände eines Kraftfahrzeuges in stationärer und instationärer Fahrbewegung in Geradeausfahrt und Kurvenfahrt
Österr. Bundesministerium für Bauten und Technik
Straßenforschung Heft 23, 1974

|5| P. Riekert; T. Schunk: Zur Fahrmechanik des gummibereiften Kraftfahrzeugs
Ing. Archiv 11 (1940)

|6| H. Springer: Untersuchungen der allgemeinen ebenen Bewegung eines luftbereiften und nicht angetriebenen PKW
Dissertation, Tech. Universität Wien, 1971

|7| P. Lugner: The influence of the structure of automobile models and tire characteristics on the theoretical results of steady-state and transient vehicle performance
Proceedings of the 5th VSD - 2nd IVTAM symposium "The dynamics of vehicles on roads and on tracks" 1977

|8| W. Schielen; E. Kreuzer: Rechnergestütztes Aufstellen der Bewegungsgleichungen gewöhnlicher Mehrkörpersysteme
Ingenieur-Archiv 46, 1977

|9| R. Gnadler: Umfassendes Ersatzsystem zur theoretischen Untersuchung der Fahreigenschaften von vierrädrigen Kraftfahrzeugen
AI, 1971, Heft 9

|10| K. Rompe: Zum Lenkverhalten von Kraftfahrzeugen bei stationärer und instationärer Kreisfahrt im Grenzbereich
Dis. TU Hannover 1972

|11| W. Gengenbach: Experimentelle Untersuchungen von Reifen auf nasser Fahrbahn
ATZ 70 (1968)

|12| J. Bukovics; P. Lugner: Eine Möglichkeit der Beschreibung von Einzelradaufhängungen im Hinblick auf fahrdynamische Untersuchungen
AI, 1980, Heft 2

|13| F. Böhm: Zur Mechanik des Luftreifens
Habilitationsschrift TH Stuttgart, 1966

|14| K. Meier-Dörnberg; B. Strackerjan: Prüfstandsversuche und Berechnung zur Querdynamik von Luftreifen
AI, Heft 4, 1977

|15| H. B. Pacejka: Tire factors and vehicle handling
Intern. Journal of Vehicle Design, Vol. 1,1 1979

HORIZONTAL MOTION OF AUTOMOBILES, VEHICLE HANDLING, MEASUREMENT METHODS AND EXPERIMENTAL RESULTS

Dr.-Ing. Adam Zomotor
Daimler-Benz AG
Postfach 2 02
7000 Stuttgart 60

INTRODUCTION

In addition to the subjective evaluation of vehicle handling, measurements of vehicle dynamic are more and more used as an aid in the development of vehicles. The general desire to retain the feeling of skilled test engineers in the shape of reproducible data for the purpose of obtaining the means for comparative improvements resulted especially during the past years in an increased application of measuring techniques. Based on this development, a number of test methods were established which cover important driving situations for vehicle evaluation. The same assignment and last but not least the cooperation between manufacturers of vehicles and the manufacturers of measuring instruments, as well as between institutes, resulted worldwide in a certain standardization of measuring methods and measuring instruments. The successful support of research and development by means of measurement requires that the measuring results of different variations are available as quickly as possible. Meeting such a demand assumes the use of computers for data input and evaluation. The following is a description of the measuring instruments now in general use for vehicle measurements.

MEASURING INSTRUMENTS FOR VEHICLE MEASUREMENTS

To cover the variables relevant to vehicle dynamic the manufacturers of measuring instruments, to a great extent in cooperation with the manufacturers of vehicles, have been developed special measuring data transducers and evaluation instruments.

Longitudinal and Lateral Acceleration

In most cases accelerations and decelerations in a horizontal level are measured with an accelerometer mounted on a stable platform, Fig. 1. Stabilizing excludes any influence

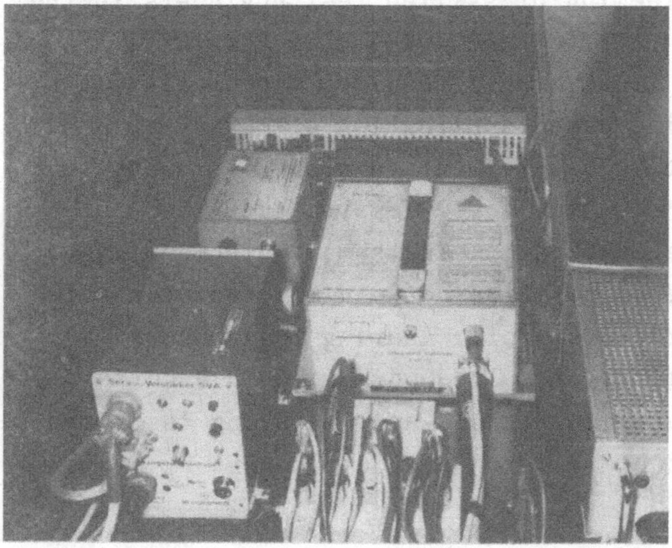

Fig. 1:
Stable Platform (Novotechnik)

on measuring results by components of gravity due to both the vehicle roll angle and pitch angle.

The measuring direction remains horizontal. However, influences caused by a road surface inclination have to be taken into consideration.

Accelerations can also be measured with transducers mounted directly on the sprung mass of the vehicle. In this case its output has to be corrected for the component of gravity on the accelerometer axis due to both the vehicle roll angle and pitch angle.

Forward and Lateral Velocity

Forward and lateral vehicle velocity are mainly measured with non-contact speed sensors. The Leitz sensors Correvit L and Q, Fig. 2 used for this purpose are working on account of optical correlation method with spatial frequency filtering. The surface structure of the road is reproduced on a grating

Fig. 2: Optical Speed Sensors on the Vehicle
Leitz Correvit L and Q

and the passing light is collected by a photographical recorder. The frequency of the received signal is proportional to the velocity at which the picture field is moved normal to the grid lines.

In another often used method the forward velocity is determined by the spin velocity of the wheel. For this purpose, an inductive impulse transducer is used in combination with a toothed disk which rotates with the wheel. A brake disk, for example, is suitably prepared for this purpose, Fig. 3. The impulses are then processed by a frequency to voltage converter.

Fig. 3:
Impuls Transducer and toothed brake disk for measuring forward velocity

Yaw Angle, Yaw Velocity

For yaw angle measuring a directional gyro stabilized in a vertical plane may be used, Fig. 4. The turning of the gyro housing attached to the vehicle in relation to gyro is obtained by a potentiometer. A built-in differentiator can also be used to derive the yaw velocity.

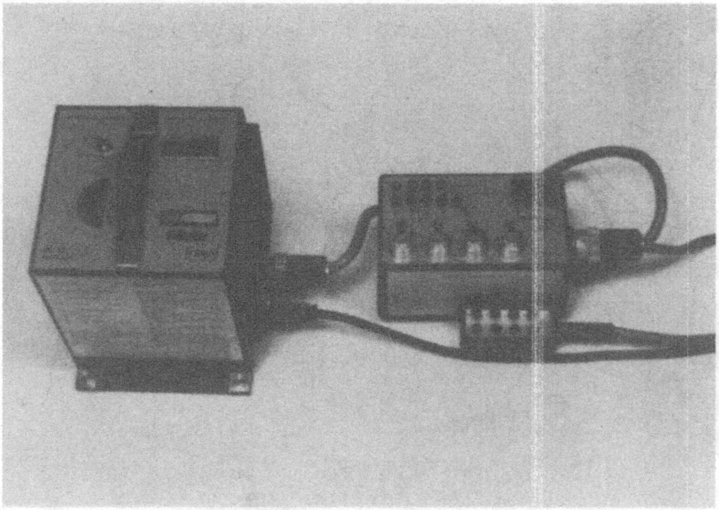

Fig. 4: Directional Gyro for obtaining yaw angle
and yaw velocity (Novotechnik)

Another possibility for direct measurement of the yaw velocity is the spring restrained rate gyro, Fig. 5. The control current for restraining the gyro is proportional to the yaw velocity.

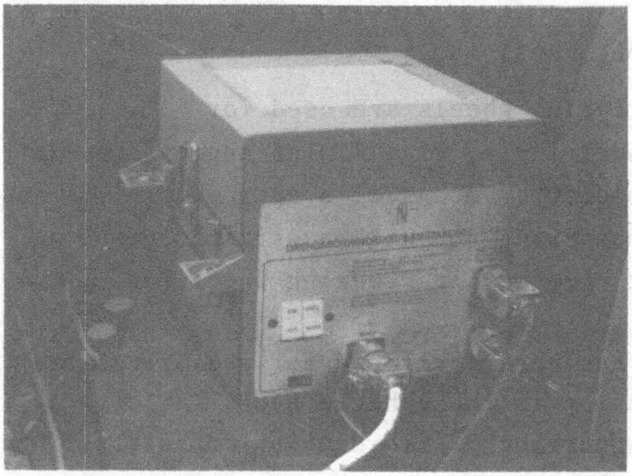

Fig. 5:
Spring Restrained
Rate Gyro (Novotechnik)

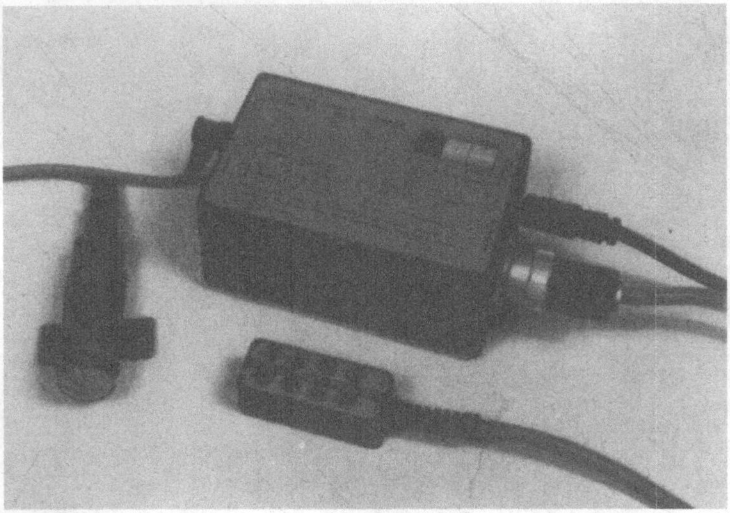

Fig. 6: Rate Sensor (Novotechnik)

For direct measurement of the yaw velocity the so-called rate sensor is often used, Fig. 6. The principle is based on the fact that during the rotation of the housing a gas jet is diverted by the Corriolis force. This diversion is proportional to the angular velocity. The small dimensions and the robustness of the instrument are of advantage. There are no sensitive bearings in contrast to the gyro.

Steering Wheel Angle, Front Wheel Angle

Special measuring steering wheels are used for measuring the steering wheel angle, Fig. 7. The angles are transferred to a potentiometer by means of a gear wheel transmission. A torsion measuring hub with a strain gages bridge is integrated in the measuring steering wheel for measuring the torque.

A number of devices has been developed for measuring the front wheel angle in relation to the vehicle body. The device for measuring the front wheel lock while driving consists of

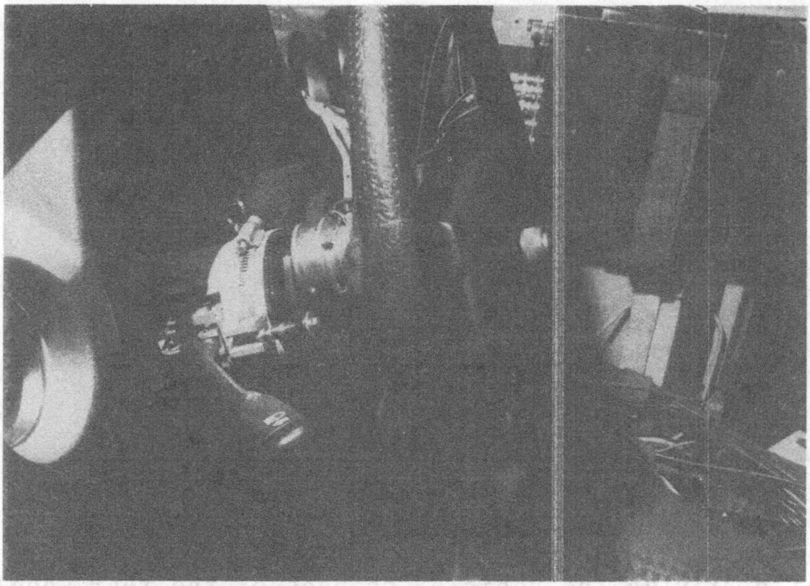

Fig. 7: Measuring Steering Wheel for obtaining steering wheel angle and torque

a lever mechanism which presses a sensor against a face plate on the wheel. The face plate is attached to the hub stud and is not rotating with the wheel. The steering angle of the front wheel is transmitted to an electric angle transducer. Changes in track width and camber are compensated by guiding on a parallelogram. This can be attached to the vehicle body by means of a frame, Fig. 8.

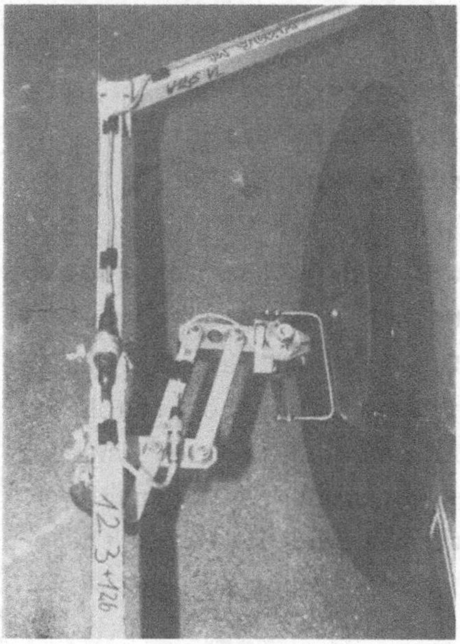

Fig. 8: Device for measuring front wheel lock angle

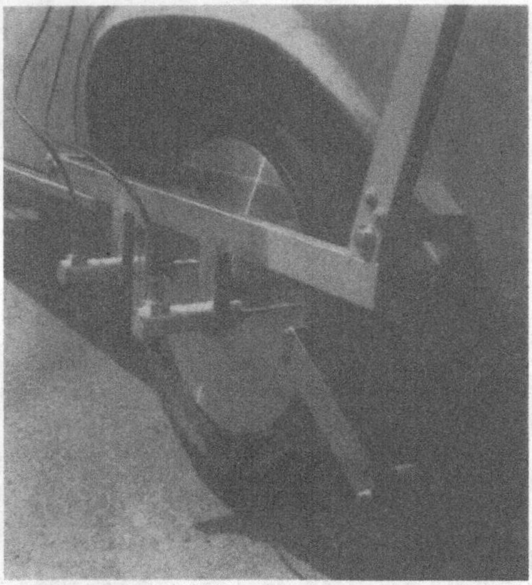

Fig. 9:
Inductive Transducers for measuring wheel toe changes on the rear axle

While driving the vehicle, changes of the rear wheel toe angle in relation to the body, can be obtained with the same device. The face plate is rotatably mounted on the rear wheel rim and held by means of an arm in relation to the vertical axis, so that the face plate is not rotating with the wheel. Since the angles on the rear axle are only small, simple inductive transducers can here also be used, Fig. 9.

Sideslip Angle, Slip Angle

Sideslip and slip angles can be computed from the measured forward and lateral velocity or can be directly determined by a rotatably suspended trolley wheel. The measuring instruments can be attached to the vehicle body for measuring the Sideslip angle, see Fig. 2. For measuring the slip angle, these measuring instruments are attached to the wheel. For this purpose, the optical forward and lateral speed sensor Fig. 10 may be used or a trolley wheel may be attached to the hub stud, Fig. 11. The rotatably mounted trolley wheel arm adjusts itself in direction of the movement. A potentiometer measures the angle in relation to the wheel plane, the

Fig. 10:

Leitz Sensors on the wheel for measuring slip angle

slip angle. A second potentiometer permits determining the camber, that is the inclination of the wheel plane to the road surface.

Fig. 11: Trolley Wheel for measuring slip angle (Eng. School Offenburg)

Roll Angle, Pitch Angle

The stable platform (refer to Fig. 1) is also suitable for measuring the roll or pitch angle of the vehicle. The swivel movement of the instrument housing firmly attached to the vehicle as compare to the always horizontally stabilized platform is obtained by a potentiometer. Another possibility is the non-contact measuring of the ground distance at three points of the vehicle by optical distance sensors, Fig. 12. Roll and pitch angle can be computed from the ground distances and the geometric dimensions of the locating points of the three sensors on the vehicle body.

Brake Pressure

The oil pressure in the braking system is obtained by means of a pressure transducer, which is installed at the master cylinder output.

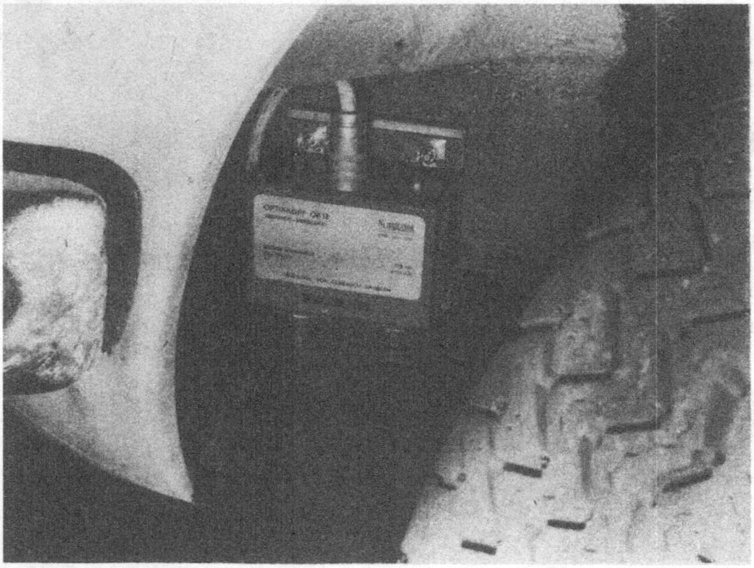

Fig. 12: Optical Distance Sensor for obtaining roll and pitch angle (Novotechnik)

RECORDING AND EVALUATION METHODS

In practical measuring techniques, a few methods and instruments for recording and evaluation of pertinent data have been extracted. Depending on assignment and equipment the following methods can be applied in general:

Direct Recorder

The most simplified method is the use of a directly writing oscillograph for recording if few measuring variables and short measuring periods are involved, Fig. 13. The time functions are immediately displayed while measuring. Evaluations are made by measuring amplitudes manually. The method is less suited for fast vibrations where very many data are coming up. In such cases it will nevertheless be useful to observe basic connections if the signals recorded on magnetic tape are made visible on an oscillograph.

Fig. 13: Directly Recording Oscillographs
Visicorder (Honeywell)

FM Magnetic Tape

For long measuring periods and quick processes, the recording with frequency modulation on a tape recorder has proven its worth, Fig. 14. The frequency modulated analog tapes are subsequently digitalized for evaluation by a computer. The conversion of the tapes requires additional time and computer capacity, for this reason the method should be used where results are not required immediately. The number of channels is limited.

With an 1/2" tape seven tracks and one voice track are available. A disadvantage is that fluctuations of tape speed will show up as measuring faults and therefore very accurate synchronization is required. Only a few tape recorders are available to withstand any acceleration suitable for use in the vehicles; they are in addition relatively large and are consequently rather heavy.

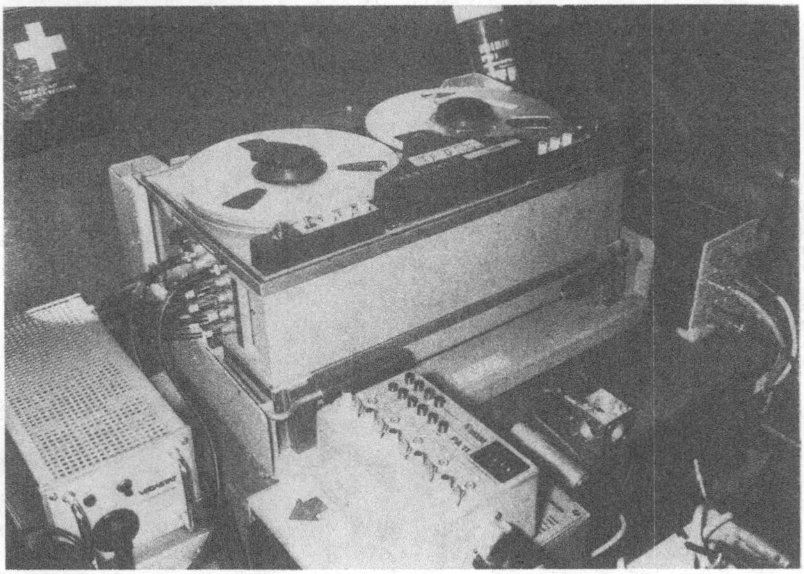

Fig. 14: FM-Tape Recorder, Frequency Modulation, (Honeywell)

Fig. 15: PCM Measuring System, Pulse-Code-Modulation, (Lennartz and Hewlett-Packard)

PCM Method

The recording of measuring signals with a PCM system (pulse-code-modulation) permits the storage of large quantities of data at low space requirements for the unit, Fig. 15. For this method regular samples are taken from analog signals and are shown as a binary digit on the magnetic tape. The time-equivalent samples of different functions are recorded on a tape by a multiplex system one after the other. This system permits the recording of eight functions on one track. Upon conversion these measuring variables can directly be evaluated by a computer. A disadvantage is that the measuring signals are not visible and that the evaluation cannot be made at the test site.

Mobile, Computer-Aided Measuring and Evaluation System

The high demands of today's practical research and the need to quickly obtain large volumes of data for immediate evaluation at the test site lead to the development of a mobile, computer-aided measuring and evaluation system. The measuring data recording system comprises the respectively required analog amplifiers, a process computer with operating terminal and a digital cassette mechanism for intermediate storage of the recorded data. The evaluation system includes a desk computer (e.g. HP 9845 B) with an additional cassette mechanism and a four-colour plotter of DIN A 3 size to issue the final diagrams. For a combination of the two systems there are basically two possibilities available, the "OFF LINE" and the "ON LINE" data connection, Fig. 16.

In most cases, the "OFF LINE" data transfer performed by a digital cassette is used in practice. For this purpose the data collecting unit with signal processing, storage and operating terminal is installed in the test car, Fig. 17 as shown in Fig. 16 too. The evaluation equipment is housed in a separate vehicle acting as a mobile computer center, Fig. 18.

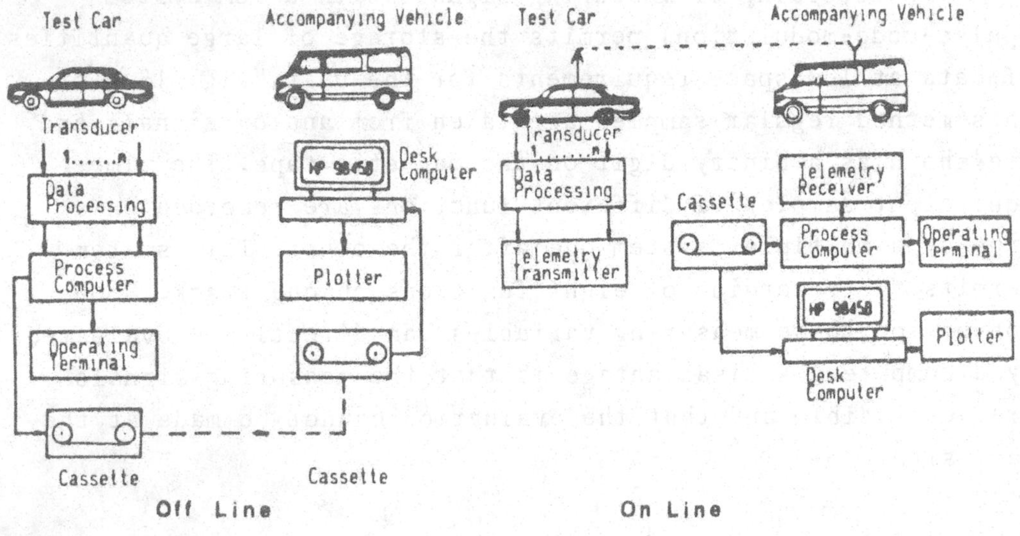

Fig. 16: Mobile Measuring and Evaluation Systems

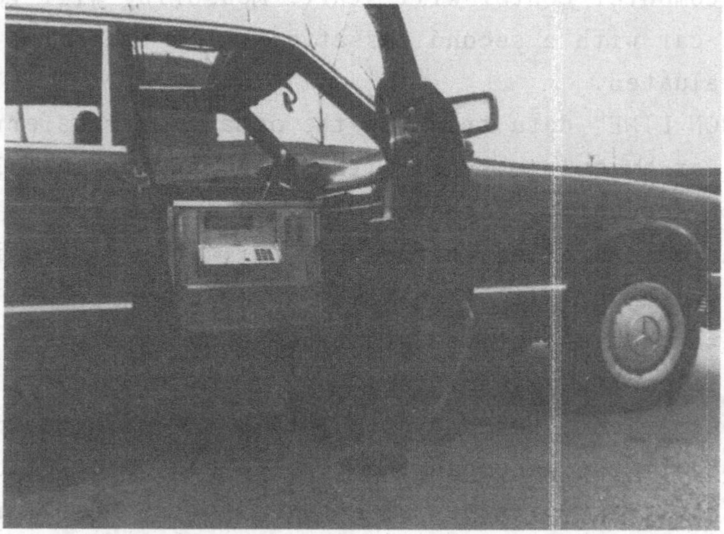

Fig. 17: Measuring Equipment VCS 102 installed in the test car

Following one or several measuring series the data cassette and the printout are taken from the test car and the evaluation

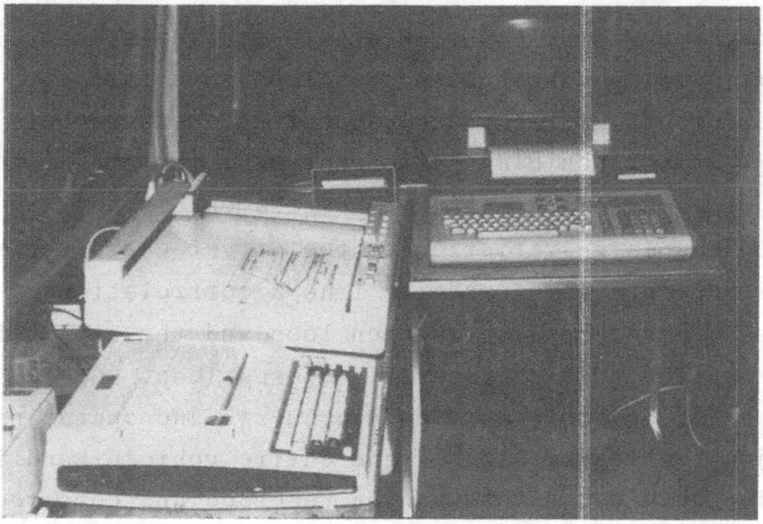

Fig. 18: Evaluation Equipment with desk computer, plotter and operating terminal installed in the accompanying vehicle

in the mobile computer center will start. Measuring will go on in the test car with a second cassette, while the first cassette is evaluated.

For the "ON LINE" data transfer the data are transferred from the test car by telemetry into the accompanying vehicle, Fig. 16. If, on account of technical reasons only a few instruments can be installed in the test car, this system is preferred. The test car will then hold only the required transducers with the respective data processing (analog amplifier and telemetry transmitter).

However, experience has shown that this measuring method can only be used if the accompanying vehicle can be positioned close to the test track with visual contact to the test car, if possible, since otherwise during telemetry transmission disturbances may occur depending on the environment.

TEST METHODS AND MEASUREMENT PROCEDURES FOR VEHICLE DYNAMICS

Today, on the basis of experience selected individual disciplines are usually examined in open loop to obtain technical measurements of the vehicle handling. Measurements in closed loop, driver/vehicle/environment have shown that the determination of an absolute measuring assessment is not possible due to the large variation of the driver characteristics. An attempt is therefore being made to find a corrrelation between the measurement results in open loop and the subjective evaluation of skilled test engineers in closed loop. Due to the large number of possible driving manoeuvres and operating conditions objective measurement of the entire vehicle handling is not possible, only subsectors can be recorded and compared for different vehicles.

Measurement of the Tractive Resistances

The rolling resistance and aerodynamic drag are primary factors for the fuel consumption of a passenger car. For this reason many attempts have been made in the past to measure these variables under realistic conditions using various methods.

One possibility is the measurement of the drive torque with the aid of highly sensitive torque measurement hubs. This allows the total tractive resistance without the losses in the drive train to be recorded. Due to the sensitivity of the measurement hubs such tests can only be accomplished on a blocked track while avoiding sharp starting, braking and turning manoeuvers.

To determine the rolling resistance the vehicle to be tested can be pulled by a second vehicle or with a cable winch, whereby the tractive forces in the tow cable are measured. The disadvantages of this method are that the measurements can only be accomplished at very low speed to eliminate the aerodynamic drag and that strong disturbances resulting from oscillation of the cable are superimposed on the measurement values. An improvement to the towing method is achieved by protecting the vehicle from the aerodynamic drag with a large trailer, Fig. 19. The tractive force between the trailer and the protected vehicle is measured with a special tow bar. The measured force corresponds to the pure rolling resistance. The measurement can also be accomplished at higher speeds. This method is, however, a large-scale affair due to the required trailer.

The coast-down test which has already been in use for some time offers the possibility to measure the entire tractive resistance and subdivide it into aerodynamic drag and rolling resistance. Experience has shown that, above all, to separate the aerodynamic and rolling resistance an extremely high degree of accuracy is required in measuring the values. The previous measuring techniques did not assure this degree of accuracy.

Fig. 19: Rolling Resistance Measurement by towing behind a trailer to eliminate Aerodynamic Drag

Only in the recent past it has become possible to accomplish such measurements with the required degree of accuracy using modern sensor and calculation technology. Various methods were also examined for the coast-down test. The method, which led to the best results will be shown here.

Up to driving speeds of approximately 150 km/h the following applies for the deceleration resulting from tractive resistances on an even track without wind:

$$a_x = A\,v_x^2 + B\,v_x + C$$

The aerodynamic drag is proportional to the square of speed:

$$W_L = m\,A \cdot v_x^2 \quad \text{where} \quad A = c_w \cdot F \cdot \frac{\rho}{2}/m$$

or solved for the drag coefficient:

$$c_w = \frac{A \cdot m}{F\ \rho/2}$$

where the air density equals

$$\rho(\mathrm{kg/m^3}) = 0.463\ \frac{P\ (\mathrm{Torr})}{273 \cdot T\ (^{\circ}\mathrm{C})}$$

The rolling resistance consists of one term which is proportional to the speed and one constant term:

$$W_R = (B\, v_x + C)\, m$$

where the speed-dependent coefficient of rolling resistance:

$$f_R\ (v) = v_x \frac{B}{g} + \frac{G}{g}$$

a_x = vehicle longitudinal acceleration
v_x = vehicle speed
m = vehicle mass
F = frontal area of the vehicle
P = air pressure
T = air temperature
g = natural gravity

The linear statement for the rolling resistance only applies to approximately 150 km/h depending upon the make of the tire.

At higher speeds terms of a higher order have also to be taken into consideration.

In coast-down tests either the deceleration or the speed can be measured. Very high requirements are placed upon the accuracy of the measurement, because all methods react very sensitively to measurement inaccuracies. In the tests the speed was measured with optical Leitz sensors and the acceleration curve ascertained through numerical differentiation.

In measuring the speed, the road surface inclination is also included (a gradient of 0.1 % results in a speed deviation of 1 m/s for a measurement duration of 100 s; coast-down time from 120 km/h to stand-still approximately 170 s).

Therefore measurement is only possible on a road with a known gradient. Correction of the measured speed:

$$v\ (t) = v_m\ (t) + \int_0^t b\ (t)\ dt$$

where

$$b\ (t) = g \cdot \sin \vartheta_s\ (t) \qquad \vartheta_s = \text{angle of inclination}$$

Starting at a fixed point each individual position, the associated angle of inclination and therefore the interfering acceleration can be determined from the road profile through integration of the measured speed. The tractive resistances are determined by direct regression. Other solution strategies gave no usable results.

In a statement according to equation (1) a polynomial regression of the second order can be accomplished for the acceleration using the speed. The regression calculation leads directly to the coefficients A, B, and C, from which the aerodynamic drag and rolling resistance W_L and W_R or the coefficients c_w and f_R (v) can be calculated. The vehicle mass, the frontal area of the vehicle and the air pressure and air temperature in terms of the air density are fully considered in the calculation of the coefficients. Measurement errors for these variables therefor result in corresponding coefficient errors. Usabe results have already been achieved with this statement, Fig. 20.

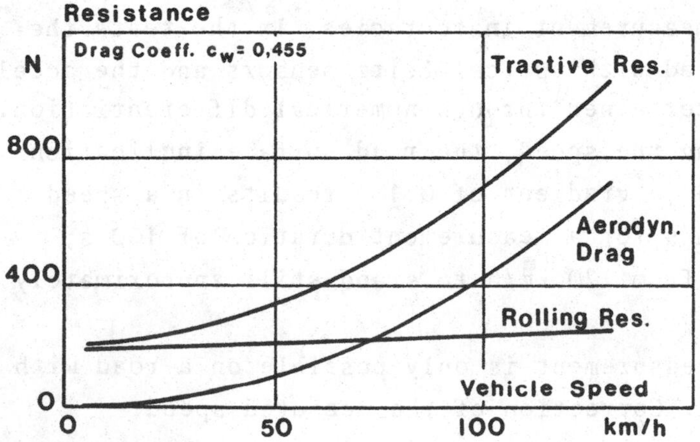

Fig. 20: Results of Coast-Down Test: Aerodynamic Drag and Rolling Resistance

Steady State Turning

One of the oldest test methods in vehicle dynamics is the steady state turning. It supplies basic comparative data for the development of vehicles. In addition to the maximum acceleration this data also include the curve of various vehicle dynamic variables using the lateral acceleration on a circular track with a constant radius. Other methods where the steering wheel angle or the driving speed remain constant are used less often because they require a very large test surface.

In the steady state turning test the lateral acceleration, steering wheel angle and moment, side slip angle and roll angle are measured. The vehicle is accelerated to the desired speed on a circular track usually with a radius between 30 and 50 m. When the steady state condition is reached, the data shall be recorded. The speeds can be run in steps or the vehicle can be accelerated in a quasi-steady state from the lowest speed up to the limit range, whereby the measured variables are recorded during the entire procedure.

The evaluation is accomplished by a computer program. The recorded tape is continuously read and checked whether the measured lateral acceleration is within certain limits; if so, all measured values are simultaneously coordinated to this lateral acceleration interval. This program allows the range from minus 10 to plus 10 m/s^2 to be evaluated in 100 classes. The class width then corresponds to 0.2 m/s^2. The mean value, standard deviation and peak values are calculated and recorded from the adjoined variables measured. These values can be given in tabular form. Presentation of the measured values as a function of the lateral acceleration is also possible. Fig. 21 shows the steering wheel angle, steering wheel torque, sideslip angle and roll angle. The lateral forces on the wheels can also be measured with special measurement hubs, Fig. 22.

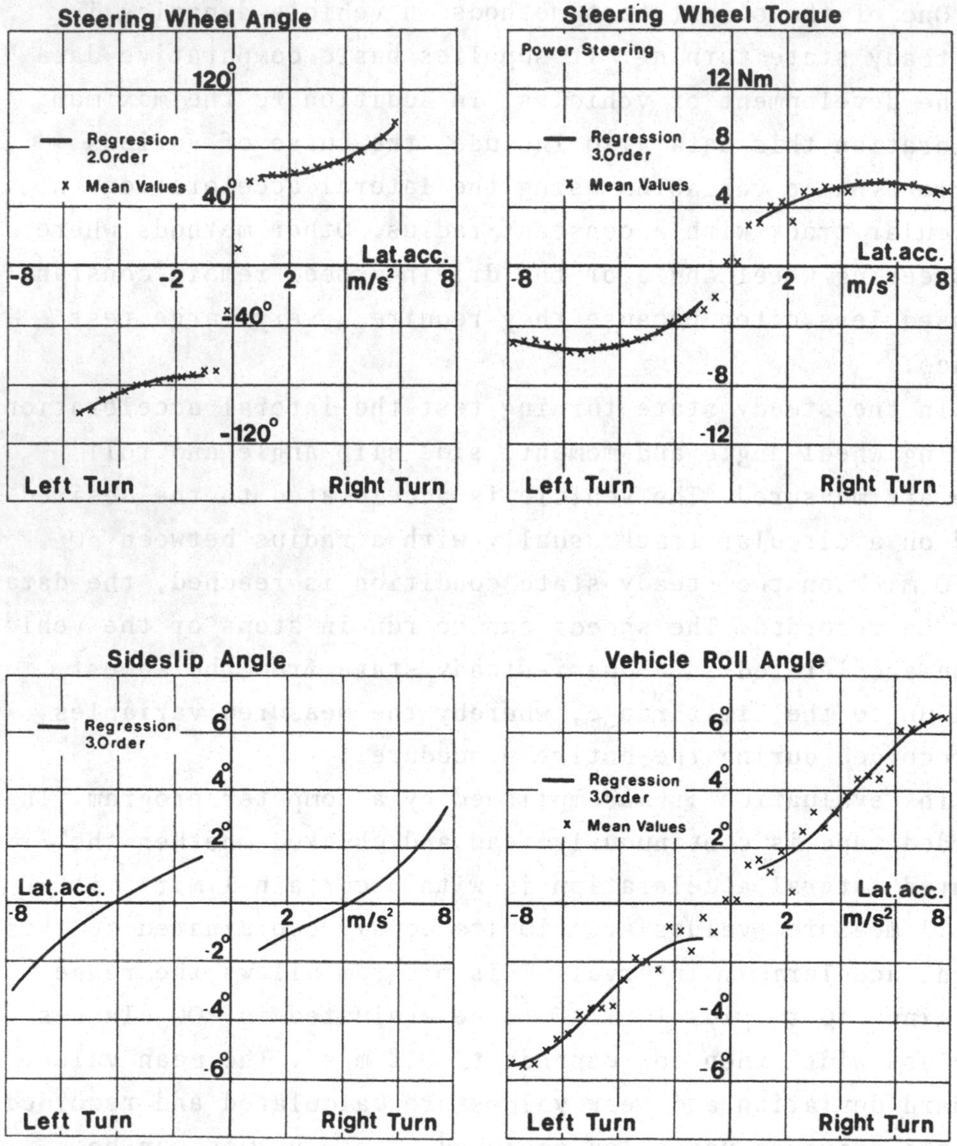

Fig. 21: Steering Wheel Angle, Steering Wheel Torque, Side Slip Angle and Roll Angle versus lateral Acceleration in Steady State Turning (Radius: 42,5 m)

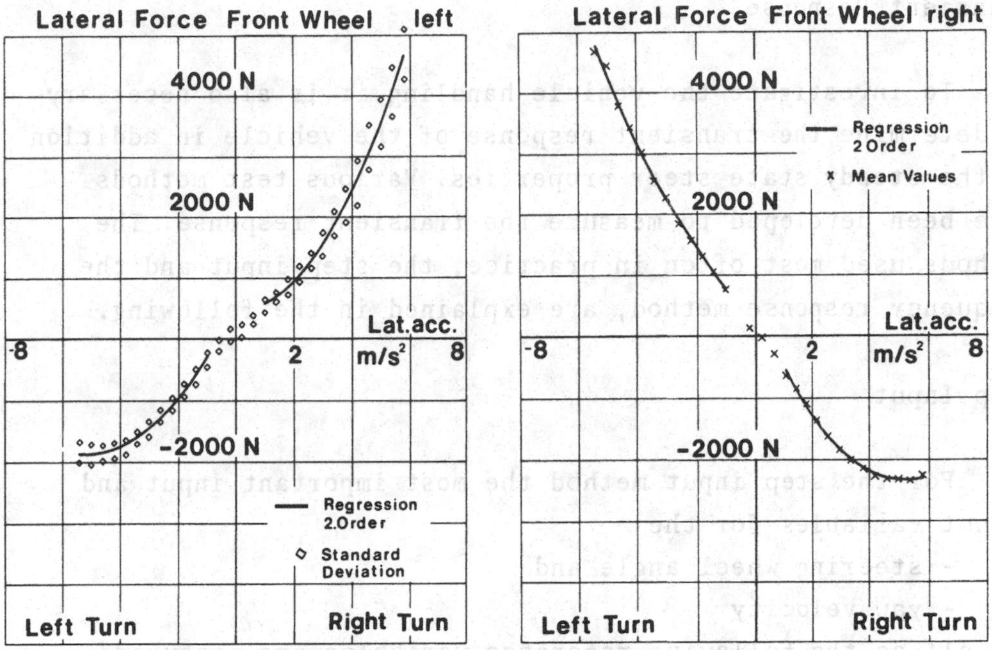

Fig. 22: Lateral Force on the front wheels in Steady State Turning (Radius: 42,5 m)

For determination of the steer properties the elasto-kinematic characteristics of the wheel suspension are used in addition. Essential information can be obtained for the steer properties from the changes in the steer angle of the individual wheels due to the lateral force and during vehicle roll. For this reason the wheel angles are recorded versus lateral acceleration on a circular path, Fig. 23.

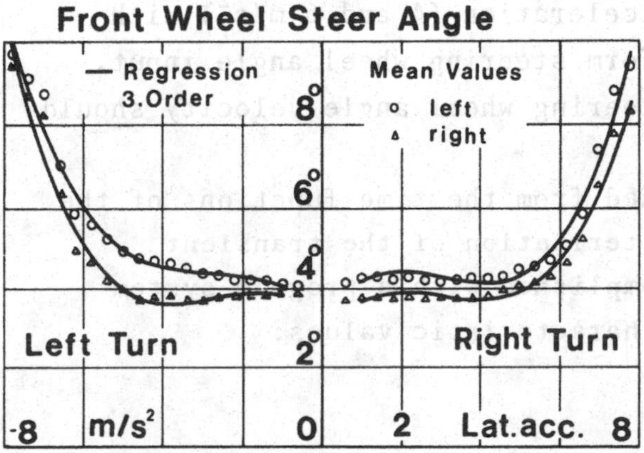

Fig. 23:

Front wheel steer angle versus lateral acceleration in steady state turning (Radius 42,5 m)

Transient Response

To investigate the vehicle handling it is also necessary to determine the transient response of the vehicle in addition to the steady state steer properties. Various test methods have been developed to measure the transient response. The methods used most often in practice, the step input and the frequency response method, are explained in the following.

Step Input

For the step input method the most important input and output variables for the

- steering wheel angle and
- yaw velocity

as well as the following reference variables are measured:

- steering wheel angle velocity
- forward velocity
- lateral acceleration.

For comparison with the subjective evaluation the

- sideslip angle and
- roll angle

are recorded in addition.

At the given speed, generally between 80 and 120 km/h, the vehicle is moved from a straight path into a circular path with a given lateral acceleration (4 and 6 m/s^2) with the quickest possible ramp form steering wheel angle input. During this procedure the steering wheel angle velocity should be between 200 and 500 $^o/s$.

Parameters are determined from the time functions of the recorded variables for characterization of the transient response. Evaluation is accomplished with a program system resulting in the following characteristic values:

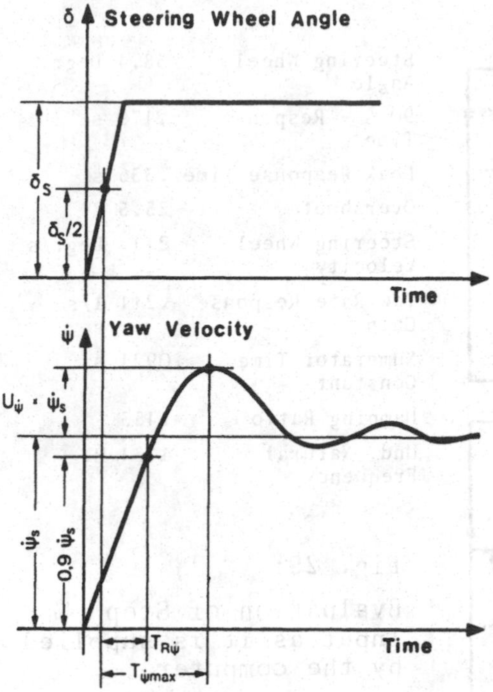

90 % - Response time of yaw velocity $T_{R\dot\psi}$

Peak response time of yaw velocity $T_{\dot\psi max}$

Maximum overshoot of yaw velocity $U_{\dot\psi}$

Yaw Rate response gain (yaw velocity divided by steady state steering wheel angle) $(\dot\psi/\delta)_s$

Fig. 24:

Definition of Several Characteristic Values for transient response test Step Input

See Fig. 24 for the definition of these parameters. The time at which the steering angle reaches 50 % of its final value is chosen as the initial point for determination of the response time.

Moreover the model time functions are determined using a simple vehicle model of the second order with an identification routine. The vehicle yaw motion is determined by the transfer function

$$\frac{\dot\psi}{\delta} = \left(\frac{\dot\psi}{\delta}\right)_s \frac{1 + T_Z \cdot s}{1 + \frac{2D}{\nu_n} s + \frac{1}{\nu_n^2} s^2}$$

Using this model the measured data can be approximated well in most cases. The model serves as a mathematical aid for evaluation of measured transfer functions. The program calculates the model parameters:

- numerator time constant T_Z
- damping ratio D
- undamped natural frequency ν_n

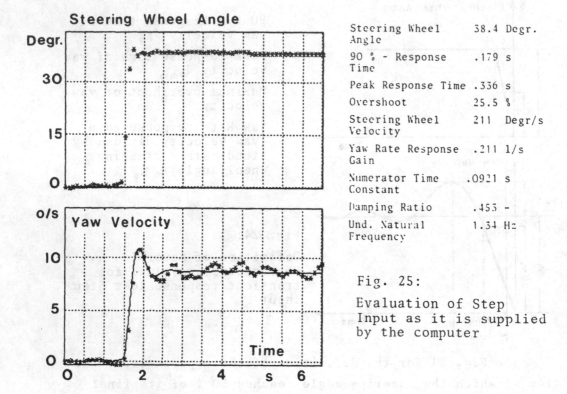

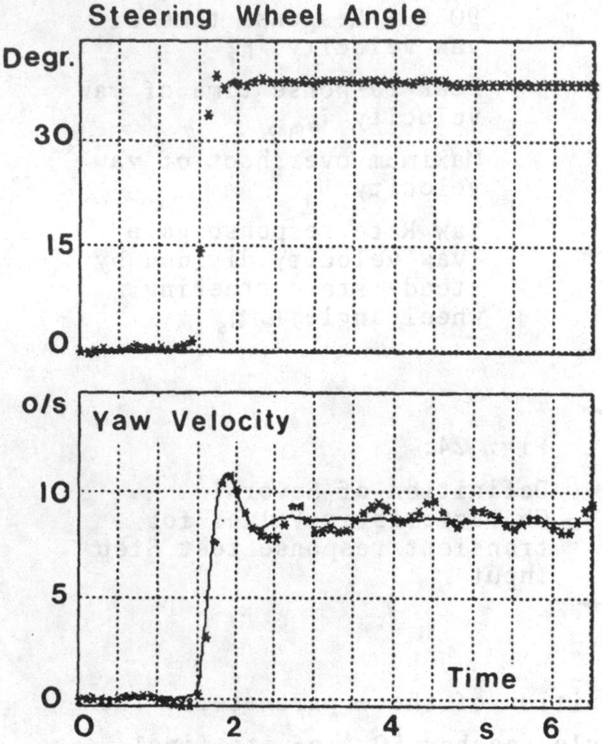

Steering Wheel Angle	38.4 Degr.
90 % - Response Time	.179 s
Peak Response Time	.336 s
Overshoot	25.5 %
Steering Wheel Velocity	211 Degr/s
Yaw Rate Response Gain	.211 1/s
Numerator Time Constant	.0921 s
Damping Ratio	.453 -
Und. Natural Frequency	1.34 Hz

Fig. 25:

Evaluation of Step Input as it is supplied by the computer

An additional evaluation variable is the vehicle factor TB, which also shows a good correlation with the subjective evaluation. The TB factor is the product of the peak response time and the steady state sideslip angle, TB = $T_{\dot{\psi}max} \cdot \beta_s$.

The measured characteristic variables and the calculated model parameters of various vehicles correlated with the subjective evaluation of numerous test drivers in comprehensive test series. It has been shown that at least two characteristic variables are required to characterize the transient response; one of these variables is the yaw rate response gain under steady state conditions. The additional characteristic values should represent the most important frequency-dependent characteristics of the dynamic vehicle behaviour.

The yaw rate response gain $(\dot{\psi}/\delta)_s$ for the yaw velocity has been clearly identified as a suitable steady state characteristic value. The yaw rate response gain shows a correlation with the subjective evaluation on a curvy path for more than 0.8. The greater $(\dot{\psi}/\delta)_s$ was, the better the vehicle was evaluated. Above all the damping D, the peak response time $T_{\dot{\psi}max}$ and the factor TB proved to be the suitable dynamic characteristic values for the evaluation of the vehicle based on the correlation. The greater the damping and the smaller $T_{\dot{\psi}max}$ and TB were, the better the evaluation of the vehicles.

To compare different vehicles in regard to their transient response based on the step input measurements the characteristic values given here, above all, must be examined more closely in addition to the time history of the individual variables.

Frequency Response

A further possibility for evaluating the transmission characteristics of a vehicle is the measurement and charting of the frequency response. Here the input and output variables are collated with one another at various sinusoidal excitation frequencies, as is usual in control technology. These transmission characteristics change according to amplitude and phase over the steer frequency.

The advantage in relation to the step input measurement is that a large test surface is not required; a long straight path with corresponding width is sufficient. At constant driving speed the steer frequency is increased in steps. It has proven practical to use an automatic steering machine for these measurements. Fig. 26 shows an electrohydraulic steering machine installed in a vehicle. The tests were performed with a driver who accelerated the test vehicle to the correct test speed and held this speed constant during the measurement procedure.

During the measurement period the driver removes his hands from the steering wheel and actuates two push buttons which start the steering machine. The steer angle input is then

Fig. 26: Steering machine for frequency response measurements installed in the vehicle

accomplished automatically according to previously set values.

The automatic mechanism is switched off by releasing the push buttons and the driver can reassume manual control of the steering wheel in order to drive the vehicle to the starting position for the next test. In these tests the data are transmitted to the accompanying vehicle by telemetry, where the test run is evaluated immediately.

The following variables are measured:

- steering wheel angle
- yaw velocity
- lateral acceleration
- sideslip angle
- roll angle

The amplitude characteristics and phase angles are calculated and plotted from the measured time history using a program. Fig. 27 shows the amplitude and phase curve for the yaw velocity and the lateral acceleration for a vehicle with

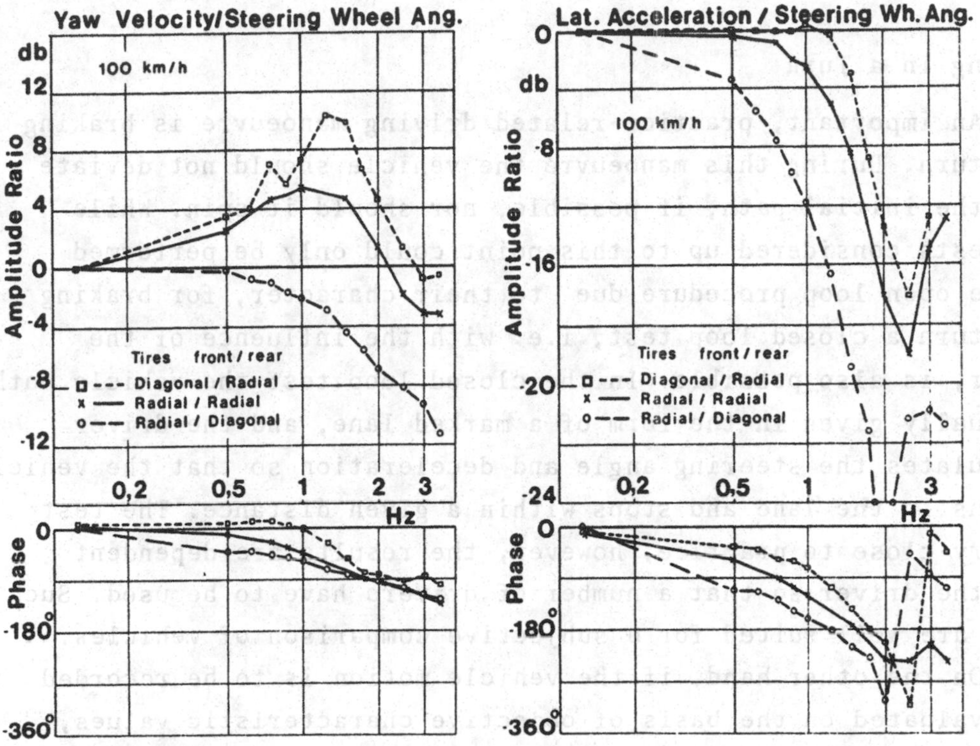

Fig. 27: Frequency Response of Yaw Velocity and Lateral Acceleration for three Several Vehicle Lay-Outs

different steer properties varied through the tires.

The yaw velocity shows a marked resonance hump at 1.5 Hz, which increases with increasing understeer. This hump also increases with the forward velocity. The differences in the vehicle lay-out can also be seen clearly from the frequency response of the lateral acceleration. The oversteer already shows a decrease of 12 dB at 1 Hz steering frequency. The high degreee of understeer is still not subject to any decrease in the lateral acceleration response at this point. On the basis of such frequency response measurements the various vehicles can be compared and evaluated in regard to their transmission characteristics.

Braking in a Turn

An important, practice-related driving manoeuvre is braking in a turn. During this manoeuvre the vehicle should not deviate from the initial path, if possible, nor should it spin. While the tests considered up to this point could only be performed in the open loop procedure due to their character, for braking in a turn a closed loop test, i.e. with the influence of the driver, is also possible. In the closed loop test the vehicle path is usually given in the form of a marked lane, and the driver manipulates the steering angle and deceleration so that the vehicle remains in the lane and stops within a given distance. The test is very close to practice, however, the results are dependent upon the driver so that a number of drivers have to be used. Such tests are well suited for a subjective comparison of vehicles.

On the other hand, if the vehicle motion is to be recorded and evaluated on the basis of objective characteristic values, the open loop procedure have to be used. The goal of this test is to determine the effect of braking upon the directional behaviour of the vehicle whose steady state turning is disrupted only by the braking procedure. The steering wheel is kept fix during the test. The following variables are measured:

- steering wheel angle
- brake pressure
- lateral acceleration
- longitudinal deceleration
- forward velocity
- yaw velocity
- sideslip angle

It has been determined that the differences between various vehicles can best be determined using an initial circle with a diameter of 100 m at an initial lateral acceleration value with a minimum of 5 m/s^2.

This corresponds to an initial vehicle speed of 80.5 km/h. The vehicle is braked at an incrementally increasing longitudinal deceleration from the steady state circular turning. During this procedure the brake pedal is actuated as quickly as possible. The brake pressure is increased this far, until all wheels lock up.

The recorded measurement variables are processed further using a computer. In this procedure the following additional variables are calculated:

- reference yaw velocity
- reference lateral acceleration
- yaw angle deviation from initial circle

The reference condition is the condition of the vehicle slowing down with the same longitudinal deceleration time history as the test vehicle without any deviation of the vehicle center of gravity from the initial circular trajectory. In addition the steady state values from the beginning of the braking procedure, mean values and maximum values as well as values at the time 1 s after braking begins are determined from the time history of the measured variables for each test. The time 1 s was therefore chosen, because this period corresponds approximately to the reaction time of the driver. When the driver after this period desires to initiate a correction the vehicle motion should still be under his control. The mean values, maximum values and the 1 second values are standardized in part to the steady state values or the reference values in order to avoid differences through dispersion in the initial conditions.

The curves for three different vehicles are shown here to illustrate the behaviour characteristics obtained from a large number of tests. Due to the fact that the vehicle handling changes at variously high degrees of deceleration, the curves are charted in relation to the deceleration. The 1 s values are graphed versus the 1 s value for the deceleration and the mean values versus the mean deceleration value.

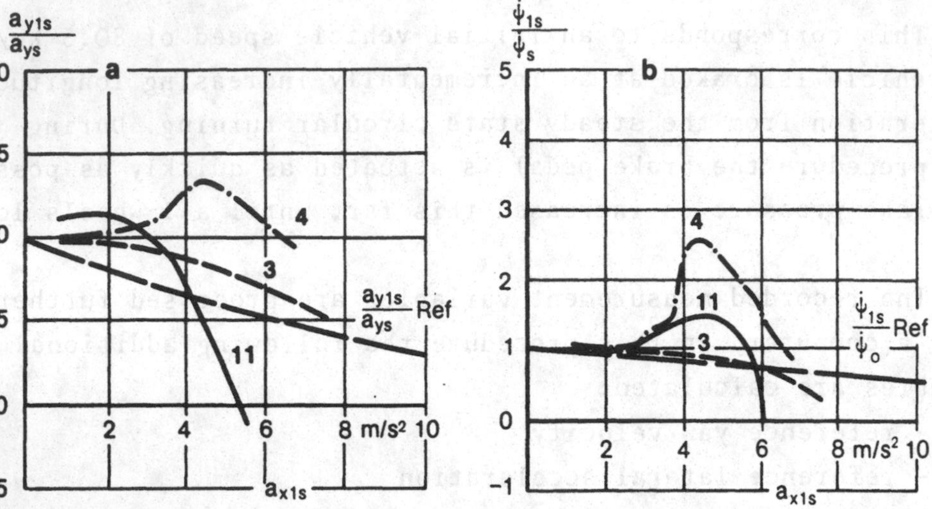

a Lateral Acceleration b Yaw Velocity

Fig. 28: Related Time Values 1 s after braking begins versus the associated time value of the longitudinal deceleration for 3 various vehicles no. 3, 4 and 11

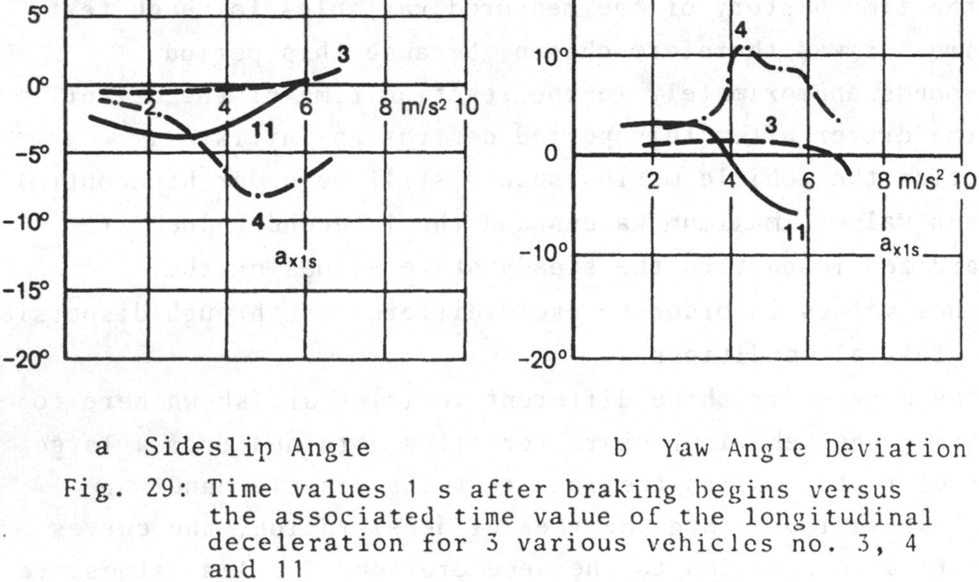

a Sideslip Angle b Yaw Angle Deviation

Fig. 29: Time values 1 s after braking begins versus the associated time value of the longitudinal deceleration for 3 various vehicles no. 3, 4 and 11

Fig. 28 and 29 show characteristic variables at the time 1 s after braking begins. Moreover in Fig. 28a and 28b the reference lines for exact keeping of the nominal circular trajectory are indicated. The presentation of the initial values based on the lateral acceleration in Fig. 28a shows that vehicle 3 comes relatively close to the desired reference curve for exact keeping of the trajectory. The relatively high values for vehicle 4 in the middle deceleration range show the entry into the circle. Vehicle 11 also allows the observation that turning into the circle at slight deceleration, while the point of intersection with the reference line indicates the begin of sliding out of the trajectory at a deceleration of approximately 4 m/s^2. The point at which the lateral acceleration crosses the zero axis at a deceleration of approximately 5.5 m/s^2 means that the limit of steerability has been reached for this vehicle. The vehicle then skids tangentially out of the circular trajectory.

The corresponding handling properties can also be seen in the curve of the initial values in relation to the yaw velocity in Fig. 28b. The related yaw motion, e.g. for vehicle 4, is approximately twice as great as the related lateral acceleration. The reason for this is that an increase in the sideslip angle in addition to a decrease in the radius driven is expressed in the value of the yaw velocity. The associated sideslip angle is charted in Fig. 29a. Fig. 29b shows the difference between the actual und the calculated yaw angle for keeping the nominal circular trajectory 1 s after braking begins. These curves can also be used for evaluation of the vehicle handling. The driver also visually recognizes the deviation in the yaw angle. A greater angle to the vehicle longitudinal axis in relation to the nominal path tangent after braking for 1 s would frighten the driver more than a smaller deviation. The mean values and maximum values of this characteristic variable supply the same information so that it is not necessary to explain them here.

CORRELATION BETWEEN MEASURED VALUES AND SUBJECTIVE EVALUATION OF THE VEHICLE HANDLING

The relationship between the characteristic values determined in the open loop procedure and the actual traffic situations as well as to the accident causes is unknown. However, the goal of vehicle development is to improve the handling characteristics so that these improvements contribute to the reduction of the number of accidents resulting from the handling characteristics and therefore increase the active safety. The manner in which the driver masters the handling characteristics of his vehicle in actual traffic situations can only be determined through subjective evaluation of several skilled drivers while driving under normal traffic conditions. The test results can often only be interpreted in terms of feelings based on experience. Therefore the correlation between objective criteria and the subjective evaluation is important. Today, correlation studies are performed for every test method used.

A questionnaire is filled in for each test procedure for subjective evaluation of the vehicle, in which the driver can evaluate the vehicles according to a certain scale. At least 7 possible grades can be given:

1 for "poor" and 7 for "very good". The selection and formulation of the questions asked play a primary roll for the success of the evaluation.

The goal of the study is to find the objective criteria from the measured variables, which are best suited for describing the subjective evaluation. The stepwise multiple regression analysis method has proven itself suitable to connect the objective and subjective data. With this method an analytic relationship is determined between the dependent variables, the subjective evaluation points and the independent measurement values, i.e. the perception of the driver is interpreted by a weighted combination of the individual test values. Since the multiple regression assumes variables independent of one another, the objective characteristic values, which correlate highly with one another and are therefore interchangeable with one another, have to be reduced. Following this data reduction only those characteristic variables are used in the calculation and compared with the subjective evaluations in the corresponding criteria, which on one hand show the highest possible correlation with the subjective evaluations and which on the other hand have a low correlation between one another.

In numerous studies it has been shown that the subjective evaluation can be described best by a linear regression model. For this reason the linear relationship

$$y = a_0 + a_1 \cdot x_1 + a_2 \cdot x_2 + \ldots + a_n \cdot x_n$$

has been assumed as a model for the stepwise multiple regression analysis. With this y is the dependent variable (evaluation point number), x_1, x_2 ... x_n are the independent variables and a_0, a_1 ... a_n are the coefficients sought. In stepwise multiple regression a regression equation results in the following manner for each calculation step by adding a new variable:

$$y = a_0 + a_1x_1$$

$$y = a_0' + a_1'x_1 + a_2x_2$$

$$y = a_0'' + a_1''x_1 + a_2'x_2 + a_3x_3$$

... etc.

In this procedure that variable is inserted into the calculation in each case, which effects the greatest possible improvement of the multiple correlation coefficients. The results are indicated by the computer in tabular form and recorded in a plotter diagram with the measurement values.

An example of the subjective evaluation of the transient response with 8 vehicles in comparison to the measurement values from the step input is shown in Fig. 30. The criterion given here is an inquiry regarding the ability of the vehicle for rapid driving over a curvy path. The table shows that the yaw response gain $\dot{\psi}/\delta$ was inserted into the regression as the first variable. The second variable inserted, the damping ratio D improved the multiple correlation coefficient from 0.821 (initial value) to 0.858. By stepwise inserting the peak response time $T_{\dot{\psi}max}$, the response time of the lateral acceleration T_{Ray} and the numerator time constant T_Z a multiple correlation coefficient of 0.897 is finally achieved.

Calculation Step	Variable inserted into the calculation	Multiple Correlation Coefficient
1	$\dot{\psi}/\delta$	0,821
2	D	0,858
3	$T_{\dot{\psi}max}$	0,882
4	T_{Ra_y}	0,891
5	T_z	0,897

Fig. 30: Stepwise Multiple Regression Analysis with 8 vehicles from the values of the subjective evaluation and the step input measurement

REFERENCES

1. Vehicle Dynamics Terminologie. SAE J 670 e, 1978

2. A. Zomotor: Ein korrelationsoptisches Verfahren zur direkten Messung von instationären Schwimm- und Schräglaufwinkeln an Kraftfahrzeugen. (An Optical Correlation Method for the Direct Measurements of Transient Sideslip and Slip Angles of Motor Vehicles). ATZ 1975, Nr. 7-8.

3. R. Rönitz, H.H. Braess, A. Zomotor: Verfahren und Kriterien zur Bewertung des Fahrverhaltens von Personenkraftwagen. (Methods and Criteria for the Evaluation of Handling and Road-Holding Properties of Passenger Cars - State-of-the-Art and Problems). Automobilindustrie 1977, Heft 1 und 3.

4. A. Zomotor: Meßverfahren bei der Auslegung des Fahrverhaltens. (Measuring Procedures for the Evaluation of Vehicle Handling). Automobilindustrie 1978, Heft 2.

5. T.P. Yasin: The Analytical Basis of Automobile Coastdown Testing. SAE 780334.

6. ISO/TC 22/SC 9: Road Vehicles - Steady State Circular Test Procedure. Draft ISO Standard, DIS 4138.2, 1980.

7. E. Bisimis, H.D. Beckmann, R. Rönitz, A. Zomotor: Lenkwinkelsprung und Übergangsverhalten von Kraftfahrzeugen. (Steer Step Input and Transient Response of Motor Vehicles). ATZ 1977, Nr. 12

8. A. Zomotor, A. Horn, K. Rompe: Bremsen in der Kurve - Untersuchung eines Testverfahrens. (Braking in a Turn - Investigation of a Testprocedure). ATZ 1980, Nr. 9.

9. H.-U. Lücke, W. Renz: Mobiles rechnergesteuertes Meßdatenerfassungs- und -auswertesystem für den Fahrzeugeinsatz. Messen und Prüfen, Juli/August 1981.

COMPLEX NONLINEAR VEHICLES UNDER STOCHASTIC EXCITATION

Werner O. Schiehlen
Institut B für Mechanik
Universität Stuttgart
Pfaffenwaldring 9, Stuttgart 80, F.R.G.

INTRODUCTION

The modeling of guideways and vehicles and the corresponding mathematical methods of solution including linear and nonlinear techniques, random response and ride quality evaluation have been presented in detail. In this paper firstly the equations of motion of a complex automobile vehicle are derived and secondly numerical results for stochastic excitation of this vehicle by guideway irregularities are presented. It will be shown in particular that the dynamical analysis of random vehicle vibrations requires an integrated investigation of guideway roughness, nonlinear vehicle dynamics and human response to vibration exposure.

SYSTEM EQUATIONS

The equations of vehicle system dynamics follow from the guideway irregularities, the vehicle itself and the passenger sensation to vibration.

Guideway Irregularities

The guideway irregularities for the right and the left trace are presented by the random processes $\zeta_r(t)$ and $\zeta_l(t)$. Thus, the spatial guideway is given by the 2x1-vector process

$$\zeta(t) = [\zeta_r(t) \quad \zeta_l(t)]^T \; . \tag{1}$$

For the description of the road characteristics it is more convenient to use the uncorrelated random processes $\zeta_M(t)$ and $\zeta_D(t)$ of the mean trace and the trace difference, respectively, resulting in the 2x1-vector process

$$v(t) = [\zeta_M(t) \quad \zeta_D(t)]^T \; . \tag{2}$$

see Bormann[1], Parkhilovski[2]. Then, it follows

$$\zeta = H v \tag{3}$$

where the 2x2-matrix

$$H = \begin{bmatrix} 1 & 1 \\ 1 & -1 \end{bmatrix} \tag{4}$$

appears. The 2x1-vector process $v(t)$ may be given by a first order shape filter

$$\dot{v} = F v + G \tilde{w} \tag{5}$$

where $F = \mathrm{diag}\,\{-f_1 \quad -f_2\}$, $G = \mathrm{diag}\,\{g_1 \quad g_2\}$ and the intensity matrix of the 2x1-white noise process $\tilde{w}(t)$ reads as $Q = q\,E$. For a rough road and a speed $v = 25$ m/s one obtains an intensity of $q = 2{,}5 \cdot 10^{-3}$ m^2/s. Now, the parameters f_1, f_2 and g_1, g_2 represent various roads, Table 1.

Parallel Traces, White Velocity Noise
$f_1 = f_2 = 0, \quad g_1 = 1, \quad g_2 = 0$
Uncorrelated Traces, White Velocity Noise
$f_1 = f_2 = 0, \quad g_1 = \frac{1}{\sqrt{2}}, \quad g_2 = \frac{1}{2}$
Correlated Traces, Colored Displacement Noise
$f_1 = 0{,}25\ \frac{1}{s}, \quad f_2 = 1{,}25\ \frac{1}{s}, \quad g_1 = \frac{1}{\sqrt{1+0{,}6}}, \quad g_2 = \frac{0{,}6}{\sqrt{1+0{,}6}}$

Table 1. Parameters of various roads

Vehicle Dynamics

The vehicle under consideration will include numerous parts as shown in Fig. 1. The model consists of 4 mass points and 7 rigid bodies subject to 35 constraints resulting in 19 degrees of freedom. Altogether 67 parameters describe the model, Table 2, and they have to be completed by geometrical distances.

Then, the linear equations of motion derived by the program NEWEUL reads as

$$M\ddot{y} + P\dot{y} + Qy = -Rw + S\xi \tag{6}$$

where $y(t)$ is the 19x1-position vector and M, P, Q are 19x19-matrices. In addition to the generalized coordinates the 2x1-vector

$$w(t) = [ST1 \quad ST2]^T \tag{7}$$

has to be introduced to describe the PI-Forces by the serial spring-dashpot-configurations at the engine,

$$\dot{w} + Ww + \dot{Y}\dot{y} = 0 \ . \tag{8}$$

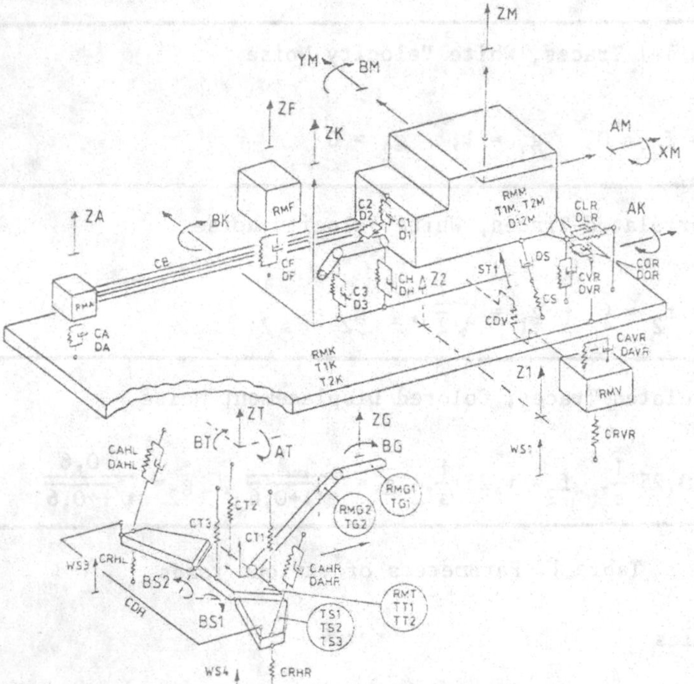

Fig. 1. Vehicle model with 19 dof

The excitation of the vehicle is restricted to the guideway irregularities. At this point, each of the wheels or wishbones, respectively, is excited independently by the 4x1-excitation vector

$$\xi(t) = [WS1 \quad WS2 \quad WS3 \quad WS4]^T . \tag{9}$$

The matrices R, S, W, $\dot{Y}$ are also time-invariant and have the corresponding dimensions.

The state equation of the vehicle reads in the linear case

$$\dot{x} = Ax + B\xi \tag{10}$$

where the 40x1-state vector

$$x(t) = [y^T(t) \quad \dot{y}^T(t) \quad w^T(t)]^T \tag{11}$$

ENGINE

RMM=293.	[KG]
T1M=9.175	[KG M M]
T2M=31.09	[KG M M]
D12M=-0.6026	[KG M M]

ENGINE MOUNT

CS=580000.	[N/M]
DS=2900.	[N S/M]
BE=0.	[RAD]
CLR=95000.	[N/M]
CLL=95000.	[N/M]
DLR=20.	[N S/M]
DLL=20.	[N S/M]
CQR=113000.	[N/M]
CQL=113000.	[N/M]
DQR=20.	[N S/M]
DQL=20.	[N S/M]
CVR=183000.	[N/M]
CVL=183000.	[N/M]
DVR=20.	[N S/M]
DVL=20.	[N S/M]
AL=0.13962634	[RAD]
CH=88000.	[N/M]
DH=25.	[N S/M]

CAR BODY

RMK=1030.	[KG]
T1K=504.	[KG M M]
T2K=1840.	[KG M M]
CAVR=10700.	[N/M]
CAVL=10700.	[N/M]
CAHR=52800.	[N/M]
CAHL=52800.	[N/M]
DAVR=1880.	[N S/M]
DAVL=1880.	[N S/M]
DAHR=5700.	[N S/M]
DAHL=5700.	[N S/M]

DRIVE SHAFT

RMG1=4.8	[KG]
RMG2=7.2	[KG]
TG1=0.78	[KG M M]
TG2=2.3	[KG M M]
C1=6380000.	[N/M]
D1=40.	[N S/M]
C2=51.	[N/M]
D2=0.00006	[N S/M]
C3=40000.	[N/M]
D3=30.	[N S/M]

EXHAUST PIPE

RMA=8.	[KG]
CB=2400.	[N/M]
CA=67600.	[N/M]
DA=210.	[N S/M]

REAR AXLE

RMT=25.	[KG]
RMS=40.2	[KG]
TT1=1.5	[KG M M]
TT2=2.	[KG M M]
TS1=0.8	[KG M M]
TS2=0.5	[KG M M]
TS3=0.4	[KG M M]
CT1=250000.	[N/M]
CT2=250000.	[N/M]
CT3=250000.	[N/M]
CRMR=175000.	[N/M]
CRML=175000.	[N/M]
CDM=5500.	[N/M]

FRONT WHEELS

RMV=42.5	[KG]
CRVR=150000.	[N/M]
CRVL=150000.	[N/M]
CDV=20000.	[N/M]

DRIVER

RMF=60.	[KG]
CF=54000.	[N/M]
DF=360.	[N S/M]

Table 2. Parameters of the vehicle model

is used. The 40x40-system matrix is then represented by

$$A = \begin{bmatrix} 0 & E & 0 \\ -M^{-1}Q & -M^{-1}P & -M^{-1}R \\ 0 & -\dot{Y} & -W \end{bmatrix} \tag{12}$$

and the 40x4-excitation matrix follows as

$$B = \left[0 \quad (M^{-1}S)^T \quad 0\right]^T \tag{13}$$

If nonlinear characteristics of springs and dashpots have to be considered, a Taylor series expansion on the corresponding forces F_S and F_D is applied:

$$F_S = F_S(\eta) = C_1\eta + C_2\eta^2 + \ldots + C_n\eta^n , \qquad (14)$$

$$F_D = F_D(\eta) = D_1\eta + D_2\eta^2 + \ldots + D_n\eta^n . \qquad (15)$$

The second and higher order terms are summarized in an additional nonlinear 40x1-vector function $f(x)$ and the state equation is extended to

$$\dot{x} = Ax + f(x) + B\xi . \qquad (16)$$

The vehicle under consideration has two axles. This means a time delay between the excitation of the front and rear wheel at each trace of the guideway. Regarding (1) and (9) one obtains

$$\xi(t) = \left[\zeta^T(t) \quad \zeta^T(t-\Delta t)\right]^T \qquad (17)$$

where the time delay follows from

$$\Delta t = \frac{L}{v} , \qquad (18)$$

L distance of axles, v speed of vehicle. Then, the excitation term in the state equation (10) or (16), respectively, reads as

$$B\,\xi(t) = \begin{bmatrix} B_1 & B_2 \end{bmatrix} \begin{bmatrix} \zeta(t) \\ \zeta(t-\Delta t) \end{bmatrix} . \qquad (19)$$

Human Response

The human sensation with respect to vibration will be discussed only for the vertical acceleration of the car body. The standard deviation $P_{\ddot{Z}}$ at an arbitrary location (U, V) in the x, y-plane of the car body reads as

$$P_{\ddot{Z}}(U,V) = P_{\ddot{Z}\ddot{K}} + V^2 P_{\ddot{A}\ddot{K}} + U^2 P_{\ddot{B}\ddot{K}}$$

$$+ 2\,V\,P_{\ddot{Z}\ddot{K}\ddot{A}\ddot{K}} - 2\,U\,P_{\ddot{Z}\ddot{K}\ddot{B}\ddot{K}} - 2\,U\,V\,P_{\ddot{A}\ddot{K}\ddot{B}\ddot{K}} \qquad (20)$$

where, e.g. $P_{\ddot{Z}\ddot{K}} = E\,\{\ddot{Z}\ddot{K}^2\}$. The acceleration variances of $\ddot{A}\ddot{K}$, $\ddot{B}\ddot{K}$, $\ddot{Z}\ddot{K}$ are to be found by the covariance analysis presented in the next chapters.

FREE VIBRATIONS OF THE VEHICLE

Free vibrations of a vehicle exist at vanishing speed or in the case of traveling on an ideal plane guideway, respectively. The eigenfrequencies of the free vibrations are also essential parameters in the comparison of the dynamical behavior of a real vehicle and its mathematical model. The computed and measured eigenfrequencies have to be consistent. The free vibrations are charactericed by the homogeneous state equation derived from (10):

$$\dot{x} = A\,x \qquad (21)$$

The eigenfrequencies ω_j, $j = 1(1)19$, are found by the solution of the eigenvalue problem

$$(A - \lambda E)\,\tilde{x} = 0 \qquad (22)$$

The numerical computation results in the eigenvalues

$$\left.\begin{aligned} \lambda_j &= \delta_k \pm i\omega_k \;, \quad j = 1(1)38, \quad k = 1(1)19 \;, \\ \lambda_j &= \delta_j \;, \quad j = 39, 40 \end{aligned}\right\} \tag{23}$$

and the corresponding eigenvectors $\tilde{x}_j$. The eigenvalues λ_{39} and λ_{40} are due to the integrals of position caused by the two spring-dashpot-configurations. The eigenvalues has been computed by Hirschberg[3]. The highest frequency $\omega_1 = 250$ Hz is due to the drive shaft, the medium frequencies $\omega_7 - \omega_{10} = 8\text{–}10$ Hz represent the wheel vibration and the lowest frequencies $\omega_{17} - \omega_{19} = 0{,}7 - 1{,}2$ Hz characterize the body vibration. The relation of the eigenfrequencies to parts of the real vehicle isn't usually unique due to the coupling phenomena. However, at least approximately the eigenfrequencies can be identified by the investigation of the eigenvectors.

COVARIANCE ANALYSIS WITH WHITE VELOCITY NOISE

For guideway irregularities characterized by white velocity noise the excitation follows from (3) and (5) with Table 1 as

$$\dot{\zeta} = H\, G\, \tilde{w} \tag{24}$$

where the 2x2-intensity matrix of the white noise process $\tilde{w}(t)$ is given by $Q = q\,E$, $q = 2{,}5 \cdot 10^{-3}\ \text{m}^2/\text{s}$. This means that the derivative $\dot{\zeta}(t)$ of the irregularity vector (1) is exactly a white noise process with the 2x2-intensity matrix

$$Q_{\dot{\zeta}} = q\, H\, G\, G^T H^T \,. \tag{25}$$

In particular for parallel traces one obtains from Table 1

$$Q_{\dot{\zeta}} = q \begin{bmatrix} 1 & 1 \\ 1 & 1 \end{bmatrix} \tag{26}$$

while for uncorrelated traces it remains

$$Q_{\dot{\zeta}} = q \begin{bmatrix} 1 & 0 \\ 0 & 1 \end{bmatrix} . \tag{27}$$

The white velocity excitation can be introduced in the state equation (10), (19) immediately after differentiation

$$\ddot{x} = A \dot{x} + B_1 \dot{\zeta}(t) + B_2 \dot{\zeta}(t-\Delta t) . \tag{28}$$

Then, the 40x40-covariance matrix P_x of the vehicle response $x(t)$ can be calculated by the algebraic Ljapunov equation[4]:

$$A P_x + P_x A^T + B_1 Q_\zeta B_1^T + B_2 Q_\zeta B_2^T$$

$$+ \phi(\Delta t) B_2 Q_\zeta B_1^T + B_1 Q_\zeta B_2^T \phi^T(\Delta t) = 0 . \tag{29}$$

Here, the 40x40-fundamental matrix

$$\phi(\Delta t) = e^{A\Delta t} = \sum_{n=o}^{\infty} \frac{(A\Delta t)^n}{n\,!} \tag{30}$$

is found by series expansion[5].

Due to (28) the variances of the accelerations required in (20) are available immediately as elements of P_x or

$$P_{\ddot{z}} = T_{\ddot{z}} P_x T_{\ddot{z}}^T , \tag{31}$$

respectively, where $T_{\ddot{z}}$ is a 1x40-transformation matrix.

The numerical results are shown in Fig. 2 for the parallel traces and in Fig. 3 for the random traces. It is obvious that the uncorrelated traces result in lower acceleration and better ride quality.

In addition to the ride quality often the ride safety is of interest. The analysis of ride safety generally cannot be performed for the white velocity noise excitation, see Rill[6]. Therefore, also the colored displacement noise excitation will be presented.

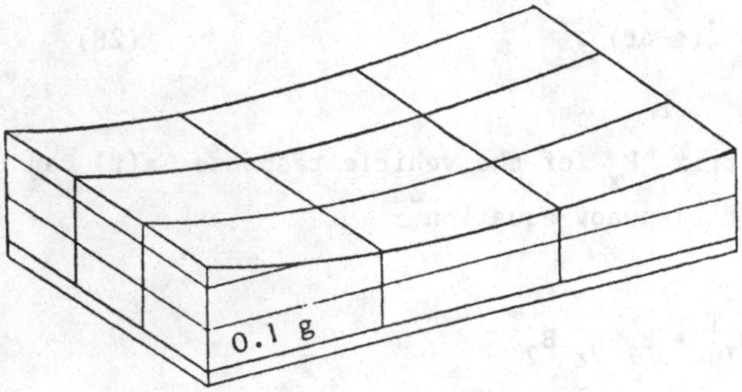

Fig. 2.
Acceleration of the car-body due to white velocity noise and parallel traces

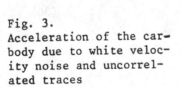

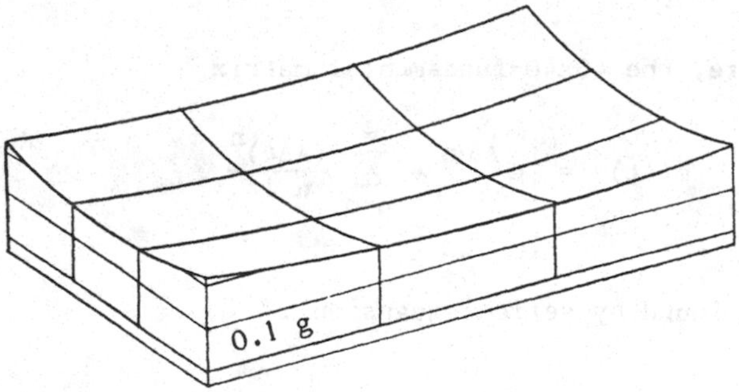

Fig. 3.
Acceleration of the car-body due to white velocity noise and uncorrelated traces

COVARIANCE ANALYSIS WITH COLORED DISPLACEMENT NOISE

Guideway irregularities modeled by colored noise result in an extended state equation where (3), (5), (10) and (19) are summarized

$$\dot{\hat{x}} = \hat{A}\,\hat{x} + \hat{B}_1\,\tilde{w}(t) + \hat{B}_2\,\tilde{w}(t-\Delta t) \tag{32}$$

with

$$\left.\begin{aligned} \hat{x} &= \begin{bmatrix} v_1 \\ v_2 \\ x \end{bmatrix}, \quad \hat{A} = \begin{bmatrix} F & 0 & 0 \\ 0 & F & 0 \\ B_1 H & B_2 H & A \end{bmatrix}, \\ \hat{B}_1 &= \begin{bmatrix} G \\ 0 \\ 0 \end{bmatrix}, \quad \hat{B}_2 = \begin{bmatrix} 0 \\ G \\ 0 \end{bmatrix}. \end{aligned}\right\} \qquad (33)$$

The parameters of the matrices F and G are listed in Table 1, the intensity remains again unchanged $q = 2{,}5 \cdot 10^{-3}$ m^2/s.

The covariance analysis via the algebraic Ljapunov equation results now in an extended 44x44-covariance matrix $\hat{P}$:

$$\hat{A}\,\hat{P} + \hat{P}\,\hat{A}^T + \hat{B}_1\,Q\,\hat{B}_1^{\,T} + \hat{B}_2\,Q\,\hat{B}_2^{\,T} + \\ + \hat{\phi}(\Delta t)\,\hat{B}_2\,Q\,\hat{B}_1^{\,T} + \hat{B}_1\,Q\,\hat{B}_2^{\,T}\,\hat{\phi}^T(\Delta t) = 0\,. \qquad (34)$$

The ride safety is characterized by the variances P_{Fi} of the wheel load variation F_i, $i = 1(1)4$, e.g.

$$P_{F1} = CRVR^2 \left[P_{Z1} - 2\,P_{Z1WS1} + P_{WS1} \right]. \qquad (35)$$

Generally, the wheel load variation reads as

$$P_{Fi} = \hat{T}_{Fi}\,\hat{P}\,\hat{T}_{Fi}^{\,T}\,, \quad i = 1(1)4\,, \qquad (36)$$

where $\hat{T}_{Fi}$ is a 1x44-transformation matrix.

In contrary to (31), the accelerations are not available in the covariance matrix $\hat{P}$. However, the accelerations can be computed if they are expressed via the state equation (32) as functions of the state vector x(t). Then, it remains

$$P_{\ddot{Z}} = \hat{T}_{\ddot{Z}} \hat{P} \hat{T}_{\ddot{Z}}^T \tag{37}$$

with another 1x44-transformation matrix $\hat{T}_{\ddot{Z}}$.

The numerical result published by Kreuzer and Rill[7], is shown in Fig. 4. There is only very little difference in the accelerations compared with Fig. 2. However, the dynamical wheel loads are now available. It turns out that the rear wheels are subject to larger dynamical loads than the front wheels.

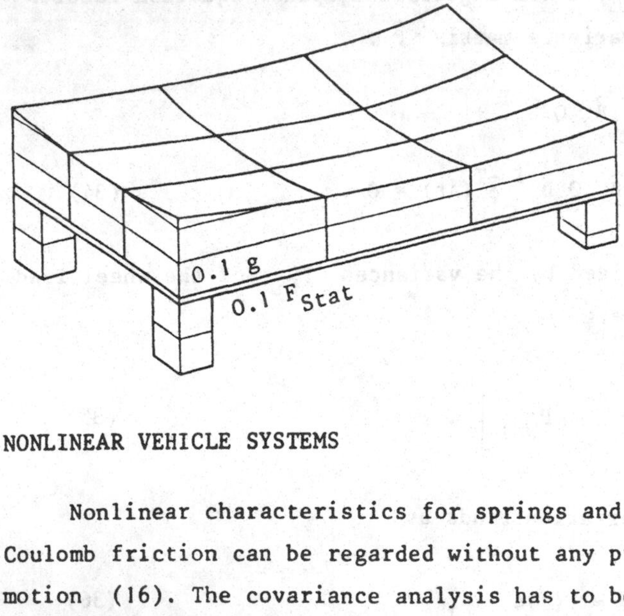

Fig. 4.
Acceleration of the car-body and wheel loads due to colored displacement noise and correlated traces

NONLINEAR VEHICLE SYSTEMS

Nonlinear characteristics for springs and dashpots inclusive Coulomb friction can be regarded without any problem in the equations of motion (16). The covariance analysis has to be extended by an iteration procedure as shown in the lecture on mathematical methods in vehicle dynamics. Some numerical results[8], however for a simple vehicle, are also known. In particular on rough roads there is an essential influence of the nonlinearities on ride quality as well as on ride safety.

REFERENCES

1 Bormann, V., Messungen von Fahrbahnunebenheiten paralleler Fahrspuren und Anwendung der Ergebnisse, *Vehicle System Dynamics* 7, 65-81, 1978.

2 Parkhilovski, J.G., Investigation of the Probability Characteristics of the Surfaces of Distributed Types of Roads. *Avtomobilnaya Promyshlennost* 8, 18-22, 1968.

3 Hirschberg, W., *Vergleich verschiedener Fahrzeugmodelle an Hand charakteristischer Kenndaten der Systemantwort bei stationärer stochastischer Erregung*, Studienarbeit STUD-1, Institut B für Mechanik, Stuttgart, 1981.

4 Müller, P.C., Popp, K. and Schiehlen, W.O., Covariance Analysis of Nonlinear Stochastic Guideway-Vehicle-Systems, in: *Dynamics of Vehicles on Roads and on Tracks*, Ed. by H.-P. Willumeit, Swets & Zeitlinger, Lisse, 337-351, 1980.

5 Müller, P.C. und Schiehlen, W.O., *Lineare Schwingungen*, Akad. Verlagsges., Wiesbaden, 1976.

6 Rill, G., Grenzen der Kovarianzanalyse bei Weißem Geschwindigkeitsrauschen. *Z. angew. Math. Mech.*, to appear.

7 Kreuzer, E. and Rill. G., Vergleichende Untersuchung von Fahrzeugschwingungen an räumlichen Ersatzmodellen. *Ing.-Arch.*, to appear.

8 Müller, P.C., Popp, K. und Schiehlen, W.O., Berechnungsverfahren für stochastische Fahrzeugschwingungen, *Ing.-Arch.* 49, 235-254, 1980.

PRINCIPLE CONCEPTS OF HIGH-SPEED TRAFFIC USING WHEEL/RAIL TECHNOLOGY

P.H. MEINKE, H. ÖRLEY
M.A.N.-Neue Technologie
Dachauerstraße 667
D-8000 München 50

1. INTRODUCTION

One main feature of human life is the necessity to transport people and goods from one point to another. Therefore much emphasis is laid on the development of efficient and comfortable systems of transportation.

Planning a journey, a passenger today has the possibility to decide between several transportation systems. For larger distances, e.g. more than 600 km, only the airplane guarantees acceptable travelling times.

In the range from 60 km to 600 km, which represents typical distances within Western Europe, the suitable transportation system may be a wheel/rail-system running on high-speed.

For smaller distances (up to 60 km) one may use different means of transportation according to the circumstances (subway, suburban railway, bus, taxi, private car).

Figure 1 illustrates the network discussed above.

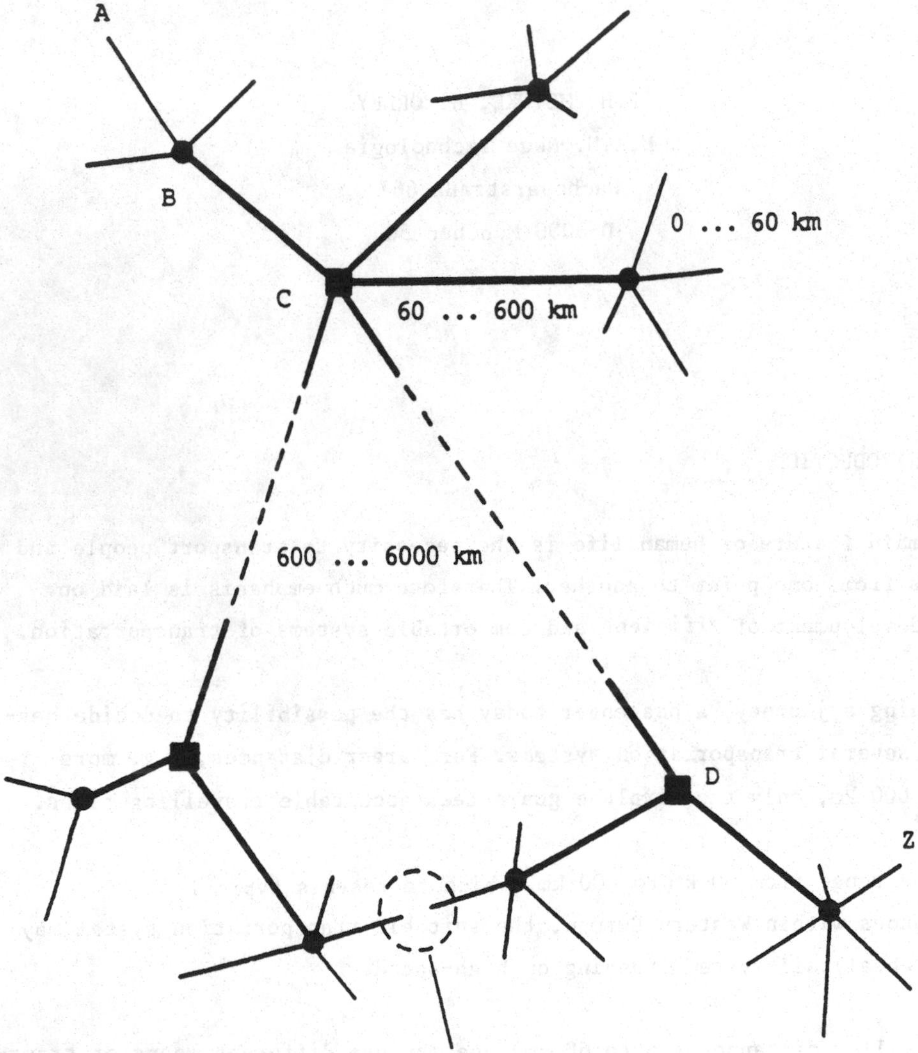

Figure 1 : Network of different means of transportation

2. TRAIN SPEED

One important step in drawing up a concept for high-speed trains is the analysis of the improvement of travelling time by speed increase.

The answer to this question depends on the structure of the population of the considered region or country. For example, in Japan, France and Italy the network is dominated by one main line, whereas in West Germany the network is very complex with comperatively small distances between the train stops[1] (cp. the IC-network in Figure 2).

In Figure 3 the distribution curve of distances in the West German IC-network is given; the mean value of distance is at about 70 km. If international connections within Western Europe are taken into account, the mean value increases up to approximately 100 km (cp. Figure 4).

Based upon the two assumptions, namely

- halt distance: 70 km
- accel ration or deceleration : less 0.7 $^{m}/s^{2}$
  (due to ride comfort criterias)

Figure 5 shows the differences in travelling time as a function of maximum speed. Although this is a very simple investigation, it gives an idea for the 'optimal' operating velocity of high-speed trains, which may be assumed in the range from 250 km/h to 270 km/h. More detailed considerations lead to similar results.

By this is not surprising, that the French TGV, introduced in 1981, is running at an operating speed of 260 km/h.

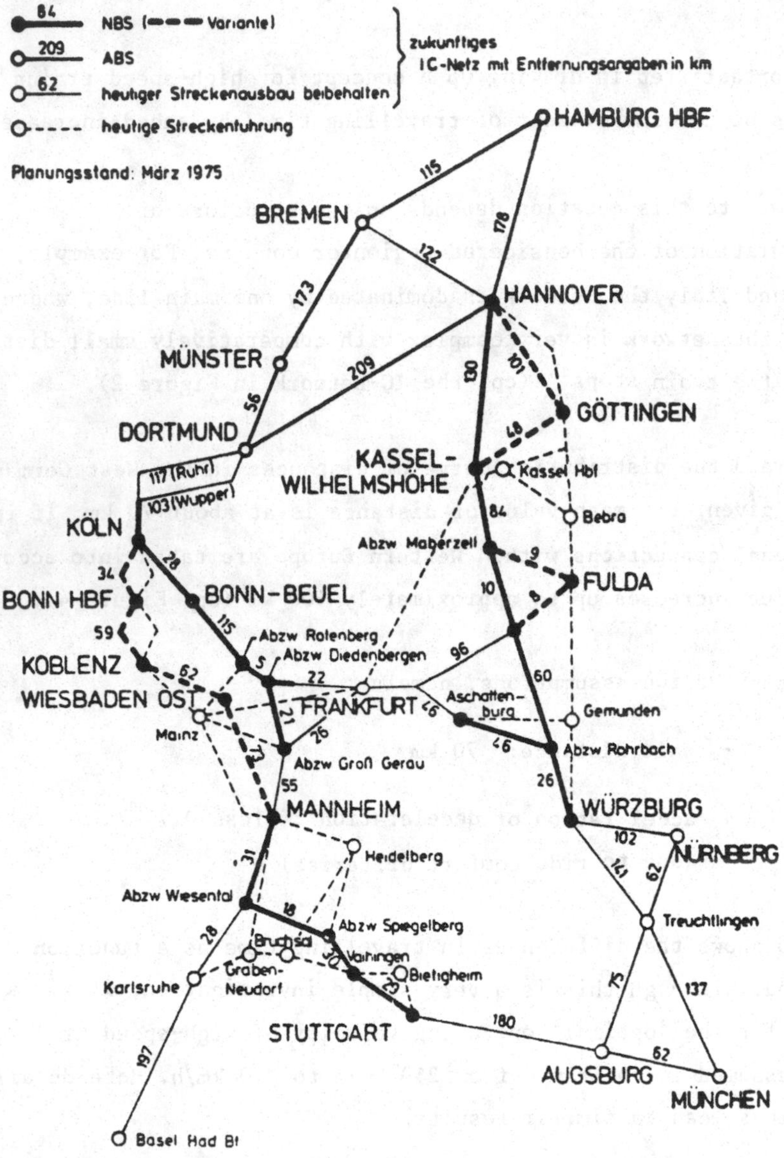

Figure : German IC-network (proposal for 1985)
(source: [2])

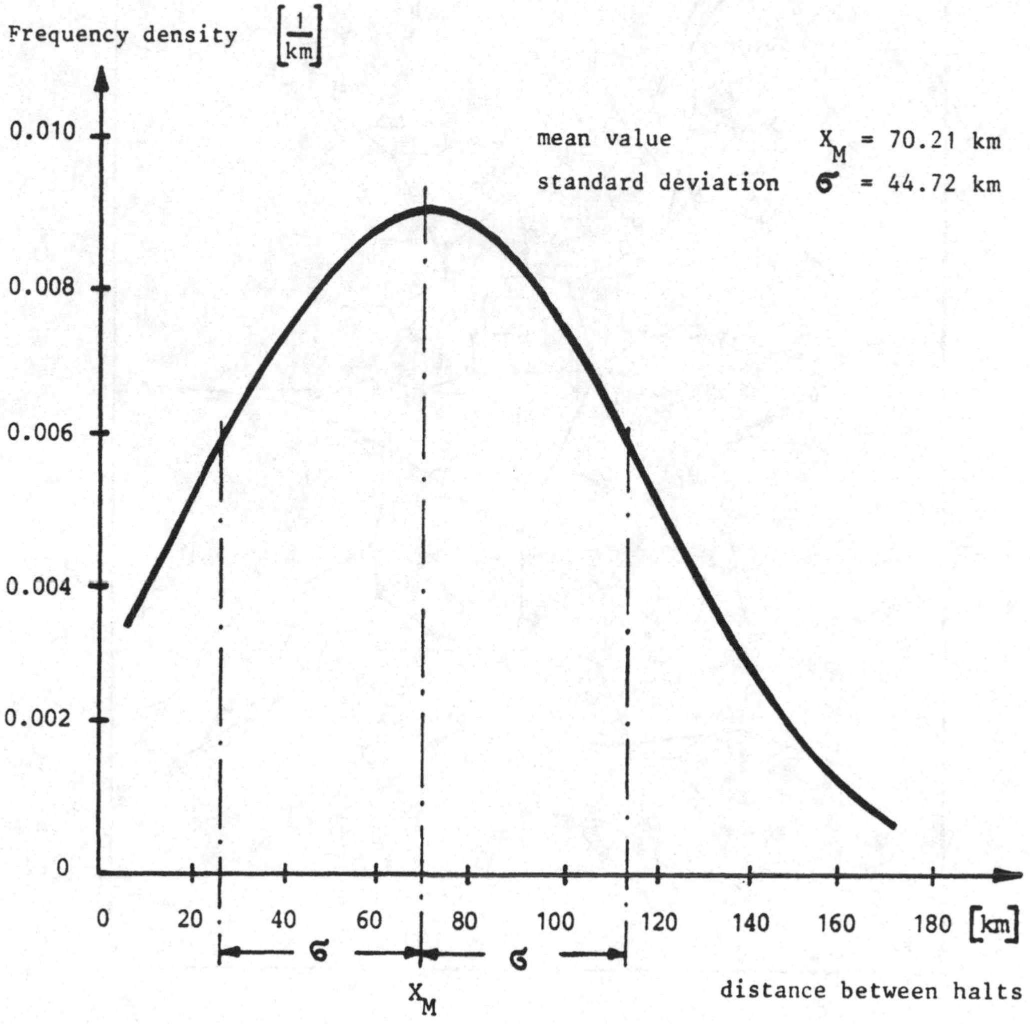

Figure 3 : Distribution curve of distance between halts within the German IC-network

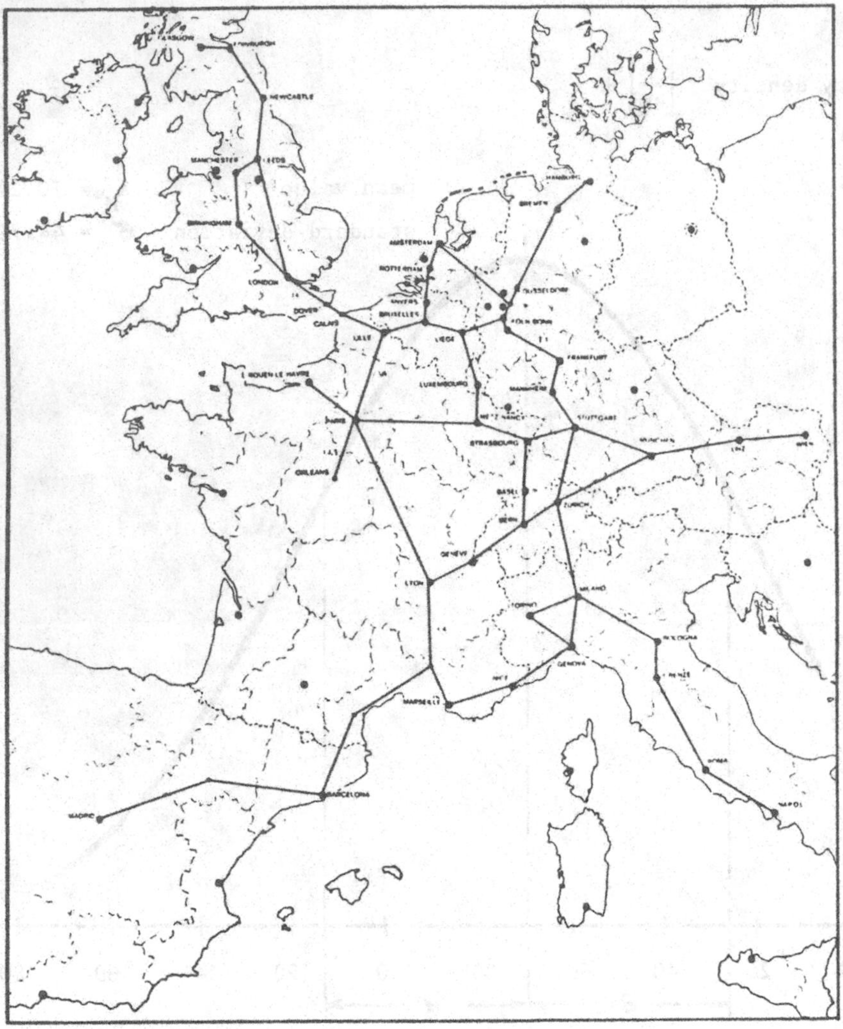

Figure 4 : Proposed European high-speed train network according to 'The future of European passenger transport', OECD, Paris 1977

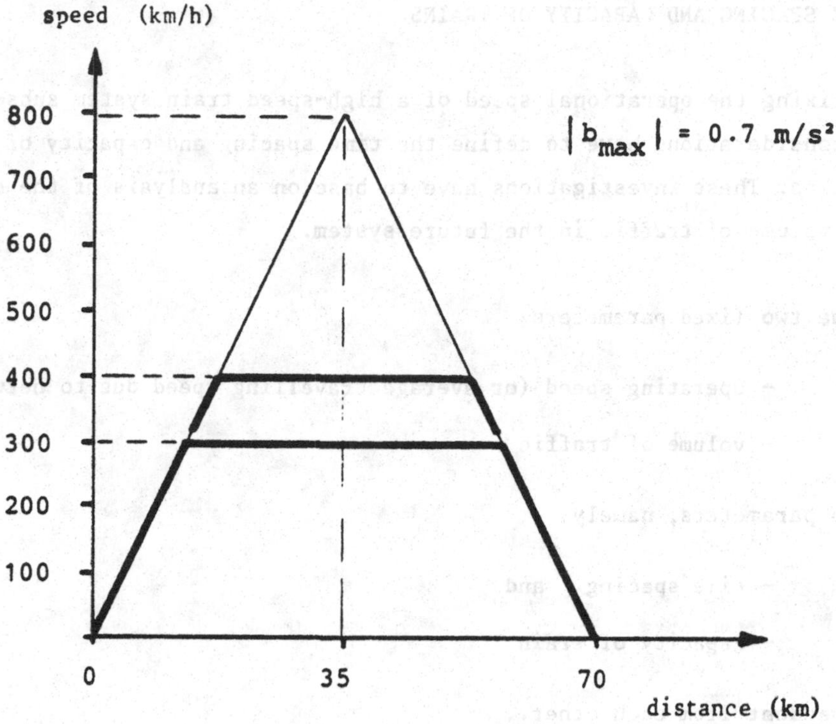

maximum speed (km/h)	travelling time (minutes)
800	10,5
400	13,1
300	16,0

Figure 5 : Speed/distance - relation for different maximum speeds

3. TIME SPACING AND CAPACITY OF TRAINS

After fixing the operational speed of a high-speed train system subsequent considerations have to define the time spacing and capacity of the trains. These investigations have to base on an analysis of the expected volume of traffic in the future system.

With the two fixed parameters

- operating speed (or average travelling speed due to network)
- volume of traffic

the two parameters, namely

- time spacing and
- capacity of train

are dependent from each other.

The decision on time spacing of trains is influenced by two characteristics:

- time spacing should be better than the IC-system operating today; the high-speed train system will become more attractive due to reduced waiting time.
- very short time spacing produces difficulties with train operation in the network; furthermore, this may result in a very small train size and therefore also in an unprofitable one.

With regard to the West German IC-network detailed investigations have shown that the values for time spacing of high-speed trains should be in the range from 20 to 40 minutes. As conclusion a train capacity of about 400 seats would be the optimal one.

4. CONFIGURATION OF TRAINSET

Next step in defining a concept for a high-speed train is the design of the configuration of a trainset, which satisfies the above introduced conditions.

There exist a lot of train configurations and hence several ways to come to a decision on an optimal one. In the following one way is illustrated taking the development of a high-speed train in West Germany as an example.

That concept of a high-speed train is based on investigations [2] carried out by a group consisting of 'Deutsche Bundesbahn' and German industry during the years 1974 - 1976.

The study starts by defining fundamental elements of trainsets (cp. Figures 6 and 7). By combining these elements a great variety of configurations can by found (cp. Figure 8).

Taking into account technical and economocal points of view ten configurations were selected out of this list (cp. Figure 9).

Doing a cost-benefit-analysis the number of configurations can be reduced furthermore. For the future German high-speed train three concepts have been selected (cp. Figure 10):

- T4: two power cars with 6 axles each and coaches.
- T8: two power cars with 4 axles each and coaches (+ booster, if necessary for higher performance).
- W7: train consisting of motor coaches, coupled two by two.

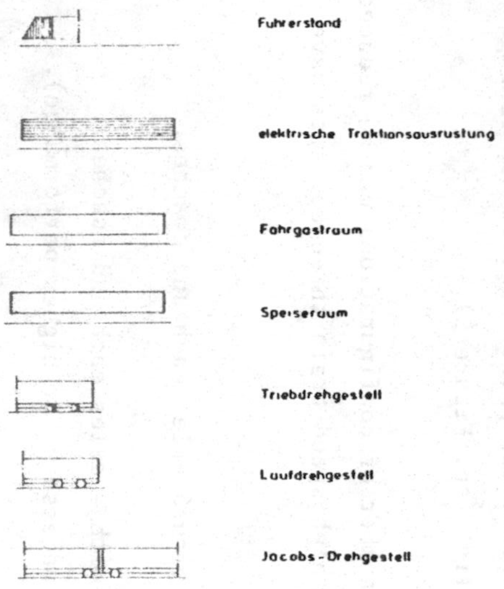

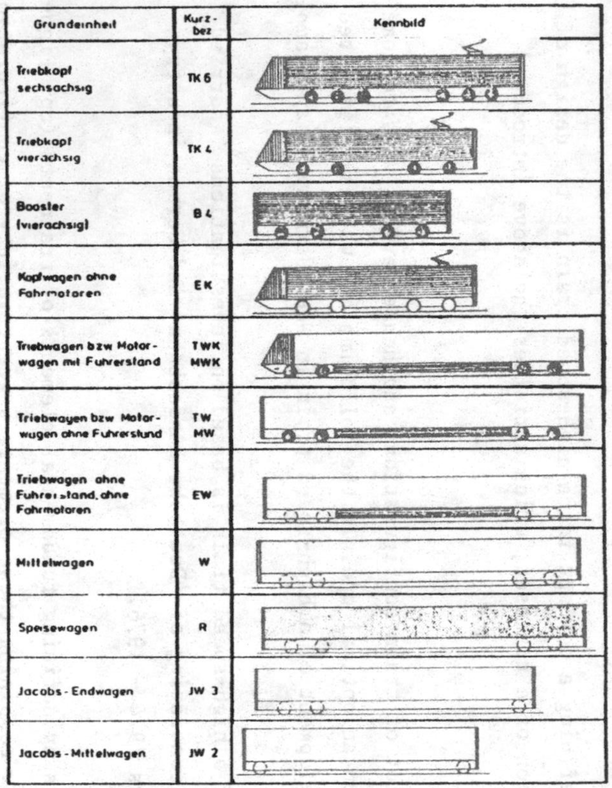

Figures 6 and 7 :

Fundamental elements of trainsets (source: [2])

Zug-kurz-bez	Zugzusammensetzung	Länge [m]	Gesamtmasse besetzt [t]	Sitzplätze	P_{erf} [MW]	$P_{üb}$ [MW]
1	2	3	4	5	6	7
T 2	TK4 + 8 W + R + TK4	273	552	432	10,32	8,83
T 4	TK6 + 8 W + R + TK6	281	624	432	10,92	13,24
T 6	TK8 + 8 W + R + TK8	305	696	432	11,87	17,66
T 8	TK4 + B4 + 8 W + R + TK4	289	624	432	11,10	13,24
T 9	TK4 + B4 + 8 W + R + B4 + TK4	305	696	432	11,87	17,66
T 11	TK6 + B4 + 8 W + R + TK6	297	696	432	11,70	17,66
T 12	TK6 + B4 + 8 W + R + B4 + TK6	313	768	432	12,47	22,07
T 14	TK6 + B6 + 8 W + R + TK6	301	732	432	12,00	19,86
T 15	TK6 + B6 + 8 W + R + B6 + TK6	321	840	432	13,07	26,49
W 1	TWK + 6 W + R + TWK	237	446	414	8,91	7,85
W 2	TWK + TW + 5 W + R + TWK	237	465	414	9,02	11,77
W 3	TWK + TW + 4 W + R + TW + TWK	237	484	414	9,14	15,69
W 4	TWK + 2 TW + 2 W + R + 2 TW + TWK	237	522	414	9,36	23,54
W 5	TWK + 3 TW + R + 3 TW + TWK	237	560	414	9,59	31,39
W 7	MWK — EW + 2 (MW — EW) + R + EW — MWK	237	560	414	9,59	15,69
W 8	MWK — EW — MW + MW — EW — MW + R + MW — EW — MWK	264	624	468	10,56	23,54
W 9	MWK — EW — MW + MW — EW + R + MW — EW				9,59	19,62
M 1	TK4 -- MW + 6 W + R + MW — TK4					16,68
M 2	TK4 — 2 MW + 4 W + R					

Figure 8 : List and main parameters of train configurations (source: [2])

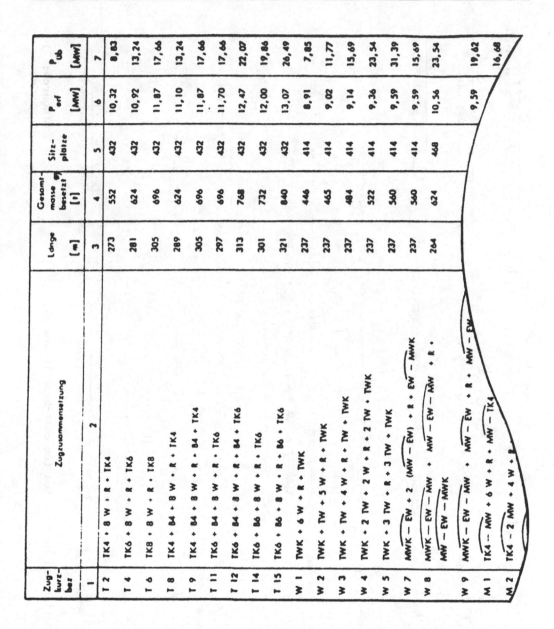

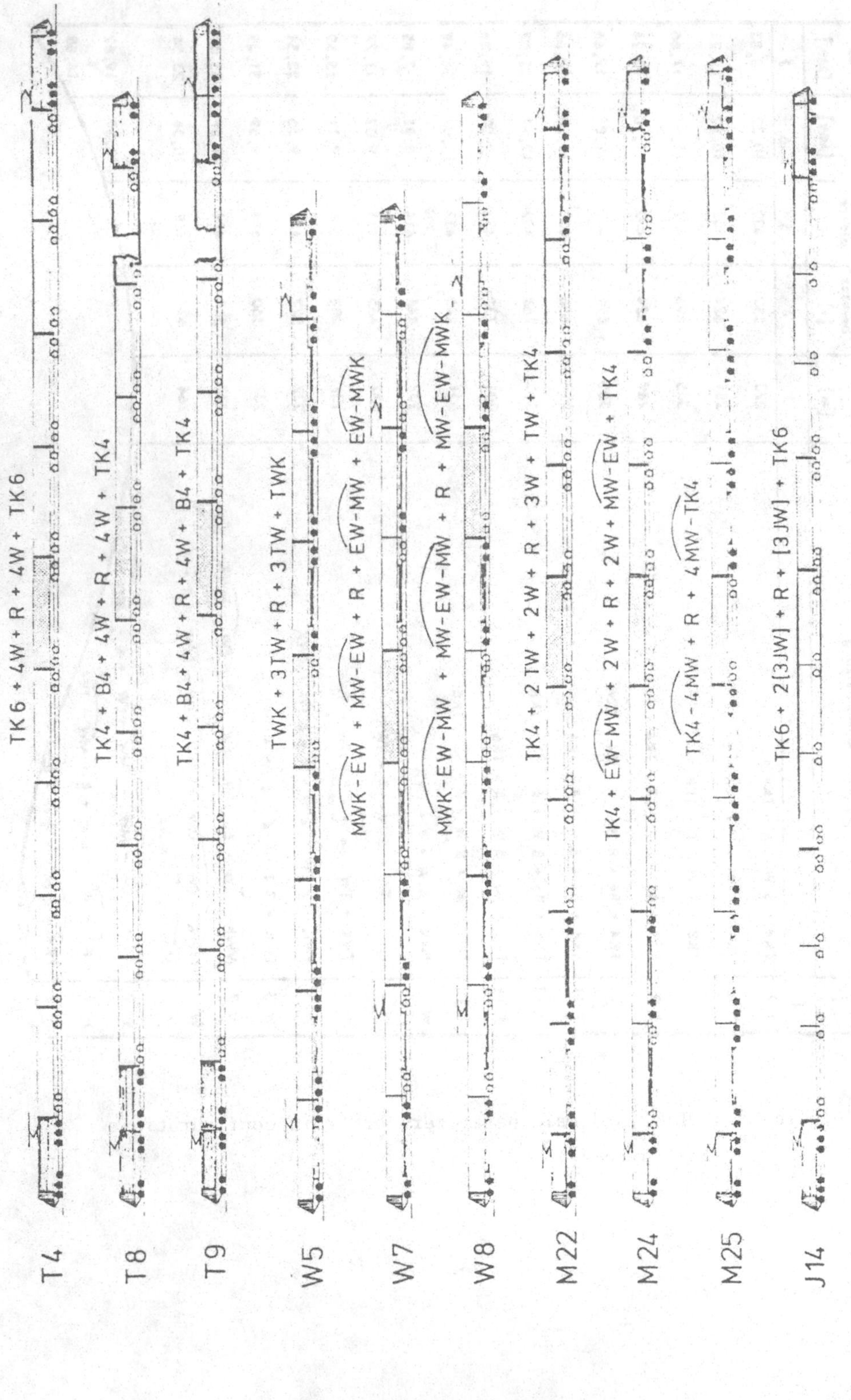

Figure 9 : Configurations selected for cost-benefit-analysis (source: 2)

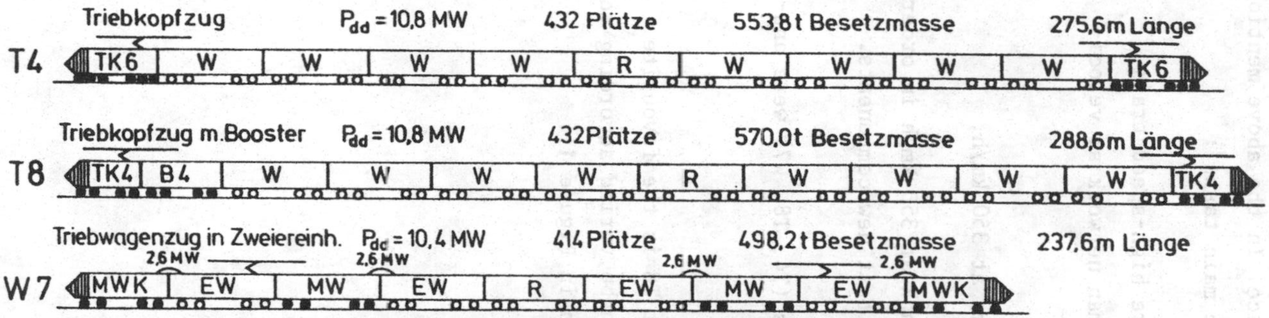

Techn. Daten der Grundeinheiten (Stahlbauweise, Drehstromtechnik)

Bezeichnung d. Grundeinheit	LüP [m]	Achsstand im Drehgestell [m]	Radsatzkraft [kN]	Drehgestell-masse [t]	Gesamtmasse [t]
TK 6	19,0	1,8 / 2,3	155,0	18,9	93,0
TK 4	17,5	2,6	168,5	13,0	67,4
B 4	16,0	2,6	168,5	13,0	67,4
MWK	26,4	2,65	140,7	10,25	56,28
EW	26,4	2,5	140,3	5,7	55,51
MW	26,4	2,65	143,3	10,25	57,31
W	26,4	2,5	100,8	5,4	40,34
R	26,4	2,5	112,7	5,4	45,1

DB	Zusammenstellung der wichtigsten techn. Daten der Konfigurationen T4, T8 u. W7 der Studie "Projektdefinition Hochgeschwindigkeitstriebfahrzeug des Gemeinschaftsbüros"	BZA München Dez 107 1010.00.62 12.3.80

Figure 10 : Configuration of trainset T4, T8, W7 (source: [3])

The final selection of the configuration came up by a particular analysis of special requirements the trainset has to face. In the above mentioned German concept the trainset must fulfill three main tasks[3]:

- demonstration of an attractive future high-speed train, running on certain parts of the German network at velocities up to 300 km/h,
- proving the system and all subsystems at 350 km/h,
- experimental operation on a test track at 350 km/h in order to prove theoretical results and to test new components.

In regard to this requirements the three trainsets (T4, T8, W7) were analysed again by a cost-benefit-analysis.

From this evaluation it turns out that the concept T8 is the favourite one. Within this configuration T8 the trainset can be varied according to the three different tasks mentioned above as shown in Figure 11.

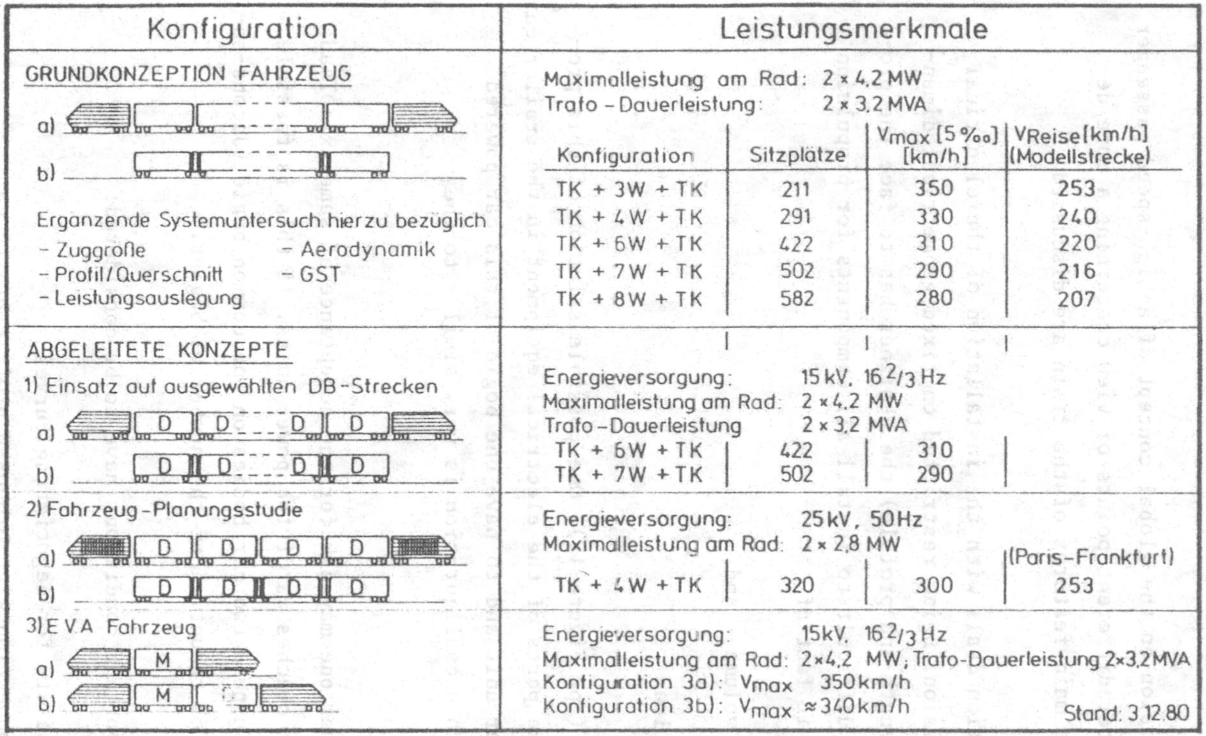
Konfiguration
Leistungsmerkmale
GRUNDKONZEPTION FAHRZEUG
a)
b)
Ergänzende Systemuntersuch hierzu bezüglich
- Zuggroße
- Aerodynamik
- Profil/Querschnitt
- GST
- Leistungsauslegung
Maximalleistung am Rad: 2 × 4,2 MW
Trafo - Dauerleistung: 2 × 3,2 MVA
Konfiguration | Sitzplätze | Vmax [5 ‰] [km/h] | VReise [km/h] (Modellstrecke)
TK + 3W + TK | 211 | 350 | 253
TK + 4W + TK | 291 | 330 | 240
TK + 6W + TK | 422 | 310 | 220
TK + 7W + TK | 502 | 290 | 216
TK + 8W + TK | 582 | 280 | 207
ABGELEITETE KONZEPTE
1) Einsatz auf ausgewählten DB-Strecken
a) D D D D
b) D D D D
Energieversorgung: 15 kV, 16 2/3 Hz
Maximalleistung am Rad: 2 × 4,2 MW
Trafo-Dauerleistung 2 × 3,2 MVA
TK + 6W + TK | 422 | 310
TK + 7W + TK | 502 | 290
2) Fahrzeug-Planungsstudie
a) D D D D
b) D D D D
Energieversorgung: 25 kV, 50 Hz
Maximalleistung am Rad: 2 × 2,8 MW
TK + 4W + TK | 320 | 300 | (Paris-Frankfurt) 253
3) E V.A Fahrzeug
a) M
b) M
Energieversorgung: 15kV, 16 2/3 Hz
Maximalleistung am Rad: 2×4,2 MW; Trafo-Dauerleistung 2×3,2 MVA
Konfiguration 3a): Vmax 350 km/h
Konfiguration 3b): Vmax ≈ 340 km/h
Stand: 3.12.80

Figure 11 : Configuration for the prototyp of a future German high-speed train (source: [5])

5. DETAILED DESIGN OF PRINCIPLE TRAIN SUBSYSTEMS

After finding a decision on the global concept of a high-speed passenger train, in this chapter different points of view concerning a more detailed design of the main features of the train are discussed.

The first consideration deals with the installation of the electrical equipment for propulsion. Being restricted on fixed geometrical dimensions (power unit length and profile) the designer has to face the problem that it can be difficult to install all components for propulsion into the power cars in fact of

- great volume and
- heavy load

of these components (cp. Figure 12). One possible solution of this problem is to mount some parts of the electrical equipment in the trailer car adjacent to the power unit and to have one bogie of this car powered[4] (cp. Figure 13). Such a configuration is very similar to the French TGV.

Going one step further one may ask for the acceptance of same axle loads at the bogies of the coaches as of the power units. If this is feasible, the possibility to suspend two car bodies on one common bogie (Jacobs-solution) is offered for reduction of the number of bogies.

In this case two important conditions have to be considered:

- suitability for negotiating curves (in connection with proposed profile of the cars; cp. Figure 15)
- limitation of axle load,

which leads to a reduction of the length of the coaches and by this to a decrease of axle load, too (cp. Figures 16 and 17).

In Western Germany the discussion on this topic is not yet finished, but preliminary results of considerations[5] show that

- the Jacobs-solution does not diminish the length or the mass of the trainset
- the Jacobs-solution offers advantages for
  * aerodynamic resistance (reduced number of bogies)
  * running behavior, ride comfort and lineside noise
  * design of passenger gangway between the coaches.

First evaluations of these two configurations (cars with two bogies and Jacobs-solution) did not result in an unequivocal statement, which is the better one.

Therefore the efforts in Germany are nowadays enforced to gain an opinion, which of the two concepts is the better one, especially with regard to the situation in Western Germany.

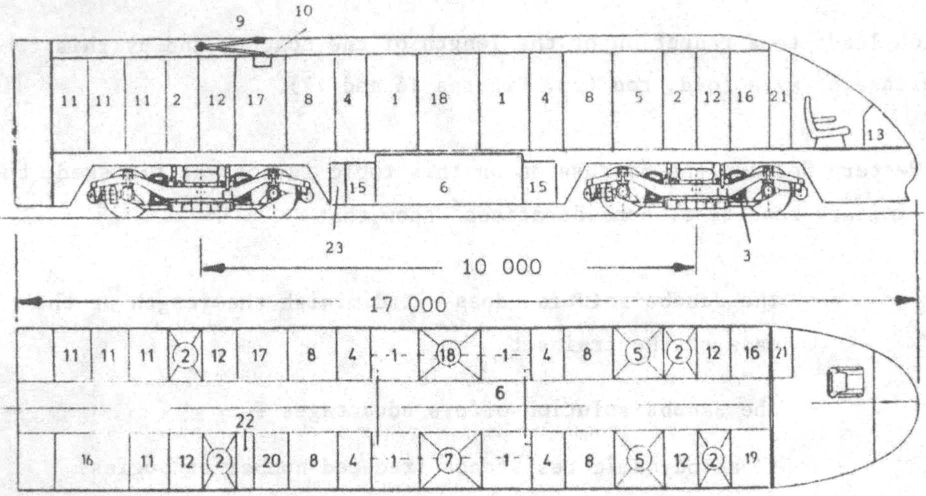

Figure 12 : Power unit with mounted propulsion equipment

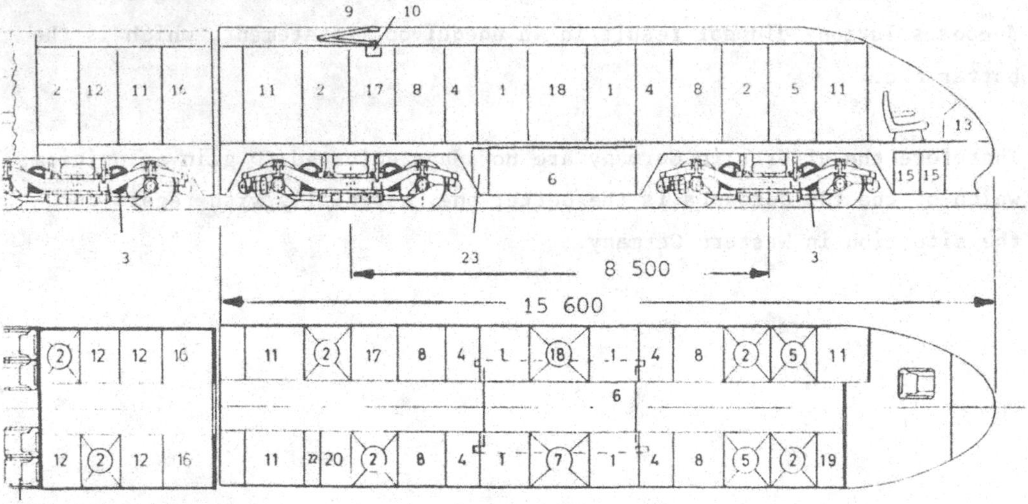

Figure 13 : Power unit (parts of the propulsion equipment mounted in the adjacent trailer car)

(Abbreviations in these figures: cp. Figure 14)

Nummer	Komponente/Bezeichnung
1	Zwischenkreis (Stützkondensatoren)
2	Fahrmotorlüfter mit Motorvordrossel
3	Fahrmotor und Antriebe (Getriebe)
4	Zwischenkreis (Saugkreisdrosseln)
5	Luftpresser (Kompressoren)
6	Transformator mit Drosseln
7	Ölkühler für Trafo und Stromrichter
8	Stromrichter
9	Stromabnehmer
10	Trennschalter
11	Hilfsbetriebe, Wechselrichter
12	Elektronische Stromversorgung, Netzgeräte
13	Fahrpult Steuerung, Regelung, Führung, Überwachung, Telefon, LZB, Lautsprecher
14	Dachausrüstung
15	Batterien
16	Hilfsbetriebe, Lade- und Schaltgeräte
17	Schüzte, Kabel, Verteiler, Schalter
18	Bremswiderstand
19	Schaltschrank Nachrichtentechnik, Meßtechnik, Elektronik für Hydraulik und Pneumatik, Sifa, Indusi, LZB, AFB, etc.
20	Nebenbetriebe, Schütze, Steuerungen, Verteiler
21	Zugbiblothek, Kleiderfach, Sicherungen
22	Hochspannungseinspeisung
23	Hauptschalter

Figure 14 : Elements of propulsion system

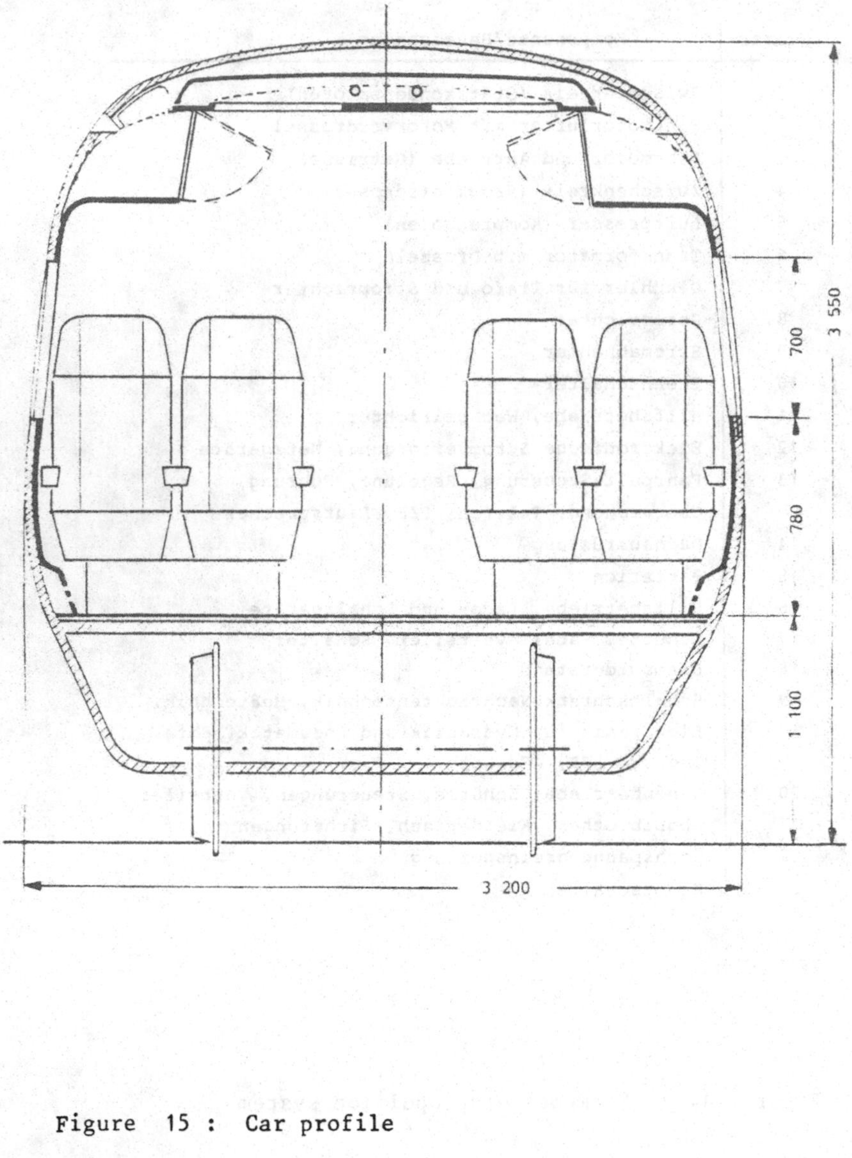

Figure 15 : Car profile

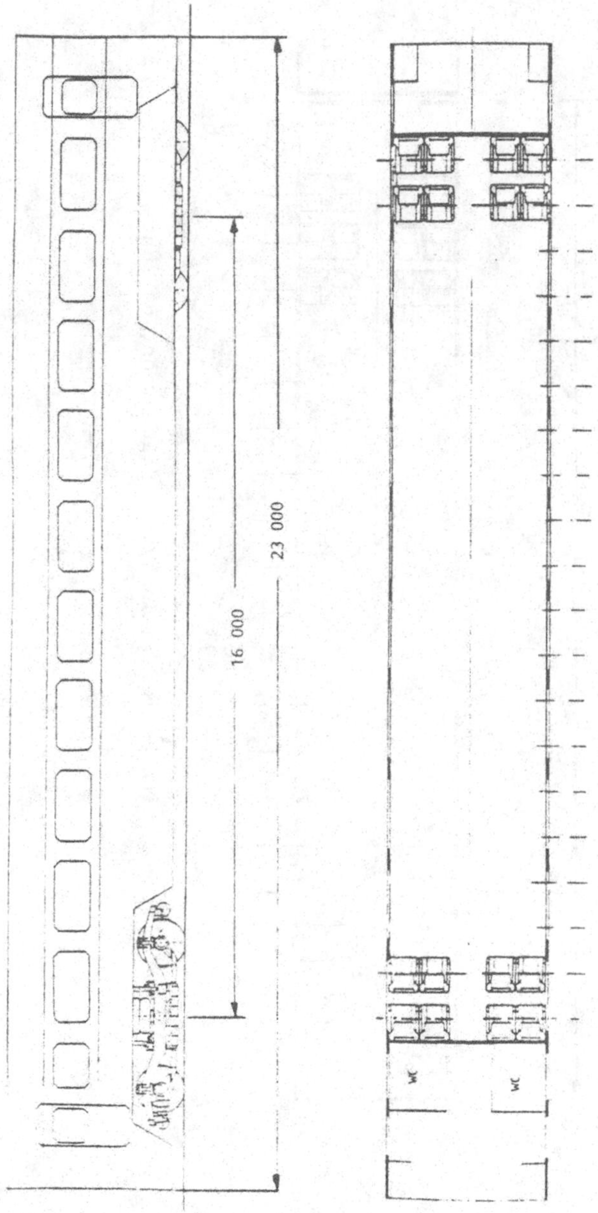

Figure 16 : Coach with two conventional bogies, 80 seats

Figure 17 : Coach with bogies in Jacobs-configuration, 64 seats

6. REFERENCES

1. Rappenglück, Fahrzeugtechnik für den Hochgeschwindigkeits-Personenverkehr, Eisenbahntechnische Rundschau, 7/8, 525, 1980

2. Gemeinschaftsbüro Hochgeschwindigkeits-Triebfahrzeug, Studie 'Hochgeschwindigkeits-Triebfahrzeug', München, 1976

3. Forschungsgemeinschaft Rad/Schiene, Arbeitsgemeinschaft Rheine-Freren, Ergebnisbericht der Konzeptphase Rad/Schiene-Versuchs- und Demonstrationsfahrzeug, 1980

4. Mauer, Mielcarek, Reichel, Örley, R/S-Planungsstudie: Voruntersuchungen und Fahrzeugentwurf von Zugsystemen mit Jakobslaufwerken, M.A.N.-Neue Technologie, München, 1981

5. Kurz, R/S-Versuchs- und Demonstrationsfahrzeug: Ausschöpfung des Forschungs- und Entwicklungspotentials für die Bahn der Zukunft, Statusseminar VII, Spurgeführter Fernverkehr-Rad/Schiene-Technik-, Titisee, 14-1, 1982

THE LATERAL BEHAVIOUR OF RAILWAY VEHICLES

A.D. de Pater
Delft University of Technology
Mekelweg 2
2628 CD Delft, Netherlands

1. INTRODUCTION

This paper will deal with aspects of the basic theory of the lateral movement of a railway vehicle. As it will be shown later on, it is desirable to investigate the motion both on a *tangent track* as in a *curve* with a *constant radius* and *cant*.

1.1. General remarks

We shall enunciate the subjects in the inductive way and we shall begin with the case of the *single wheelset* and then extend the system to a *vehicle with two wheelsets*, after which the *vehicle with two bogies* will be discussed. In the enunciation many results of a recent (not yet finished) investigation by the author in collaboration with "M.A.N./Neue Technologie" at Munich have been inserted; on the contrary, not so many references to results of other investigators will be given.

The equations of motion of the system, consisting of a track and a vehicle with a number of wheelsets, often admit a *linearisation* and most of the theory will be developped for such linearised systems; in the last chapters some aspects of the more general non-linear theory will be discussed.

1.2. The number of degrees of freedom of the system

We consider the *track*, the *wheelsets* and the *frame* as rigid. Then, the system, consisting only of one single wheelset, presents four *degrees of freedom*: the position of the wheelset is given by six *coordinates* and because each wheel contacts the rail usually in one point, there are two constraints. Two of the four coordinates determine the *symmetrical movement* and the two other ones (the *lateral displacement* and the *yaw angle*) the *lateral movement*.

In the same way we find that a two-axled vehicle has $3 \times 6 - 2 \times 2 = 14$ degrees of freedom and seven coordinates determine the symmetrical motion; seven other ones (the lateral displacements of the body and the two wheelsets, their yaw angles and the rolling angle of the body) the lateral motion.

In case of a vehicle with one *main body* and two two-axled bogies we have $7 \times 6 - 2 \times 4 = 34$ degreees of freedom. Now 17 coordinates determine the lateral motion: the lateral displacements (7) of the wheelsets, the *bogie frames* and the main body, their yaw angles (7) and the rolling angles (3) of the bogie frames and the main body.

1.3. Linearisation

We mentioned already that we shall mainly occupy ourselves with linearised systems. In such systems the differential equations for the *symmetrical motion* are completely uncoupled from the equations for the *lateral motion*, so that we can leave the symmetrical motion out of account. Further on, the differential equations for the lateral motion are *inhomogeneous equations* with constant right-hand terms, which only are zero when the vehicle is moving on a tangent track rather than on a curve. The *particular solution* of the equations (which is independent of time and which is zero on a tangent track) determines the *fundamental movement* of the vehicle. The general solution of the corresponding homogeneous equations determines its *parasitic movement*: when the parasitic movement decreases, the fundamental movement is *stable*; when it increases, the fundamental movement is *unstable*.

1.4. Restrictions

In this paper we have no possibility to discuss all the effects which influence the lateral motion.

We shall leave out of consideration the *gyroscopic effects of the wheelsets*. It can be shown that these effects only have influences at speeds from about 500 km/h, whereas conventional railway vehicles, with which we are concerned, are not faster than 300 to 400 km/h (although in determining root loci and such-like we often shall consider the case of infinite speed).

Another effect of minor importance is the so-called *gravitational stiffness*. It occurs in a wheelset with *hollow (concave) tyres*: in case of a lateral displacement the direction of the normal force in each contact point will change in such a way that the resultant of the two normal forces will have a *lateral component* which has a *restoring effect* and thus *stabilizes* the lateral motion. However, because of the *spin effect* in a contact point (see chapter 2) also a tangential force in tangential direction will occur. In the central position of the wheelset the resultant of the two tangential forces will have no lateral component but it does have in case of a lateral deviation. It can be shown that this lateral force has a *destabilizing influence*, which always is about 80% of the stabilizing influence of the resultant of the normal forces. Thus, the complete gravitational stiffness is small in comparison with the effects which we shall take into account and in neglecting it we are on the safe side.

We shall also leave out of consideration the *elasticity of the wheelset*; this seems to be admissable, let alone the torsional elasticity of the shaft, connecting the two wheels, the influence of which can sometimes be important. Further investigation about this aspect is necessary.

At last we should mention that we restrict ourselves to the case of a curve with a constant radius r_o and a constant cant angle ϕ_o, the vehicle moving with the *equilibrium speed* V_o. This means that the speed is such that the component of the gravity force which is parallel to the track in lateral direction, is equal to the component in this direction of the *centrifugal force*:

$$V_o = \sqrt{gr_o \, tg\phi_o}. \qquad (1.4\text{-}1)$$

2. THE FORCE IN A CONTACT POINT BETWEEN RAIL AND WHEEL

In the railway literature until about 1940 generally Amontons-Coulomb law was used as starting point for calculations in the field of railway dynamics. According to this law we can write

$$\left.\begin{array}{ll} T_x = \mu N \dfrac{W_{tx}}{W_t}, \quad T_y = \mu N \dfrac{W_{ty}}{W_t} & \text{for } W_t \neq 0, \\ T = \sqrt{T_x^2 + T_y^2} < \mu N & \text{for } W_t = 0, \end{array}\right\} \qquad (2\text{-}1)$$

where we have called

- N : the normal force in the contact point;
- T_x, T_y : the components in x and y directions of the tangential force in the contact point;
- μ : the coefficient of friction;
- W_t : the relative velocity in the contact point;
- W_{tx}, W_{ty} : the components in x and y directions of W_t.

The x direction is the direction of the motion of the wheelset, the y direction is perpendicular to it and is equally situated in the tangential plane in the contact point; the y direction is to the right-hand side of the track. The velocities W_{tx}, W_{ty} are in the directions x and y, whereas the forces T_x, T_y are opposite to x and y.

As Carter [1] has shown about 1920 for the two-dimensional case and Kalker [2] in 1967 for the three-dimensional case, the relations (1) are only valid when there is no angular velocity of the wheel about the normal in the contact point and, moreover, the velocity W_t is not small as compared with the speed V of the vehicle. In reality, there is not a single contact point but a *contact area*, the dimension of which can be determined by means of Hertz' theory when the elastic constants of rail and wheel are similar. In each point of the contact area there is a *normal pressure* p_n and a *tangential pressure* p_t with components p_{tx}, p_{ty}. The normal pressure is distributed over the contact area in a special way. In the contact area locally the relations (1) hold when we replace N, T_x, T_y by p_n, p_{tx}, p_{ty}. The points where the velocity is zero, belong to the *adhesion region* and the points where this is not the case belong to the *slip region*.

In points in the neighbourhood of the contact area the elastic deformation of the rail and the wheel now have to be taken into account; at a certain distance of this area the deformation can be neglected. Thus we can define an *apparent relative tangential velocity* W_t with components W_{tx}, W_{ty}, calculated for the centre of the contact area in the supposition that the rail and the wheel would still be rigid. In the same way we can define an *apparent normal angular velocity* ω_n. Now we define the quantities υ, υ_x, υ_y and ϕ by

$$\upsilon = \frac{W_t}{V}, \quad \upsilon_x = \frac{W_{tx}}{V}, \quad \upsilon_y = \frac{W_{ty}}{V}, \quad \phi \quad \frac{\omega_n}{V}. \qquad (2\text{-}2)$$

We call υ the *creep* and ϕ the *spin*; note that the creep is dimensionless whereas the spin has the dimension of an inverted length. The creep usually is of the order of 10^{-3} to 10^{-2}.

There are two limit eases: that of *complete sliding*, in which there is only a slip region, and that of *vanishing creep and spin*, in which the slip is reduced to a very narrow slip, situated along the trailing edge of the contact area.

For the two-dimensional case, i.e. the case of two parallel cylinders rolling over each other, the contact area is a strip, limited by two straight lines, parallel to the cylinder axes. There is one slip region (near the trailing edge) and one adhesion region (near the leading edge), separated by a third straight line. The total tangential force T is represented in fig. 2.1 and is equal to

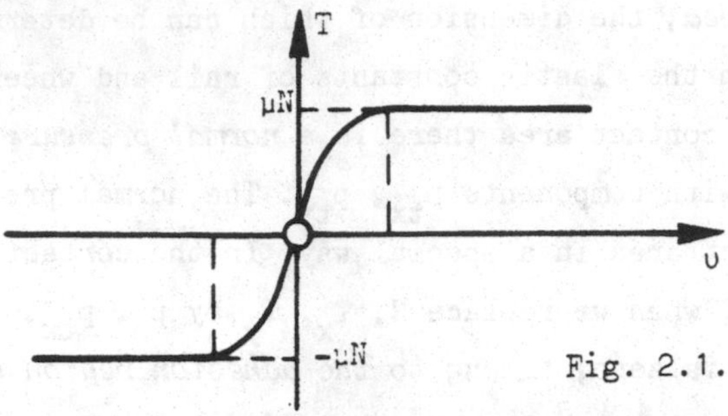

Fig. 2.1. The tangential force as a function of the slip in the two-dimensional case.

$$T = \mu N \frac{\rho \upsilon}{2\mu a}\left(1 - \frac{\rho \upsilon}{8\mu a}\right) \text{ for } 0 \leqq \upsilon \leqq \frac{4\mu a}{\rho}, \; T = \mu N \text{ for } \upsilon > \frac{4\mu a}{\rho}, \quad (2\text{-}3)$$

where a is the half width of the contact area and $\rho^{-1} = 0{,}25(R_1^{-1} + R_2^{-1})$, R_1 and R_2 being the cylinder radii.

For the three-dimensional case the situation is much more complicated. Within the contact area there can be more than one adhesion region and more than one slip region. Only in the case of vanishing slip and spin the relation between the tangential forces and the slip and spin quantities is a linear one and we have

$$T_x = G\,c^2\,C_{11}\,\upsilon_x, \; T_y = G\,c^2(C_{22}\,\upsilon_y + C_{23}\,c\phi), \; c = \sqrt{ab}, \quad (2\text{-}4^{a-c})$$

G being the shear modulus, a and b the half-axes of the contact ellipse and C_{11}, C_{22}, C_{23} dimensionless constants which only depend on Poisson's ratio ν: see Carter [1]. We also can write

$$T_x = K_x\,\upsilon_x, \quad T_y = K_y\,\upsilon_y + K_z\phi. \quad (2\text{-}5)$$

In the case of small conicities, to which we shall restrict ourselves in the next chapters, we may put $K_z = 0$. A very rough value for K_x and K_y is

$$K_x = K_y = 150\ N, \quad (2\text{-}6)$$

N being again the normal force in the contact point.

3. THE SINGLE WHEELSET, MOVING ON A TANGENT TRACK

Many aspects of the lateral motion of railway vehicles as far as its stability is concerned, can be already shown by considering a single wheelset, pulled along a tangent track by a vehicle frame which we suppose to perform only a fundamental movement, purely parallel to the track. The displacement of the wheelset is described by the longitudinal displacement of the frame and two parasitic motions: the lateral displacement and the yaw angle.

3.1. Geometrical and kinematical considerations

We consider the single wheelset which is shown in fig. 3.1-1. In order to distinguish it from other wheelsets and from the vehicle body, we add the index i to all quantities referring to wheelsets in general, i having the values 1, ... n. In the case of a two-axled vehicle we have n = 1. A second index, j, is used to make distinction between the two rails and we put j = 1 for the right hand rail and j = 2 for the left-hand rail.

A coordinate system (o_i^*, x_i^*, y_i^*, z_i^*) is accompanying the wheel-

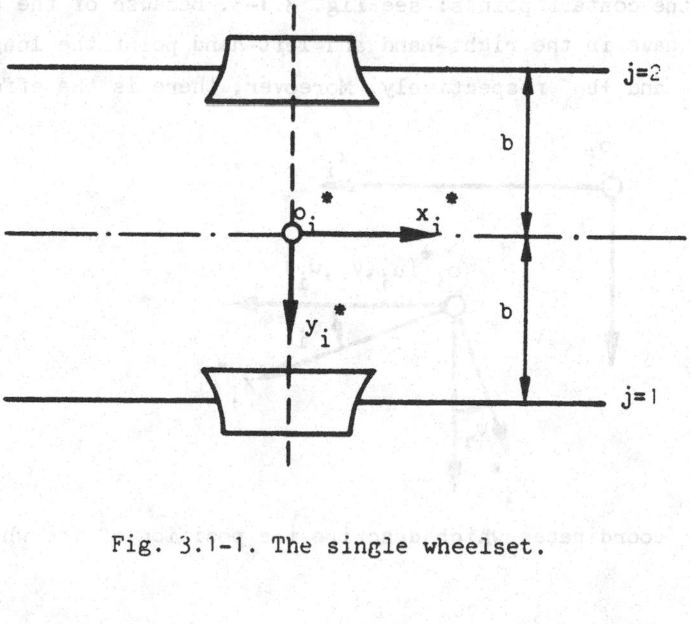

Fig. 3.1-1. The single wheelset.

set in such a way that always the origin o_i^* is coinciding with the mass centre, the axis $o_i^* x_i^*$ is horizontal and the axis $o_i^* y_i^*$ is coinciding with the axis of revolution. Now the position of the wheelset is described by means of a second coordinate system, (o_i, x_i, y_i, z_i), which is purely translating: the origin o_i is situated in such a way that, when the systems (o_i, x_i, y_i, z_i) and $(o_i^*, x_i^*, y_i^*, z_i^*)$ coincide, the wheelset is in its central position with regard to the track. The origin o_i describes a straight line in the longitudinal plane of symmetry of the track. We assume the velocity of o_i to be constant and equal to V.

The coordinates of o_i^* with respect to the system (o_i, x_i, y_i, z_i) are u_i, v_i and w_i. The rotation of the system $(o_i^*, x_i^*, y_i^*, z_i^*)$ can be described by the rotation about the axis $o_i x_i$ (the rolling angle) and the rotation ψ_i about the axis $o_i z_i$ (the yaw angle), when we restrict ourselves to linear terms in the rotations. The rotation of the wheelset about its axis of revolution is equal to $-V/r + \dot{\chi}_i$, χ_i being a third small angle and r being the radius of the wheels in the central position of the wheelset. We shall renounce the symmetrical movement with the components u_i, w_i and χ_i; moreover, ϕ_i is proportional to the lateral displacement v_i.

In the case of a tangent track it is not difficult to calculate the velocities in the contact points: see fig. 3.1-3. Because of the angular velocity $\dot{\psi}_i$ we have in the right-hand and left-hand point the longitudinal velocities $-b\dot{\psi}_i$ and $+b\dot{\psi}_i$ respectively. Moreover, there is the effect of

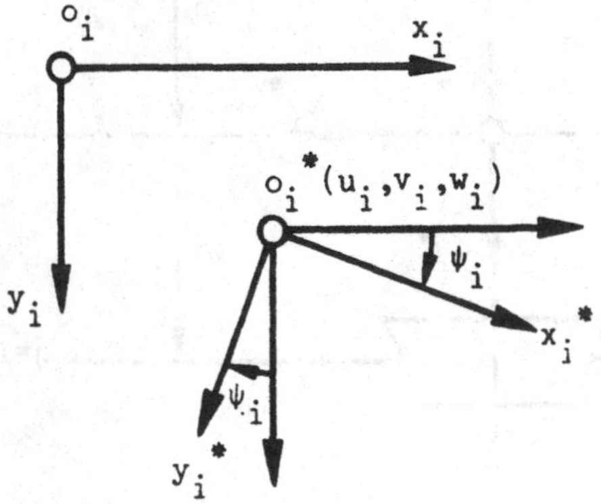

Fig. 3.1-2. The coordinates which describe the position of the wheelset.

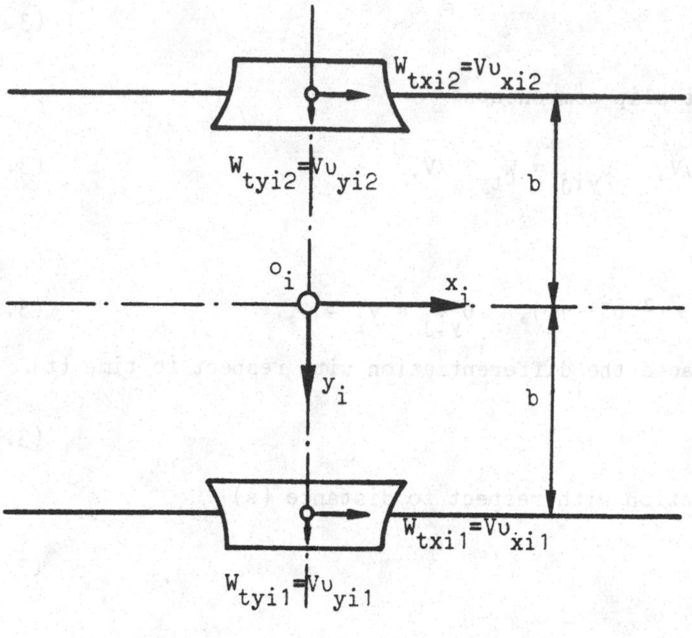

Fig. 3.1-3. The velocities in the contact points.

the change of the wheel radius: in the contact point it will have the value $\mp\gamma_o v_i \rho^*$, γ_o being the tyre conicity in the central position and ρ^* the quantity

$$\rho^* = \frac{r_y^*}{r_y^* - r_y}, \qquad (3.1\text{-}1)$$

where r_y is the radius of the rail profile and r_y^* that of the tyre profile (both in the central position of the wheelset). Note that in a combination of signs the upper sign holds for $j = 1$ and the lower one for $j = 2$

Altogether, we find for the two components of the lateral tangential velocity

$$W_{txij} = \mp(b\dot{\psi}_i + \gamma_o \rho^* V r^{-1} v_i), \qquad (3.1\text{-}2^a)$$

$$W_{tyij} = \dot{v}_i - V\psi_i. \qquad (3.1\text{-}2^b)$$

We introduce the parameter

$$\Gamma = \sqrt{\frac{\gamma_o b}{r} \rho^*} \qquad (3.1\text{-}3)$$

and, moreover, the slip components

$$\upsilon_{xij} = W_{txij}/V, \quad \upsilon_{yij} = W_{tyij}/V; \qquad (3.1\text{-}4)$$

then we find

$$\upsilon_{xij} = \mp(b\psi_i' + \Gamma^2 b^{-1} v_i), \quad \upsilon_{yij} = v_i' - \psi_i. \qquad (3.1\text{-}5^{a-b})$$

Here we have replaced the differentiation with respect to time (t)

$$\dot{} = \frac{d}{dt} \qquad (3.1\text{-}6^a)$$

by the differentiation with respect to distance (s)

$$' = \frac{d}{ds}, \qquad (3.1\text{-}6^b)$$

where

$$s = Vt. \qquad (3.1\text{-}7)$$

For being able to determine the kinetic energy of the wheelset we find the components of the translational velocity of the wheelset with respect to the system (o_i, x_i, y_i, z_i) and the corresponding components of its rotatory velocity:

$$V_{xi} = V + \dot{u}_i = V, \quad V_{yi} = \dot{v}_i, \quad V_{zi} = \dot{w}_i = 0, \qquad (3.1\text{-}8^a)$$

$$\omega_{xi} = \dot{\phi}_i = 0, \quad \omega_{yi} = -Vr^{-1} + \dot{\chi}_i = -Vr^{-1}, \quad \omega_{zi} = \dot{\psi}_i. \qquad (3.1\text{-}8^b)$$

Here again we put u_i, w_i and χ_i equal to zero. Moreover, ϕ_i appears to be equal to

$$\phi_i = -\frac{\gamma_o}{b} v_i \qquad (3.1\text{-}9)$$

and thus can be omitted for low values of γ_o.

3.2. Dynamical considerations

In fig. 3.2-1 the longitudinal and lateral contact forces are shown. They are related with the longitudinal and lateral slip components and accord-

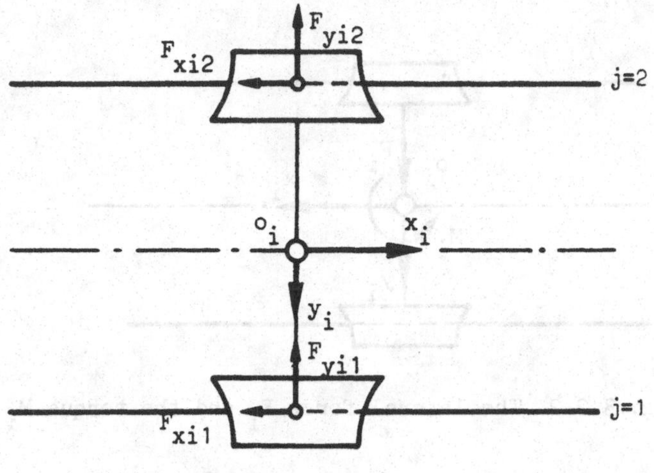

Fig. 3.2-1. The contact forces.

ing to chapter 2 we can write

$$F_{xij} = K_{xij}\,\upsilon_{xij}, \quad F_{yij} = K_{yij}\,\upsilon_{yij}, \qquad (3.2\text{-}1)$$

K_{xij} and K_{yij} being the creep coefficients. Introducing the creep coefficients per wheelset K_{xi}, K_{yi}, we have

$$K_{xij} = \tfrac{1}{2} K_{xi}, \quad K_{yij} = \tfrac{1}{2} K_{yi} \quad (j=1,2), \qquad (3.2\text{-}2)$$

so that (1) and (3.1-5^{a-b}) yield

$$F_{xij} = \mp \tfrac{1}{2} K_{xi}(b\psi_i' + \Gamma^2 b^{-1} v_i), \quad F_{yij} = \tfrac{1}{2} K_{yi}(v_i' - \psi_i). \qquad (3.2\text{-}3^{a-b})$$

Further on, we introduce the total lateral force F_i and the total torque M_i, applied at the wheelset; their signs are positive as in fig. 3.2-2. Comparing fig. 3.2-2 with fig. 3.2-1 yields

$$F_i = -(F_{yi1} + F_{yi2}), \quad M_i = (F_{xi1} - F_{xi2})b, \qquad (3.2\text{-}4)$$

so that we obtain from (3^{a-b}):

$$F_i = -K_{yi}(v_i' - \psi_i), \qquad (3.2\text{-}5^{a})$$

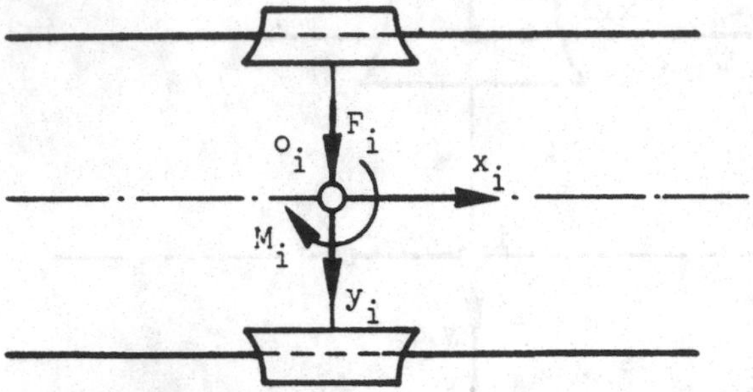

Fig. 3.2-2. The lateral force F_i and the torque M_i.

$$M_i/b = -K_{xi}(b\psi_i' + \Gamma^2 v_i). \qquad (3.2\text{-}5^b)$$

We suppose that the wheelset is connected with a purely translating main frame by means of longitudinal and lateral springs. The lateral distance of the longitudinal springs is b_i and the spring rigidities are c_{xi} and c_{yi} respectively. Then the system has the potential energy

$$U = \tfrac{1}{2}(c_{yi}\, v_i^2 + c_{xi}\, b_i^2\, \psi_i^2) \qquad (3.2\text{-}6)$$

as far as the lateral motion is concerned.

The kinetic energy for the lateral motion turns out to be equal to

$$T = \tfrac{1}{2}\, m_i\, \dot{v}_i^2 + \tfrac{1}{2}\, J_i\, \dot{\psi}_i^2. \qquad (3.2\text{-}7)$$

Now Lagrange's equations read

$$\frac{d}{dt}\frac{\partial T}{\partial \dot{q}_k} + \frac{\partial U}{\partial q_k} = Q_k \;\; (q_1 = v_i,\; q_2 = \psi_i), \qquad (3.2\text{-}8)$$

where

$$Q_1 = F_i, \quad Q_2 = M_i. \qquad (3.2\text{-}9)$$

Introducing the vector $\bar{q}_i$ in such a way that

$$^t\bar{q}_i = (v_i,\; b\psi_i), \qquad (3.2\text{-}10)$$

we find for the equations of motion

$$\bar{M}\,\ddot{\bar{q}}_i + \bar{C}\,\bar{q}_i = \bar{f}_i, \qquad (3.2\text{-}11)$$

where

$$^t\bar{f}_i = (F_i,\ M_i/b), \qquad (3.2\text{-}12)$$

$$\bar{M} = \begin{pmatrix} m_i & 0 \\ 0 & J_i/b^2 \end{pmatrix}, \quad \bar{C} = \begin{pmatrix} c_{yi} & 0 \\ 0 & c_{xi}b_i{}^2/b^2 \end{pmatrix}. \qquad (3.2\text{-}13)$$

By means of (5^{a-b}) and (12) we can write

$$\bar{f}_i = -\bar{K}\,V^{-1}\,\dot{\bar{q}}_i - \bar{B}\,\bar{q}_i, \qquad (3.2\text{-}14)$$

$\bar{K}$ and $\bar{B}$ being 2×2 matrices:

$$\bar{K} = \begin{pmatrix} K_{yi} & 0 \\ 0 & K_{xi} \end{pmatrix}, \quad \bar{B} = \begin{pmatrix} 0 & -K_{yi}/b \\ \Gamma^2 K_{xi}/b & 0 \end{pmatrix}. \qquad (3.2\text{-}15)$$

Combining (11) and (14) yields

$$\bar{M}\,\ddot{\bar{q}}_i + \bar{K}\,V^{-1}\,\dot{\bar{q}}_i + (\bar{C} + \bar{B})\bar{q}_i = \bar{0}. \qquad (3.2\text{-}16)$$

We have

$$T = \tfrac{1}{2}\,{}^t\dot{\bar{q}}_i\,\bar{M}\,\dot{\bar{q}}_i, \quad U = \tfrac{1}{2}\,{}^t\bar{q}_i\,\bar{C}\,\bar{q}_i; \qquad (3.2\text{-}17)$$

moreover, we introduce the power

$$P = {}^t\bar{f}_i\,\dot{\bar{q}}_i. \qquad (3.2\text{-}18)$$

In agreement with (11) we have the energy balance

$$\frac{d}{dt}(T + U) = P. \qquad (3.2\text{-}19)$$

For the power P of the forces $\bar{f}_i$ we find by means of (18), (14), (15) and (10):

$$P = -K_{xi}\,V^{-1}(b\dot{\psi}_i + \Gamma^2\,V\,v_i/b)b\dot{\psi}_i - K_{yi}\,V^{-1}(\dot{v}_i - V\,\psi_i)\dot{v}_i. \qquad (3.2\text{-}20)$$

The power P is equal to the power P_{in} of the contact forces and the power P_{ex} of the tractive effort L which yields a constant speed V:

$$\left. \begin{aligned} P &= P_{ex} + P_{in}, \\ P_{in} &= -V \sum_{j=1}^{2} (F_{xij}\, \upsilon_{xij} + F_{yij}\, \upsilon_{yij}), \quad P_{ex} = L\, V. \end{aligned} \right\} \qquad (3.2\text{-}21)$$

By means of (3^{a-b}) and $(3.1\text{-}5^{a-b})$ we find

$$P_{in} = -K_{xi}\, V^{-1}(b\dot{\psi}_i + \Gamma^2\, V\, v_i/b)^2 - K_{yi}\, V^{\,1}(\dot{v}_i - V\, \psi_i)^2. \qquad (3.2\text{-}22)$$

The tractive effort L is of order 2 in the displacement quantities and cannot be determined by means of the equations of motion. But from (20)-(22) we deduce

$$P_{ex} = K_{xi}(b\dot{\psi}_i + \Gamma^2\, V\, v_i/b)\Gamma^2\, v_i/b - K_{yi}(\dot{v}_i - V\, \psi_i)\psi_i. \qquad (3.2\text{-}23)$$

The power P_{in} is negative definite. The sign of the total power P and that of the power P_{ex} of the tractive effort can only be found by solving the equations of motion. So it is very well possible that the lateral motion is unstable. We shall discuss this feature in section 3.9 more in detail.

3.3. Introduction of reduced quantities and parameters

In the investigation of the behaviour of the solution of the equation of motion is is convenient to introduce reduced coordinates and parameters.

We define the reference length ℓ by

$$\ell = \Gamma\, b, \qquad (3.3\text{-}1)$$

with Γ (3.1-3). We call K_1 the creep coefficient in y direction and κ the ratio of the creep coefficients:

$$K_{xi} = \kappa K_1, \quad K_{yi} = K_1. \qquad (3.3\text{-}2)$$

Now we are able to define the reference velocity

$$W = \sqrt{\frac{K_1 \ell}{m_i}} \qquad (3.3\text{-}3)$$

and to reduce the time t and the velocity V:

$$\underline{t} = Wt/\ell, \quad \underline{V} = V/W. \tag{3.3-4}$$

For reducing the coordinates we define a second reference length, ℓ^*:

$$\ell^* = \frac{m_i\, g\, \ell}{K_1} = \frac{g\, \ell^2}{W^2} \tag{3.3-5}$$

and we put

$$\underline{\bar{q}} = \bar{q}/\ell^*, \quad \underline{v}_i = v_i/\ell^*, \quad \underline{\psi}_i = b\psi_i/\ell^*. \tag{3.3-6}$$

At last we reduce the rigidities and the moment of inertia:

$$C_x = c_{xi}\, \ell\, b_i^{\,2}/K_1 b^2, \quad C_y = c_{yi}\, \ell/K_1, \tag{3.3-7}$$

$$\Psi_1 = J_i/m_i b^2. \tag{3.3-8}$$

Then we can write

$$\underline{\bar{f}}_i = \bar{f}_i/m_i g, \tag{3.3-9}$$

$$\underline{\bar{M}} = \bar{M}/m_i = \begin{pmatrix} 1 & 0 \\ 0 & \Psi_1 \end{pmatrix}, \quad \underline{\bar{K}} = \bar{K}/K_1 = \begin{pmatrix} 1 & 0 \\ 0 & \kappa \end{pmatrix}, \tag{3.3-10}$$

$$\underline{\bar{C}} = \bar{C}\ell/K_1 = \begin{pmatrix} C_y & 0 \\ 0 & C_x \end{pmatrix}, \quad \underline{\bar{B}} = \bar{B}\ell/K_1 = \begin{pmatrix} 0 & -\Gamma^{-1} \\ \kappa\Gamma & 0 \end{pmatrix}. \tag{3.3-11}$$

Now the equations of motion (3.2-11) and (3.2-16) reduce to

$$\underline{\bar{M}}\, \ddot{\underline{\bar{q}}}_i + \underline{\bar{C}}\, \underline{\bar{q}}_i = \underline{\bar{f}}_i, \tag{3.3-12a}$$

$$\underline{\bar{M}}\, \ddot{\underline{\bar{q}}}_i + \underline{\bar{K}}\, \underline{V}^{-1}\, \dot{\underline{\bar{q}}}_i + (\underline{\bar{C}} + \underline{\bar{B}})\underline{\bar{q}}_i = \bar{0} \tag{3.3-12b}$$

and the formula (3.2-14) for the force $\bar{f}_i$ to

$$\underline{\bar{f}}_i = -\underline{\bar{K}}\, \underline{V}^{-1}\, \dot{\underline{\bar{q}}}_i - \underline{\bar{B}}\, \underline{\bar{q}}_i. \tag{3.3-12c}$$

Here the differentiation $\cdot$ now means a differentiation with respect to the reduced time $\underline{t}$. We can also differentiate with respect to the reduced distance $\underline{s}$:

$$\underline{s} = s/\ell; \tag{3.3-13}$$

(12^c) then becomes

$$\bar{\underline{f}}_i = -\bar{\underline{K}}\ \bar{\underline{q}}_i' - \bar{\underline{B}}\ \bar{\underline{q}}_i. \qquad (3.3\text{-}14)$$

Putting

$$\underline{F}_i = F_i/m_i g, \quad \underline{M}_i = M_i/m_i gb \qquad (3.3\text{-}15)$$

we can write

$$^t\bar{\underline{f}}_i = (\underline{F}_i, \underline{M}_i). \qquad (3.3\text{-}16)$$

In railway practice the parameters κ and Ψ_1 will always be in the neighbourhood of 1, whereas $\underline{V}$, C_x and C_y can have widely varing values.

3.4. The stability of the fundamental motion. The characteristic equations, the eigenvalues and the eigenvectors

The fundamental motion of the wheelset agrees with the stationary solution $\bar{\underline{q}}_i = \bar{0}$ of the equations of motion (3.2-16). We shall investigate the stability of this solution.

We do so from the equations ($3.3\text{-}12^{a-c}$). Their solution can be written as

$$\bar{\underline{f}}_i = \bar{x}\ e^{p\underline{t}}, \quad \bar{\underline{q}}_i = \bar{y}\ e^{p\underline{t}}. \qquad (3.4\text{-}1)$$

Substituting this into ($3.3\text{-}12^a$) yields the relations

$$\bar{Z}\ \bar{y} = \bar{x}, \quad \bar{Z} = \bar{\underline{C}} + \bar{\underline{M}}\ p^2, \qquad (3.4\text{-}2)$$

$\bar{Z}$ being an impedance, whereas ($3.3\text{-}12^c$) leads to

$$\bar{x} = -(\bar{\underline{K}}\ p\ \underline{V}^{-1} + \bar{\underline{B}})\bar{y}. \qquad (3.4\text{-}3)$$

The characteristic equation becomes

$$|\bar{\underline{M}}\ p^2 + \bar{\underline{K}}\ \underline{V}^{-1}\ p + \bar{\underline{C}} + \bar{\underline{B}}| = 0 \qquad (3.4\text{-}4)$$

and by means of ($3.3\text{-}10^{a-b}$) we find

$$\begin{vmatrix} p^2 + \underline{V}^{-1}\ p + C_y & -\Gamma^{-1} \\ \kappa\Gamma & \Psi_1 p^2 + \kappa \underline{V}^{-1}\ p + C_x \end{vmatrix} = 0, \tag{3.4-5}$$

viz.

$$(p^2 + \underline{V}^{-1}\ p + C_y)(\Psi_1\ p^2 + \kappa\ \underline{V}^{-1}\ p + C_x) + \kappa = 0. \tag{3.4-6}$$

The eigenvector can be determined by means of the equation

$$(p^2 + \underline{V}^{-1}\ p + C_y)y_v - \Gamma^{-1}\ y_\psi = 0. \tag{3.4-7}$$

We put

$$y_v = 1 \tag{3.4-8^a}$$

so that

$$y_\psi = \Gamma(p^2 + \underline{V}^{-1}\ p + C_y). \tag{3.4-8^b}$$

From (2) and (3.3-10^{a-b}) we find for the force amplitudes

$$x_v = p^2 + C_y, \quad x_\psi = -\frac{\kappa\Gamma(\Psi_i\ p^2 + C_x)}{\Psi_i p^2 + \kappa\underline{V}^{-1} p + C_x}. \tag{3.4-9}$$

Besides the impedance $\bar{Z}$ and the force amplitude $\bar{x}$ we introduce the quantities $\bar{Z}^*$ and $\bar{x}^*$ in such a way that

$$\bar{Z}^* = \underline{\bar{M}}\ p^2 + \underline{\bar{K}}\ p + \underline{\bar{C}}, \quad \bar{x}^* = \bar{Z}^*\ \bar{y}. \tag{3.4-10}$$

Then (3) should be replaced by

$$\bar{x}^* = -\underline{\bar{B}}\ \bar{y} \tag{3.4-11}$$

and (9) by

$$x_v{}^* = p^2 + \underline{V}^{-1}\ p + C_y, \quad x_\psi{}^* = -\Gamma. \tag{3.4-12}$$

3.5. The case of pure rolling

For $\underline{V} = 0$ we find from (3.4-5) that two roots p_k are zero, whereas

$$p_k{}^* = \underline{V}^{-1}\ p_k \tag{3.5-1}$$

is determined by

$$(p^* + C_y)(\kappa p^* + C_x) + \kappa = 0. \tag{3.5-2}$$

In agreement with (3.1-7) we can write

$$\underline{s} = \underline{V}\,\underline{t}, \tag{3.5-3}$$

so that (3.4-1) reduces to

$$\bar{\underline{f}}_i = \bar{x}\, e^{p^* \underline{s}}, \quad \bar{\underline{q}}_i = \bar{y}\, e^{p^* \underline{s}}. \tag{3.5-4}$$

The relations ($3.4\text{-}8^a$)-(3.4-9) and (3.4-12) become

$$y_v = 1,\ y_\psi = \Gamma(p^* + C_y); \quad x_v = C_y,\ x_\psi = C_x\Gamma; \quad x_v^* = p^* + C_y,\ x_\psi^* = \kappa\Gamma. \tag{3.5-5^{a-c}}$$

An important special case is that in which $C_x = C_y = 0$; then (5^b) reduces to

$$x_v = x_\psi = 0. \tag{3.5-6}$$

This is the case of *pure rolling*: rolling without sliding. For the eigenvalues, (2) yields

$$p_k^* = \pm j \quad (k=1,2). \tag{3.5-7}$$

The eigenvectors (5^a) become

$$y_v = 1, \quad y_\psi = \pm j\Gamma \tag{3.5-8}$$

so that because of (4):

$$\bar{f}_i = \bar{0}, \quad \bar{q}_i = A\begin{pmatrix}\cos \underline{s}\\ -\Gamma \sin \underline{s}\end{pmatrix} + A^*\begin{pmatrix}\sin \underline{s}\\ \Gamma \cos \underline{s}\end{pmatrix} \tag{3.5-9}$$

The motion, described by (9), is sinusoidal and the wavelength is equal to

$$\underline{\lambda} = 2\pi, \tag{3.5-10}$$

so that, according to (3.1-7) and (3.3-1):

$$\lambda = 2\pi\ell = 2\pi\sqrt{\frac{br}{\gamma_o \rho^*}}. \tag{3.5-11}$$

For $\rho^* = 1$ this result was indicated by Klingel as early as 1887.

3.6. The case in which $\kappa = 1$, $\Psi_1 = 1$. The root loci

At the end of section 3.3 we mentioned already that in railway practice usually $\kappa \approx 1$, $\Psi_1 \approx 1$. For the case $\kappa = 1$, $\Psi_1 = 1$ the calculations become much more simple and in the present and the next section we shall restrict ourselves to this case.

We have to solve the characteristic equation (3.4-6)

$$(p^2 + \underline{V}^{-1} p + C_x)(p^2 + \underline{V}^{-1} p + C_y) + \kappa = 0. \qquad (3.6\text{-}1)$$

Putting

$$q = \tfrac{1}{2} \underline{V}^{-1} \qquad (3.6\text{-}2)$$

we find that the roots of (1) are the roots of the equation

$$p^2 \mp 2qp - r_k^{\,2} e^{2j\phi_k} = 0, \qquad (3.6\text{-}3)$$

where

$$r_k^{\,2} e^{2j\phi_k} = -\tfrac{1}{2}(C_x + C_y) \pm \sqrt{\tfrac{1}{4}(C_x - C_y)^2 - 1} \quad (k=1,2). \qquad (3.6\text{-}4)$$

We now have to distinguish between two cases.

Case a:

$$|C_x - C_y| < 2. \qquad (3.6\text{-}5^a)$$

Here

$$r_k^{\,2} \cos 2\phi_k = -\tfrac{1}{2}(C_x + C_y),\; r_k^{\,2} \sin 2\phi_k = \pm\sqrt{1 - \tfrac{1}{4}(C_x - C_y)^2} \quad (k=1,2), \qquad (3.6\text{-}6)$$

so that

$$r_k^{\,2} = r_o^{\,2} = \sqrt{1 + C_x C_y}, \quad \mathrm{tg}2\phi_k = \mp \frac{\sqrt{4 - (C_x - C_y)^2}}{C_x + C_y} \quad (k=1,2) \qquad (3.6\text{-}7^a)$$

and the four roots are equal to

$$p_{1,2} = -q + \sqrt{q^2 - \tfrac{1}{2}(C_x + C_y) \pm j\sqrt{1 - \tfrac{1}{4}(C_x - C_y)^2}}, \qquad (3.6\text{-}8^a)$$

$$p_{3,4} = -q - \sqrt{q^2 - \tfrac{1}{2}(C_x + C_y) \mp j\sqrt{1 - \tfrac{1}{4}(C_x - C_y)^2}}. \qquad (3.6\text{-}8^b)$$

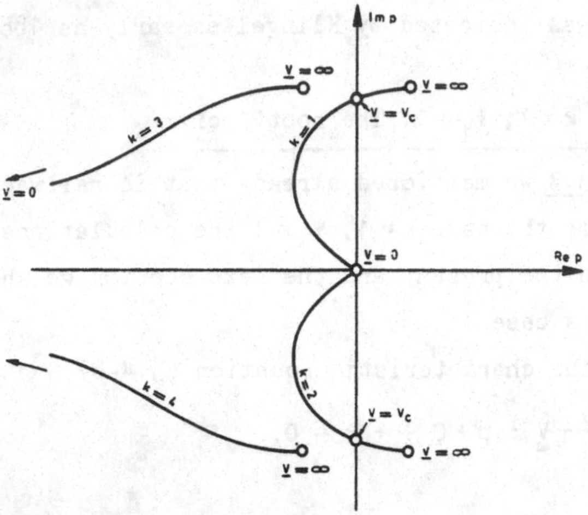

Fig. 3.6-1[a]. The root loci for $|C_x - C_y| < 2$.

This situation has been represented in fig. 3.6-1[a].

For $q = 0$ ($\underline{V} = \infty$) all four roots remain complex. For $q \to \infty$ ($\underline{V} \to 0$) we find, inspecting equation (3):

$$p_{1,2} = -\tfrac{1}{2}(C_x + C_y)\underline{V} \pm j\, \underline{V} \sqrt{1 - \tfrac{1}{4}(C_x - C_y)^2}, \tag{3.6-9^a}$$

$$p_{3,4} = -\underline{V}^{-1} + \tfrac{1}{2}(C_x + C_y)\underline{V} \pm j\, \underline{V} \sqrt{1 - \tfrac{1}{4}(C_x - C_y)^2}. \tag{3.6-9^b}$$

Case b:

$$|C_x - C_y| > 2. \tag{3.6-5^b}$$

Now we have

$$\phi_k = \phi_o = \frac{\pi}{2}, \quad r_k{}^2 = \tfrac{1}{2}(C_x + C_y) \mp \sqrt{\tfrac{1}{4}(C_x - C_y)^2 - 1} \quad (k=1,2). \tag{3.6-7^b}$$

For the four roots we can write

$$p_{1,2} = -q \pm j \sqrt{r_1{}^2 - q^2}, \quad p_{3,4} = -q \pm j \sqrt{r_2{}^2 - q^2} \text{ for } 0 \leqq q \leqq r_k, \tag{3.6-10^a}$$

$$p_{1,2} = -q \pm \sqrt{q^2 - r_1{}^2}, \quad p_{3,4} = -q \pm \sqrt{q^2 - r_2{}^2} \text{ for } r_k \leqq q \leqq \infty. \tag{3.6-10^b}$$

See fig. 3.6-1[b]: for $0 \leqq q \leqq r_k$ the roots are situated on circles with radii r_k and for $r_k \leqq q \leqq \infty$ they are real and negative.

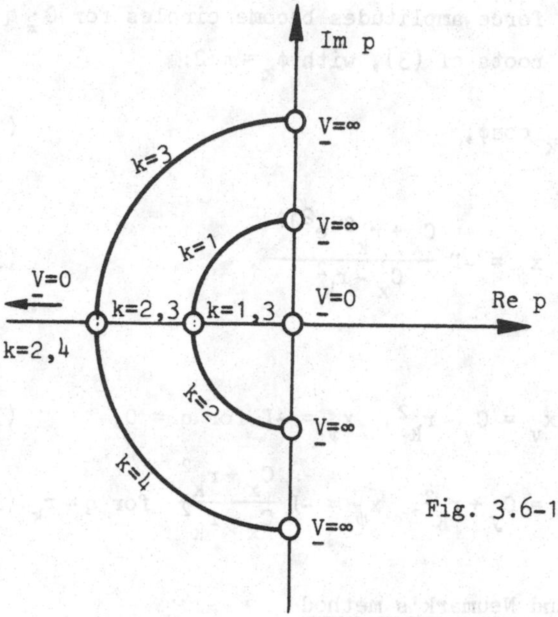

Fig. 3.6-1[b]. The root loci for $|C_x - C_y| > 2$.

Now for $q = 0$ ($\underline{V} = \infty$) all four roots become purely imaginary, whereas for $q \to \infty$ ($\underline{V} \to 0$) we obtain

$$p_{1,3} = -\tfrac{1}{2}(C_x + C_y)\underline{V} \pm \underline{V}\sqrt{\tfrac{1}{4}(C_x - C_y)^2 - 1}, \qquad (3.6\text{-}11^a)$$

$$p_{2,4} = -\underline{V}^{-1} + \tfrac{1}{2}(C_x + C_y)\underline{V} \pm \underline{V}\sqrt{\tfrac{1}{4}(C_x - C_y)^2 - 1}. \qquad (3.6\text{-}11^b)$$

We also investigate the eigenvectors. From (3.4-8[a-b]) we deduce

$$y_v = 1, \quad y_\psi = \Gamma(C_y + r_k^2\, e^{2j\phi_k}) \quad (k=1,2), \qquad (3.6\text{-}12)$$

$$y_\psi = \Gamma\{-\tfrac{1}{2}(C_x - C_y) \pm j\sqrt{1 - \tfrac{1}{4}(C_x - C_y)^2}\} \text{ for case a,} \qquad (3.6\text{-}13^a)$$

$$y_\psi = \Gamma\{-\tfrac{1}{2}(C_x - C_y) \pm \sqrt{\tfrac{1}{4}(C_x - C_y)^2 - 1}\} \text{ for case b.} \qquad (3.6\text{-}13^b)$$

Thus each of the loci of the displacement amplitudes reduces to one single point. The same holds for the amplitudes of the vector $\bar{x}^*$; (3.4-12) yields

$$x_v^* = C_y + r_k^2\, e^{2j\phi_k}, \quad x_\psi^* = -\Gamma \quad (k=1,2), \qquad (3.6\text{-}14)$$

$$x_v^* = -\tfrac{1}{2}(C_x - C_y) \pm j\sqrt{1 - \tfrac{1}{4}(C_x - C_y)^2} \text{ for case a,} \qquad (3.6\text{-}15^a)$$

$$x_v^* = -\tfrac{1}{2}(C_x - C_y) \pm \sqrt{\tfrac{1}{4}(C_x - C_y)^2 - 1} \text{ for case b.} \qquad (3.6\text{-}15^b)$$

In case b the loci of the force amplitudes become circles for $0 \leq q \leq r_k$. Here we can write for the roots of (3), with $\phi_k = \pi/2$:

$$p = r_k\, e^{j\phi}, \quad q = -r_k \cos\phi, \tag{3.6-16}$$

so that

$$x_v = C_y + r_k^2\, e^{2j\phi}, \quad x_\psi = -\Gamma\, \frac{C_x + r_k^2\, e^{2j\phi}}{C_x - r_k^2}. \tag{3.6-17}$$

Special cases are:

$$\phi = \frac{\pi}{2}, \ p = j\, r_k, \ x_v = C_y - r_k^2, \ x_\psi = -\Gamma \text{ for } q = 0, \tag{3.6-18a}$$

$$\phi = \pi, \ p = -r_k, \ x_v = C_y + r_k^2, \ x_\psi = -\Gamma\, \frac{C_x + r_k^2}{C_x - r_k^2} \text{ for } q = r_k. \tag{3.6-18b}$$

3.7. Hurwitz' criterion and Neumark's method

The results of section 3.6 indicate that for low speeds the motion is asymptotically stable, whereas it is often unstable at high speeds. The stability can be investigated by means of Hurwitz' criterion: when

$$A_0 > 0,\ A_1 > 0,\ D_2 > 0,\ D_3 > 0,\ \ldots\ D_{n-1} > 0,\ A_n > 0, \tag{3.7-1}$$

where A_k is the coefficient of p^{n-k} in the characteristic equation and the determinants $D_2, D_3, \ldots D_{n-1}$ are determined by

$$D_2 = \begin{vmatrix} A_1 & A_3 \\ A_0 & A_2 \end{vmatrix}, \quad D_3 = \begin{vmatrix} A_1 & A_3 & A_5 \\ A_0 & A_2 & A_4 \\ 0 & A_1 & A_3 \end{vmatrix}, \ \ldots \tag{3.7-2}$$

(with $A_k = 0$ for $k < 0$ and for $k > n$), the motion is asymptotically stable.

For the characteristic equation (3.4-6) we have

$$\left.\begin{aligned} &A_0 = \Psi_1,\ A_1 = (\Psi_1 + \kappa)\underline{V}^{-1},\ A_2 = C_2 + \Psi_1\, C_y + \kappa \underline{V}^{-2}, \\ &\quad A_3 = (C_x + \kappa C_y)\underline{V}^{-1},\ A_4 = C_x C_y + \kappa, \end{aligned}\right\} \tag{3.7-3}$$

$$D_2 = \kappa \underline{V}^{-1}\, \{C_x + \Psi_1\, C_y + (\Psi_1 + \kappa)\underline{V}^{-2}\}, \tag{3.7-4a}$$

$$D_3 = \{\kappa(C_x + \kappa C_y)(C_x + \Psi_1 C_y) - (C_x C_y + \kappa)(\Psi_1 + \kappa)^2\}\underline{V}^{-2} + \kappa(C_x + \kappa C_y)(\Psi_1 + \kappa)\underline{V}^{-4}. \qquad (3.7\text{-}4^b)$$

For

$$\kappa(C_x + C_y)(C_x + \Psi_1 C_y) - (C_x C_y + \kappa)(\Psi_1 + \kappa)^2 > 0 \qquad (3.7\text{-}5)$$

the motion is always asymptotically stable; when this condition is not satisfied, there is a critical speed $\underline{V}_c$ such that the motion is asymptotically stable for $\underline{V} < \underline{V}_c$ and unstable for $\underline{V} > \underline{V}_c$.

For $\kappa = \Psi_1 = 1$ the condition (5) reduces to

$$|C_x - C_y| > 2 \qquad (3.7\text{-}6)$$

and the critical speed $\underline{V}_c$ is equal to

$$\underline{V}_c = \sqrt{\frac{2(C_x + C_y)}{4 - (C_x - C_y)^2}} \qquad (3.7\text{-}7)$$

The critical speed $\underline{V}_c$ is the speed for which a root locus crosses the imaginary axis. Neumark's method boils down to drawing a diagram with one of the parameters of the system as the abscissa and another one as the ordinate and to find the locus of the points of intersection, mentioned before. Ennunciating this method we shall restrict ourselves again to the case $\kappa = \Psi_1 = 1$.

In a point on the imaginary axis we can put $p = j\omega$. Thus the characteristic equation (3.4-6) breaks up into the two real equations

$$\left.\begin{aligned} &\omega^4 - (C_x + C_y + \underline{V}^{-2})\omega^2 + C_x C_y + 1 = 0, \\ &\omega\underline{V}^{-1}(-2\omega^2 + C_x + C_y) = 0. \end{aligned}\right\} \qquad (3.7\text{-}8)$$

We eliminate ω^2: this yields the relation (7) with $\underline{V} = \underline{V}_c$, which we also can write as

$$u = (2 - v^2)\underline{V}^2 \sqrt{2} \qquad (3.7\text{-}9)$$

when we put

$$u = \tfrac{1}{2}(C_x + C_y)\sqrt{2}, \quad v = \tfrac{1}{2}(-C_x + C_y)\sqrt{2}. \qquad (3.7\text{-}10)$$

In the (u, v) plane the curves $\underline{V}$ = const. are parabolae: see fig. 3.7-1. The transition to the (C_x, C_y) plane is readily made. Now a crossing of the imaginary axis means that two asymptotically stable roots become unstable. Thus, taking in fig. 3.7-2 a certain value of $\underline{V}$, we find that for higher values of u four roots are asymptotically stable, whereas for lower values of u two roots are asymptotically stable and two are unstable. We can distinguish between these regions by indicating them by the symbols (4,0) and (2,2) respectively.

3.8. The case of double roots for infinite speed

From the figures 3.7-1 and 3.7-2 we may conclude that for infinite speed and $\kappa = \Psi_1 = 1$ the characteristic equation (3.4-6) has two double roots when

$$|C_x - C_y| = 2. \qquad (3.8\text{-}1)$$

In this case (3.4-6) reduces to

$$p^4 + (C_x + C_y)p^2 + C_x C_y + 1 = 0, \qquad (3.8\text{-}2)$$

so that

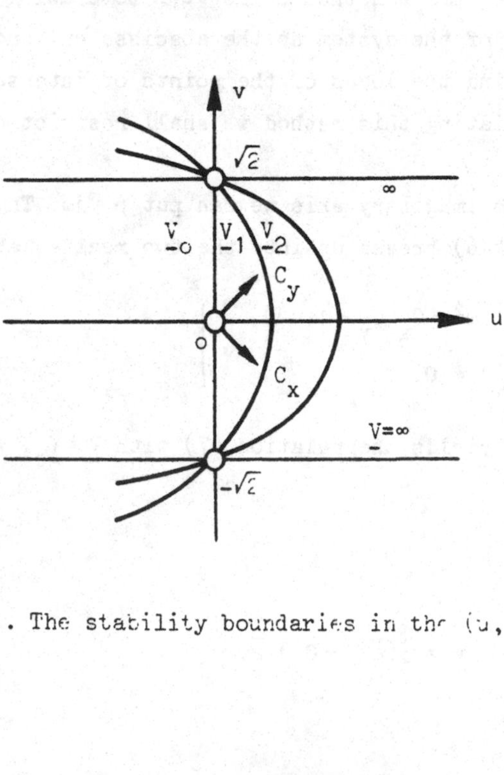

Fig. 3.7-1. The stability boundaries in the (u, v) plane.

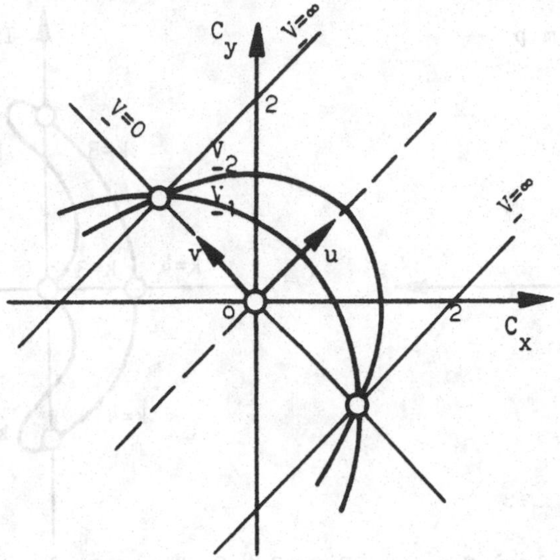

Fig. 3.7-2. The stability boundaries in the (C_x, C_y) plane.

$$p^2 = -\tfrac{1}{2}(C_x + C_y) \pm \tfrac{1}{2}\sqrt{(C_x - C_y)^2 - 4} \tag{3.8-3}$$

and we find indeed that, when (1) holds, double roots occur.

When $|C_x - C_y| > 2$, the root loci are stable for all values of $\underline{V}$. The condition that the characteristic equation has double roots for $\underline{V} = \infty$ seems to give us an easy method to find the values of the parameters which ensure such stability.

However, sometimes the situation is more complicated. In the case $\kappa = \Psi_1$ the root loci have the form of fig. 3.8-1: all root loci are asymptotically stable for $\underline{V} < \infty$. But this is not the case when $\kappa \neq \Psi_1$: then the root loci for $k = 1,2$ are unstable for high but finite speeds. The investigation of this more general case is left as an exercise to the reader. Anyhow, when κ and Ψ_1 do not differ very much from unity, the unstable parts of the root loci in fig. 3.8-2 for $k = 1,2$ are insignificant.

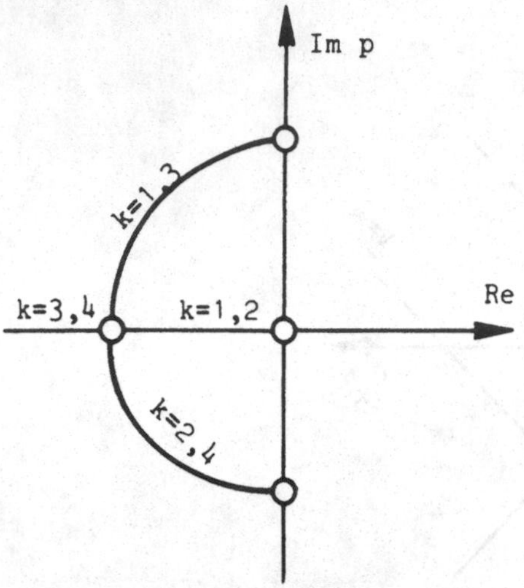

Fig. 3.8-1. The root loci for $\kappa = \Psi_1$ when there are double roots for infinite speed.

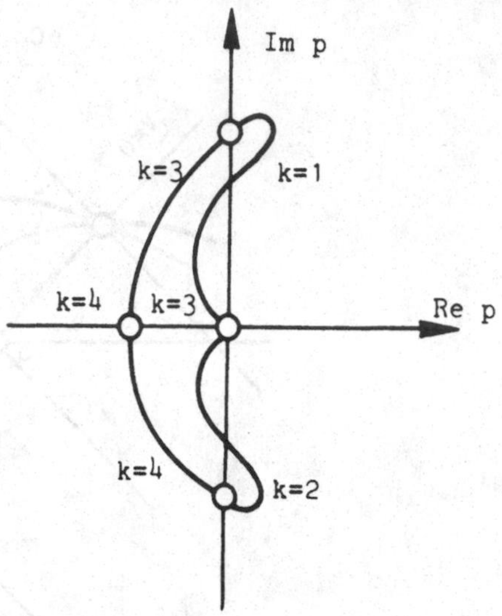

Fig. 3.8-2. The root loci for $\kappa \neq \Psi_1$ when there are double roots for infinite speed.

4. THE TWO-AXLED VEHICLE

The results, obtained in chapter 3, have only academic interest. In reality, a purely translating motion of the vehicle body cannot be prescribed and the elastic interconnection between wheelsets and body must be designed in such a way that the complete system is stable. We shall investigate this in the present chapter, together with the behaviour in curves.

4.1. Geometrical and kinematical considerations

We consider a symmetrical vehicle, consisting of a body and two wheelsets. The displacements are described by means of the three coordinate systems (o, x, y, z), (o_1, x_1, y_1, z_1) and (o_2, x_2, y_2, z_2) and the wheelbase is equal to 2a.

The system now has six degrees of freedom and the coordinates are: the lateral displacement v of the mass centre o^* of the body, its rotation ψ about a vertical axis and the coordinates v_i, ψ_i of the wheelsets,

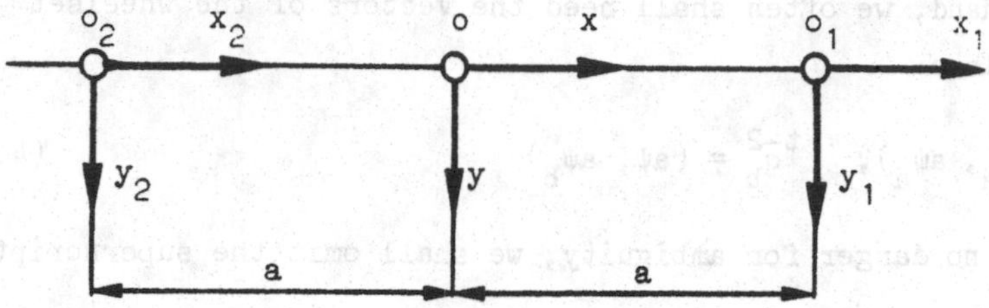

Fig. 4.1-1. The three reference coordinate systems.

mentioned in section 3.1. We also introduce the coordinate system (o^*, x^*, y^*, z^*) in such a way that in the central position of the vehicle on a tangent track it coincides with the system (o, x, y, z).

When the vehicle is in a curve with the constant radius r_o and cant ϕ_o, we assume that the point o remains in the centre of the track and that the axis oy contains the curve centre.

In the forthcoming calculations it will turn out that it is advantageous to replace v_i and ψ_i by the new coordinates

$$v_a = \tfrac{1}{2}(v_1 + v_2),\ \psi_a = \tfrac{1}{2}(\psi_1 - \psi_2);\ v_b = \tfrac{1}{2}(v_1 - v_2),\ \psi_b = \tfrac{1}{2}(\psi_1 + \psi_2), \quad (4.1\text{-}1)$$

so that

$$v_i = v_a \pm v_b, \quad \psi_i = \pm \psi_a + \psi_b \quad (i=1,2). \quad (4.1\text{-}2)$$

We shall also write

$$^t\bar{q} = (v,\ v_a,\ a\psi_a;\ a\psi,\ v_b,\ a\psi_b), \quad (4.1\text{-}3)$$

$$^t\bar{q}_a^{\,3} = (v,\ v_a,\ a\psi_a), \quad ^t\bar{q}_b^{\,3} = (a\psi,\ v_b,\ a\psi_b). \quad (4.1\text{-}4)$$

The motions which are related with $\bar{q}_a$ and $\bar{q}_b$, we call the "a motion" and "b motion" respectively. Usually, a coupling between them does exist.

When the rolling displacement ϕ of the vehicle body is considered, it can be incorporated into the a motion, so that in that case

$$^t\bar{q}_a^{\,4} = (v,\ \varphi,\ v_a,\ a\psi_a \quad (4.1\text{-}5)$$

On the other hand, we often shall need the vectors of the wheelset coordinates alone:

$$ {}^{t}\underline{q}_a^2 = (v_a,\ a\psi_a), \qquad {}^{t}\underline{q}_b^2 = (a\psi,\ a\psi_b). \tag{4.1-6}$$

When there is no danger for ambiguity, we shall omit the superscripts 2, 3 and 4.

In a curve, the axes $o_i y_i$ are not purely radial. This means that in the expression (3.1-5^b) for υ_{yij} the angle ψ_i has to be replaced by $\psi_i \mp a\, r_o^{-1}\cos\phi_o$. Moreover, because of the difference in length of the inner rail (j=1) and the outer rail (j=2) we have to replace $\psi_i{}'$ by $\psi_i{}' + r_o^{-1}\cos\phi_o$ in (3.1-5^a). Altogether, we now obtain for the creep components

$$\upsilon_{xij} = \mp\,(b\, r_o^{-1}\cos\phi_o + b\,\psi_i{}' + \Gamma^2\, b^{-1}\, v_i), \tag{4.1-7^a}$$

$$\upsilon_{yij} = a_i\, r_o^{-1}\cos\phi_o + v_i{}' - \psi_i, \tag{4.1-7^b}$$

with

$$a_i = \pm\, a. \tag{4.1-8}$$

4.2. The elastic interconnection between wheelsets and body

For the time being we assume the body coordinates v and ψ to be zero. When the forces $(F_i,\ M_i)$ are applied at the wheelsets, they undergo displacements $(v_i,\ \psi_i)$. In the most general case, when the wheelsets are also connected with each other, the relation between the forces and the displacements can be written in the form

$$\begin{pmatrix} F_1 \\ M_1/a \\ F_2 \\ M_2/a \end{pmatrix} = \begin{pmatrix} c_{11} & c_{12} & c_{13} & c_{14} \\ c_{21} & c_{22} & c_{23} & c_{24} \\ c_{31} & c_{32} & c_{33} & c_{34} \\ c_{41} & c_{42} & c_{43} & c_{44} \end{pmatrix} \begin{pmatrix} v_1 \\ a\psi_1 \\ v_2 \\ a\psi_2 \end{pmatrix}, \tag{4.2-1}$$

the matrix being symmetrical because of Maxwell's theorem.

As the vehicle is assumed to be symmetrical with respect to a lateral vertical plane, the relation (1) must also hold when we interchange F_1 for F_2, M_1 for $-M_2$, v_1 for v_2 and ψ_1 for $-\psi_2$:

$$\begin{pmatrix} F_2 \\ -M_2/a \\ F_1 \\ -M_1/a \end{pmatrix} = \begin{pmatrix} c_{11} & c_{12} & c_{13} & c_{14} \\ c_{21} & c_{22} & c_{23} & c_{24} \\ c_{31} & c_{32} & c_{33} & c_{34} \\ c_{41} & c_{42} & c_{43} & c_{44} \end{pmatrix} \begin{pmatrix} v_2 \\ -a\psi_2 \\ v_1 \\ -a\psi_1 \end{pmatrix} ; \qquad (4.2\text{-}2)$$

this relation can be rewritten as

$$\begin{pmatrix} F_1 \\ M_1/a \\ F_2 \\ M_2/a \end{pmatrix} = \begin{pmatrix} c_{33} & -c_{34} & c_{32} & -c_{32} \\ -c_{43} & c_{44} & -c_{41} & c_{42} \\ c_{13} & -c_{14} & c_{11} & -c_{12} \\ -c_{23} & c_{24} & -c_{21} & c_{22} \end{pmatrix} \begin{pmatrix} v_1 \\ a\psi_1 \\ v_2 \\ a\psi_2 \end{pmatrix} . \qquad (4.2\text{-}3)$$

Comparing (1) and (3) we find the relations

$$c_{11} = c_{33},\ c_{22} = c_{44},\ c_{12} + c_{34} = 0,\ c_{14} + c_{32} = 0, \qquad (4.2\text{-}4)$$

which hold in addition to the ordinary symmetry relations $c_{12} = c_{21}$, etc.

The relation (1) can be considerably simplified by introducing the quantities $v_{a,b}$ and $\psi_{a,b}$ (4.1-1) and the corresponding force quantities

$$F_a = F_1 + F_2,\ M_a = M_1 - M_2,\ F_b = F_1 - F_2,\ M_b = M_1 + M_2. \qquad (4.2\text{-}5)$$

Thus we also have

$$F_i = \tfrac{1}{2}(F_a \pm F_b), \quad M_i = \tfrac{1}{2}(\pm M_a + M_b). \qquad (4.2\text{-}6)$$

Then we can write

$$\begin{pmatrix} F_a \\ M_a/a \\ F_b \\ M_b/a \end{pmatrix} = \begin{pmatrix} c_1 & c_2 & 0 & 0 \\ c_2 & c_3 & 0 & 0 \\ 0 & 0 & c_4 & c_5 \\ 0 & 0 & c_5 & c_6 \end{pmatrix} \begin{pmatrix} v_a \\ a\psi_a \\ v_b \\ a\psi_b \end{pmatrix}, \qquad (4.2\text{-}7)$$

with

$$\left. \begin{aligned} c_1 &= 2(c_{11} + c_{13}),\ c_2 = 2(c_{12} - c_{14}),\ c_3 = 2(c_{22} - c_{24}), \\ c_4 &= 2(c_{11} - c_{13}),\ c_5 = 2(c_{12} + c_{14}),\ c_6 = 2(c_{22} + c_{24}). \end{aligned} \right\} \qquad (4.2\text{-}8)$$

In the general case the body coordinates v and ψ differ from zero, whereas on the other side a lateral force F is applied at the body mass centre and, moreover, a torque with the movement M is applied at the body in the (x, o, y) plane, F and M being in equilibrium with F_i, M_i (i=1,2), so that

$$F + F_1 + F_2 = 0,\ (F_1 - F_2)a + M + M_1 + M_2 = 0, \qquad (4.2\text{-}9)$$

by means of (5) we can reduce the equilibrium conditions to

$$F + F_a = 0,\ M/a + F_b + M_b/a = 0. \qquad (4.2\text{-}10)$$

Now we write

$$\begin{pmatrix} F \\ M/a \\ F_a \\ M_a/a \\ F_b \\ M_b/a \end{pmatrix} = \bar{C} \begin{pmatrix} v \\ a\psi \\ v_b \\ a\psi_a \\ v_b \\ a\psi_b \end{pmatrix},\quad \bar{C} = \begin{pmatrix} \bar{C}_{11} & \bar{C}_{12} \\ \bar{C}_{21} & \bar{C}_{22} \end{pmatrix}, \qquad (4.2\text{-}11)$$

$\bar{C}_{22}$ being the matrix in (7). With the first equation (10) corresponds the displacement vector

$$ {}^t\bar{q} = (1,\ 0,\ 1,\ 0,\ 0,\ 0), \qquad (4.2\text{-}12^a)$$

and with the second one the vector

$$ {}^t\bar{q} = (0,\ 1,\ 0,\ 0,\ 1,\ 1); \qquad (4.2\text{-}12^b)$$

both are eigenvectors of the matrix $\bar{C}$. So we can write

$$\begin{pmatrix} \bar{C}_{11} & \bar{C}_{12} \\ \bar{C}_{21} & \bar{C}_{22} \end{pmatrix} \begin{pmatrix} 1 & 0 \\ 0 & 1 \\ 1 & 0 \\ 0 & 0 \\ 0 & 1 \\ 0 & 1 \end{pmatrix} = \bar{0} \qquad (4.2\text{-}13)$$

and

$$\bar{C}_{21} = -\bar{C}_{22} \begin{pmatrix} 1 & 0 \\ 0 & 0 \\ 0 & 1 \\ 0 & 1 \end{pmatrix}, \quad \bar{C}_{12} = {}^t\bar{C}_{21}, \quad \bar{C}_{11} = -\bar{C}_{12} \begin{pmatrix} 1 & 0 \\ 0 & 0 \\ 0 & 1 \\ 0 & 1 \end{pmatrix}. \qquad (4.2\text{-}14)$$

Together with (7) we obtain

$$\bar{C}_{22} = \begin{pmatrix} c_1 & c_2 & 0 & 0 \\ c_2 & c_3 & 0 & 0 \\ 0 & 0 & c_4 & c_5 \\ 0 & 0 & c_5 & c_6 \end{pmatrix}, \quad \bar{C}_{21} = \begin{pmatrix} -c_1 & 0 \\ -c_2 & 0 \\ 0 & -c_{45} \\ 0 & -c_{56} \end{pmatrix}, \quad \bar{C}_{11} = \begin{pmatrix} c_1 & 0 \\ 0 & c_{456} \end{pmatrix} \qquad (4.2\text{-}15)$$

and

$$\bar{C} = \left[\begin{array}{cc|cccc} c_1 & 0 & -c_1 & -c_2 & 0 & 0 \\ 0 & c_{456} & 0 & 0 & -c_{45} & -c_{56} \\ \hline -c_1 & 0 & c_1 & c_2 & 0 & 0 \\ -c_2 & 0 & c_2 & c_3 & 0 & 0 \\ 0 & -c_{45} & 0 & 0 & c_4 & c_5 \\ 0 & -c_{56} & 0 & 0 & c_5 & c_6 \end{array}\right], \qquad (4.2\text{-}16)$$

when we put

$$c_{45} = c_4 + c_5,\ c_{56} = c_5 + c_6,\ c_{456} = c_{45} + c_{56} = c_4 + 2c_5 + c_6. \qquad (4.2\text{-}17)$$

Thus we have found that the behaviour of the elastic interconnection between body and wheelsets is determined by the six elasticity coefficients c_1, ... c_6. This result was obtained by Kar, Wormley and Hedrick [4].

The potential energy of the elastic interconnection can be given in terms of the matrix $\bar{C}$:

$$U = \tfrac{1}{2}\, {}^t\bar{q}\, \bar{C}\, \bar{q}. \qquad (4.2\text{-}18)$$

We now shall calculate the values of the coefficients for various interconnecting constructions. First of all, we investigate the classical design with longitudinal and lateral springs with rigidities $c_x/2$ and $c_y/2$ per wheelset, so that

$$F_1 = \tfrac{1}{2}\, c_y\, v_i, \quad M_i = \tfrac{1}{2}\, c_x\, b_1^2\, \psi_i, \qquad (4.2\text{-}19)$$

b_1 being the half-distance between the axle-boxes. By means of (5) and (6) we obtain

$$F_{a,b} = c_y\, v_{a,b}, \quad M_{a,b}/a = c_x\, b_1^2 (a\psi_{a,b})/a^2; \qquad (4.2\text{-}20)$$

comparing this with (6) yields

$$c_1 = c_y,\ c_2 = 0,\ c_3 = c_x\, b_1^2/a^2,\ c_4 = c_y,\ c_5 = 0,\ c_6 = c_x\, b_1^2/a^2. \qquad (4.2\text{-}21)$$

When the wheelsets are connected by means of two longitudinal springs with the total rigidity c_b (bending spring device), we can write

$$F_i = 0, \quad M_1 = -M_2 = c_b a^2(\psi_1 - \psi_2), \tag{4.2-22}$$

viz. by means of (4.1-1) and (5):

$$F_a = F_b = 0, \; M_a/a = 4\, c_b\, a\psi_a, \; M_b = 0 \tag{4.2-23}$$

and comparing this with (7) yields

$$c_1 = c_2 = 0, \; c_3 = 4\, c_b, \; c_4 = c_5 = c_6 = 0. \tag{4.2-24}$$

In the case of a shear spring device (fig. 4.2-1) with the spring rigidity c_s, we find for the spring elongation

$$-(v_1 - v_2) + a(\psi_1 + \psi_2) = -2v_b + 2a\psi_b,$$

so that

$$\left.\begin{aligned} F_1 &= -F_2 = 2c_s(v_b - a\psi_b) \\ M_1 &= M_2 = -2c_s\, a(v_b - a\psi_b), \end{aligned}\right\} \tag{4.2-25}$$

and, according to (5.1-1) and (5),

$$F_a = 0, \; -F_b = M_b/a = 4c_s(a\psi_b - v_b), \; M_a = 0. \tag{4.2-26}$$

Now (6) yields

$$c_1 = c_2 = c_3 = 0, \; c_4 = -c_5 = c_6 = 4c_s. \tag{4.2-27}$$

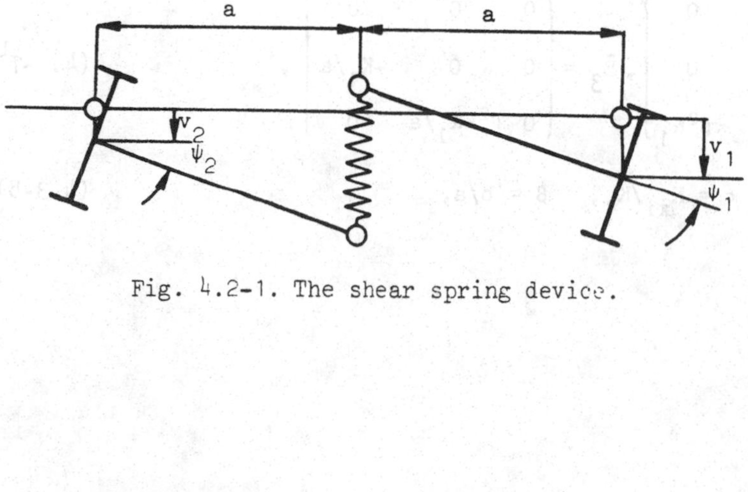

Fig. 4.2-1. The shear spring device.

4.3. Dynamical considerations

For the contact forces we can write down expressions, analogous to (3.2-5^{a-b}). By means of (4.1-7^{a-b}) we find

$$\begin{pmatrix} F_i \\ M_i/a \end{pmatrix} = -\begin{pmatrix} K_{yi} & 0 \\ 0 & K_{xi}b^2/a^2 \end{pmatrix} V^{-1} \begin{pmatrix} \dot{v}_i \\ a\dot{\psi}_i \end{pmatrix}$$

$$- \begin{pmatrix} 0 & -K_{yi}/a \\ \Gamma^2 K_{xi}/a & 0 \end{pmatrix} \begin{pmatrix} v_i \\ a\psi_i \end{pmatrix} + \begin{pmatrix} \mp K_{yi}a \\ -K_{xi}b^2/a \end{pmatrix} r_o^{-1}\cos\phi_o. \quad (4.3\text{-}1)$$

We now define the fectors $\bar{f}$, $\bar{f}_a$ and $\bar{f}_b$ by formulae, analogous to (4.1-3)-(4.1-6):

$$^t\bar{f} = (0,\ F_a,\ M_a/a;\ 0,\ F_b,\ M_b/a), \quad (4.3\text{-}2)$$

$$^t\bar{f}^3_a = (0,\ F_a,\ M_a/a),\quad {}^t\bar{f}^3_b = (0,\ F_b,\ M_b/a), \quad (4.3\text{-}3)$$

$$^t\bar{f}^2_a = (F_a,\ M_a/a),\quad {}^t\bar{f}^2_b = (F_b,\ M_b/a). \quad (4.3\text{-}4)$$

Thus by means of (4.1-1) and (4.2-5) we can write:

$$^t\bar{f} = ({}^t\bar{f}^3_a,\ {}^t\bar{f}^3_b), \quad (4.3\text{-}5)$$

$$\bar{f}_a = -2\bar{K}\, V^{-1}\, \dot{\bar{q}}_a - 2\bar{B}\, \bar{q}_b + 2\bar{g}_a,\quad \bar{f}_b = -2\bar{K}\, V^{-1}\, \dot{\bar{q}}_b - 2\bar{B}\, \bar{q}_a + 2\bar{g}_b, \quad (4.3\text{-}6^{a-b})$$

$$\bar{K}_2 = \begin{pmatrix} K_1 & 0 \\ 0 & \kappa\beta^2 K_1 \end{pmatrix},\quad \bar{B}_2 = \begin{pmatrix} 0 & -K_1/a \\ \Gamma^2\kappa K_1/a & 0 \end{pmatrix}, \quad (4.3\text{-}7^a)$$

$$\bar{K}_3 = \begin{pmatrix} 0 & 0 & 0 \\ 0 & K_1 & 0 \\ 0 & 0 & \kappa\beta^2 K_1 \end{pmatrix},\ \bar{B}_3 = \begin{pmatrix} 0 & 0 & 0 \\ 0 & 0 & -K_1/a \\ 0 & \Gamma^2\kappa K_1/a & 0 \end{pmatrix}, \quad (4.3\text{-}7^b)$$

$$K_1 = K_{yi},\quad \kappa = K_{xi}/K_1,\quad \beta = b/a, \quad (4.3\text{-}8)$$

$$\bar{g}_a = \bar{0}, \quad \bar{g}_b^2 = -\begin{pmatrix} 1 \\ \kappa\beta^2 \end{pmatrix} K_1 a r_o^{-1} \cos\phi_o . \tag{4.3-9}$$

The kinetic energy of the system turns out to be equal to

$$T = \tfrac{1}{2} m \dot{v}^2 + \tfrac{1}{2} J \dot{\psi}^2 + \tfrac{1}{2} m_1(\dot{v}_1{}^2 + \dot{v}_2{}^2) + \tfrac{1}{2} J_1(\dot{\psi}_1{}^2 + \dot{\psi}_2{}^2), \tag{4.3-10}$$

m being the body mass and J its moment of inertia about the axis o^*z^*. By means of (4.1-1) and (4.1-3) we can reduce (10) to

$$T = \tfrac{1}{2}\, {}^t\dot{\bar{q}}\, \bar{M}\, \dot{\bar{q}} \tag{4.3-11}$$

with

$$\bar{M} = \begin{pmatrix} \bar{M}_a & \bar{0} \\ \bar{0} & \bar{M}_b \end{pmatrix}, \quad \bar{M}_a = \begin{bmatrix} m & 0 & 0 \\ 0 & 2m_1 & 0 \\ 0 & 0 & 2J_1/a^2 \end{bmatrix}, \quad \bar{M}_b = \begin{bmatrix} J/a & 0 & 0 \\ 0 & 2m_1 & 0 \\ 0 & 0 & 2J_1/a^2 \end{bmatrix}. \tag{4.3-12}$$

The potential energy was already given by (4.2-18). For the energies we can also write by means of (4.1-4):

$$T = \tfrac{1}{2}\, {}^t\dot{\bar{q}}_a\, \bar{M}_a\, \dot{\bar{q}}_a + \tfrac{1}{2}\, {}^t\dot{\bar{q}}_b\, \bar{M}_b\, \dot{\bar{q}}_b, \quad U = \tfrac{1}{2}\, {}^t\bar{q}_a\, \bar{C}_a\, \bar{q}_a + \tfrac{1}{2}\, {}^t\bar{q}_b\, \bar{C}_b\, \bar{q}_b, \tag{4.3-13}$$

where, according to (4.2-16):

$$\bar{C}_a = \begin{bmatrix} c_1 & -c_1 & -c_2 \\ -c_1 & c_1 & c_2 \\ -c_2 & c_2 & c_3 \end{bmatrix}, \quad \bar{C}_b = \begin{bmatrix} c_{456} & -c_{45} & -c_{56} \\ -c_{45} & c_4 & c_5 \\ -c_{56} & c_5 & c_6 \end{bmatrix}. \tag{4.3-14}$$

Now the equations of motion read

$$\bar{M}_a \ddot{\bar{q}}_a + \bar{C}_a \bar{q}_a = \bar{f}_a, \quad \bar{M}_b \ddot{\bar{q}}_b + \bar{C}_b \bar{q}_b = \bar{f}_b, \tag{4.3-15}$$

the right-hand terms being given by (6^{a-b}), so that also

$$\begin{pmatrix} \bar{M}_a/2 & 0 \\ 0 & \bar{M}_b/2 \end{pmatrix} \begin{pmatrix} \ddot{\bar{q}}_a \\ \ddot{\bar{q}}_b \end{pmatrix} + \begin{pmatrix} \bar{K} & \bar{O} \\ \bar{O} & \bar{K} \end{pmatrix} V^{-1} \begin{pmatrix} \dot{\bar{q}}_a \\ \dot{\bar{q}}_b \end{pmatrix} + \begin{pmatrix} \bar{C}_a/2 & \bar{B} \\ \bar{B} & \bar{C}_b/2 \end{pmatrix} \begin{pmatrix} \bar{q}_a \\ \bar{q}_b \end{pmatrix} = \begin{pmatrix} \bar{g}_a \\ \bar{g}_b \end{pmatrix} . \tag{4.3-16}$$

For the energy balance the relation (3.2-19) holds again, P now being equal to the sum of the powers (3.2-20) for i=1 and i=2.

4.4. Introduction of reduced quantities

The equations of motion can be simplified again by introducing reduced coordinates and parameters.

The reference length ℓ is now defined by

$$\ell = \Gamma^{-1} a, \tag{4.4-1}$$

with Γ (3.1-3), and the reference speed W by

$$W = \sqrt{K_1 \ell/m_1} \tag{4.4-2}$$

with K_1 (4.3-8). The time t and the speed V are again reduced by means of the formulae

$$\underline{t} = Wt/\ell, \quad \underline{V} = V/W. \tag{4.4-3}$$

The second reference length, ℓ^*, is now equal to

$$\ell^* = G\ell/2K_1, \tag{4.4-1}$$

G being the total weight of the vehicle, whereas

$$\bar{\underline{q}} = \bar{q}/\ell^*, \ \bar{\underline{q}}_{a,b} = \bar{q}_{a,b}/\ell^*, \ q_v = v/\ell^*, \ q_\psi = a\psi/\ell^*, \tag{4.4-5}$$

$$q_{vi} = v_i/\ell^*, \ q_{\psi i} = a\psi_i/\ell^*, \ q_v{}^{a,b} = v_{a,b}/\ell^*, \ q_\psi{}^{a,b} = a\psi_{a,b}/\ell^*. \tag{4.4-6}$$

The rigidities and the mass parameters are reduced according to

$$C_i = c_i\ell/2K_1 \quad (i=1, \dots 6), \tag{4.4-7}$$

$$\Psi = J/ma^2, \quad \mu = 2m_1/m, \quad \Psi_1 = J_1/m_1b^2. \tag{4.4-8}$$

Now we can write

$$\underline{f} = 2\ \bar{f}/G,\ \underline{f}_{a,b} = 2\ \bar{f}_{a,b}/G,\ \underline{g}_{a,b} = 2\ \bar{g}_{a,b}/G, \tag{4.4-9}$$

$$\underline{M}_a = \bar{M}_a/2m_1 = \begin{pmatrix} \mu^{-1} & 0 & 0 \\ 0 & 1 & 0 \\ 0 & 0 & \Psi_1\beta^2 \end{pmatrix},\ \underline{M}_b = \bar{M}_b/2m_1 = \begin{pmatrix} \mu^{-1}\Psi & 0 & 0 \\ 0 & 1 & 0 \\ 0 & 0 & \Psi_1\beta^2 \end{pmatrix}, \tag{4.4-10}$$

$$\underline{C}_a = \bar{C}_a\ell/2K_1 = \begin{pmatrix} C_1 & -C_1 & -C_2 \\ -C_1 & C_1 & C_2 \\ -C_2 & C_2 & C_3 \end{pmatrix},\ \underline{C}_b = \bar{C}_b\ell/2K_1 = \begin{pmatrix} C_{456} & -C_{45} & -C_{56} \\ -C_{45} & C_4 & C_5 \\ -C56 & C_5 & C_6 \end{pmatrix}, \tag{4.4-11}$$

$$\underline{K} = \bar{K}/2K_1 = \begin{pmatrix} 1 & 0 \\ 0 & \kappa\beta^2 \end{pmatrix},\ \underline{B} = \bar{B}/2K_1 = \begin{pmatrix} 0 & -\Gamma^{-1} \\ \kappa\Gamma & 0 \end{pmatrix}. \tag{4.4-12}$$

Then the equations (4.3-15), (4.3-6^{a-b}) and (4.3-16) reduce to

$$2\ \underline{M}_a\ \ddot{\underline{q}}_a + 2\ \underline{C}_a\ \underline{q}_a = \underline{f}_a,\ 2\ \underline{M}_b\ \ddot{\underline{q}}_b + 2\ \underline{C}_b\ \underline{q}_b = \underline{f}_b, \tag{4.4-13^{a-b}}$$

$$\underline{f}_a = 2\ \underline{K}\ \underline{V}^{-1}\ \dot{\underline{q}}_a - 2\ \underline{B}\ \underline{q}_b + 2\ \underline{g}_a,\ \underline{f}_b = -2\ \underline{K}\ \underline{V}^{-1}\ \dot{\underline{q}}_b - 2\ \underline{B}\ \underline{q}_a + 2\ \underline{g}_b, \tag{4.4-14^{a-b}}$$

$$\begin{pmatrix} \underline{M}_a & \bar{0} \\ \bar{0} & \underline{M}_b \end{pmatrix}\begin{pmatrix} \ddot{\underline{q}}_a \\ \ddot{\underline{q}}_b \end{pmatrix} + \begin{pmatrix} \underline{K} & \bar{0} \\ \bar{0} & \underline{K} \end{pmatrix}\underline{V}^{-1}\begin{pmatrix} \dot{\underline{q}}_a \\ \dot{\underline{q}}_b \end{pmatrix} + \begin{pmatrix} \underline{C}_a & \underline{B} \\ \underline{B} & \underline{C}_b \end{pmatrix}\begin{pmatrix} \underline{q}_a \\ \underline{q}_b \end{pmatrix} = \begin{pmatrix} \underline{g}_a \\ \underline{g}_b \end{pmatrix}. \tag{4.4-15}$$

From (9) and (4.3-9) we find for $\underline{g}$:

$$\underline{g}_a = \bar{0},\quad \underline{g}_b = -\begin{pmatrix} 1 \\ \kappa\beta^2 \end{pmatrix}\Theta_o \tag{4.4-16}$$

with

$$\Theta_o = \frac{a\,\ell\cos\phi_o}{r_o\ell^*} = \frac{2\,K_1\,a\cos\phi_o}{G\,r_o}\ . \tag{4.4-17}$$

We also put

$$f_v^{\ a,b} = 2\ F_{a,b}/G,\ f_\psi^{\ a,b} = 2\ M_{a,b}/Ga,\ f_{vi} = 2\ F_i/G,\ f_{\psi i} = 2\ M_i/Ga, \tag{4.4-18}$$

so that

$$f_v^{\,a} = f_{v1} + f_{v2},\ f_\psi^{\,a} = f_{\psi 1} - f_{\psi 2},\ f_v^{\,b} = f_{v1} - f_{v2},\ f_\psi^{\,b} = f_{\psi 1} + f_{\psi 2} \qquad (4.4\text{-}19)$$

and

$$f_{vi} = \tfrac{1}{2}(f_v^{\,a} \pm f_v^{\,b}),\ f_{\psi i} = \tfrac{1}{2}(\pm\, f_\psi^{\,a} + f_\psi^{\,b}). \qquad (4.4\text{-}20)$$

4.5. The behaviour in curves

We now shall determine the particular solution of the equations of motion which agrees with the stationary motion of the vehicle through curves. Before doing so, we investigate how the relations reduce when the speed vanishes. In that case it is advantageous to replace the differentiation with respect to time by the differentiation with respect to distance. Then (4.4-15) becomes

$$\begin{pmatrix} \bar{\underline{M}}_a & \bar{0} \\ \bar{0} & \bar{\underline{M}}_b \end{pmatrix} \underline{V}^2 \begin{pmatrix} \bar{\underline{q}}''_a \\ \bar{\underline{q}}''_b \end{pmatrix} + \begin{pmatrix} \bar{\underline{K}} & \bar{0} \\ \bar{0} & \bar{\underline{K}} \end{pmatrix} \begin{pmatrix} \bar{\underline{q}}'_a \\ \bar{\underline{q}}'_b \end{pmatrix} + \begin{pmatrix} \bar{\underline{C}}_a & \bar{\underline{B}} \\ \bar{\underline{B}} & \bar{\underline{C}}_b \end{pmatrix} \begin{pmatrix} \bar{\underline{q}}_a \\ \bar{\underline{q}}_b \end{pmatrix} = \begin{pmatrix} \bar{\underline{g}}_a \\ \bar{\underline{g}}_b \end{pmatrix} \qquad (4.5\text{-}1)$$

and for vanishing speed $\underline{V}$ we may omit the mass term, so that we obtain the equations

$$\bar{\underline{K}}\,\bar{\underline{q}}'_a + \bar{\underline{C}}_a\,\bar{\underline{q}}_a + \bar{\underline{B}}\,\bar{\underline{q}}_b = \bar{\underline{g}}_a,\ \bar{\underline{K}}\,\bar{\underline{q}}'_b + \bar{\underline{B}}\,\bar{\underline{q}}_a + \bar{\underline{C}}_b\,\bar{\underline{q}}_b = \bar{\underline{g}}_b. \qquad (4.5\text{-}2)$$

Here $\bar{\underline{q}}_a$ and $\bar{\underline{q}}_b$ are 3×1 vectors. The first scalar equation, related with the first vector equation, reads

$$C_1\, q_v - C_1\, q_v^{\,a} - C_2\, q_\psi^{\,a} = 0 \qquad (4.5\text{-}3^a)$$

and the first scalar equation, belonging to the second vector equation, is

$$C_{456}\, q_\psi - C_{45}\, q_v^{\,b} - C_{56}\, q_\psi^{\,b} = 0. \qquad (4.5\text{-}3^b)$$

Thus we can express q_v and q_ψ in terms of the other coordinates:

$$q_v = q_v^{\,a} + C_2\, C_1^{-1}\, q_\psi^{\,a},\quad q_\psi = C_{45}\, C_{456}^{-1}\, q_v^{\,b} + C_{56}\, C_{456}^{-1}\, q_\psi^{\,b}. \qquad (4.5\text{-}4)$$

Writing out (2), we obtain

$$\begin{pmatrix} & q_v^{a'} \\ \kappa\beta^2 & q_\psi^{a'} \\ & q_v^{b'} \\ \kappa\beta^2 & q_\psi^{b'} \end{pmatrix} + \begin{pmatrix} 0 & 0 & 0 & -\Gamma^{-1} \\ 0 & C_o & \kappa\Gamma & 0 \\ 0 & -\Gamma^{-1} & C_o^* & -C_o^* \\ \kappa\Gamma & 0 & -C_o^* & C_o^* \end{pmatrix} \begin{pmatrix} q_v^a \\ q_\psi^a \\ q_v^b \\ q_\psi^b \end{pmatrix} = \begin{pmatrix} 0 \\ 0 \\ -1 \\ -\kappa\beta^2 \end{pmatrix} \Theta_o \qquad (4.5\text{-}5)$$

with

$$C_o = C_3 - C_2^2/C_1,\quad C_o^* = C_4 - C_{45}^2/C_{456} = -C_5 + C_{45}\, C_{56}/C_{456} = C_6 - C_{56}^2/C_{456}. \qquad (4.5\text{-}6)$$

We find that the behaviour at vanishing speed is only determined by two combinations of the six rigidity coefficients: the rigidities of the resultant bending and shear springs, connecting the two wheelsets.

The contact force vectors $\bar{f}_a$ and $\bar{f}_b$ become, according to $(4.4\text{-}13^{a-b})$:

$$\bar{\underline{f}}_a = 2\,\bar{\underline{C}}_a\,\bar{\underline{q}}_a,\quad \bar{\underline{f}}_b = 2\,\bar{\underline{C}}_b\,\bar{\underline{q}}_b \qquad (4.5\text{-}7)$$

and because of the transformation (4) and by means of (4.3-4) and (4.4-18) we obtain

$$\begin{pmatrix} f_v^a \\ f_\psi^a \end{pmatrix} = 2\,C_o \begin{pmatrix} 0 & 0 \\ 0 & 1 \end{pmatrix} \begin{pmatrix} q_v^a \\ q_\psi^a \end{pmatrix},\quad \begin{pmatrix} f_v^b \\ f_\psi^b \end{pmatrix} = 2\,C_o^* \begin{pmatrix} 1 & -1 \\ -1 & 1 \end{pmatrix} \begin{pmatrix} q_v^b \\ q_\psi^b \end{pmatrix}. \qquad (4.5\text{-}8)$$

The stationary motion through a curve agrees with the particular solution of (4.4-15) in which the derivatives with respect to time disappear. This also happens with the derivatives $q_v^{a'}$ etc. in the equation (5), so that

$$(C_o C_o^* + \kappa) \begin{pmatrix} q_v^a \\ q_\psi^a \\ q_v^b \\ q_\psi^b \end{pmatrix} = \begin{pmatrix} -C_o C_o^*(1+\kappa\beta^2)/\kappa\Gamma \quad -\kappa\beta^2\Gamma^{-1} \\ \kappa\Gamma \\ -C_o \\ 0 \end{pmatrix} \Theta_o. \qquad (4.5\text{-}9)$$

This gives rise to the forces (8), viz.

$$\begin{pmatrix} f_v^{\,a} \\ f_\psi^{\,a} \end{pmatrix} = \frac{2\,C_o\Theta_o}{C_oC_o^{*}+\kappa}\begin{pmatrix} 0 \\ \kappa\Gamma \end{pmatrix},\quad \begin{pmatrix} f_v^{\,b} \\ f_\psi^{\,b} \end{pmatrix} = \frac{2\,C_oC_o^{*}\Theta_o}{C_oC_o^{*}+\kappa}\begin{pmatrix} -1 \\ 1 \end{pmatrix}, \qquad (4.5\text{-}10)$$

which satisfy the equilibrium conditions (4.2-10). Thus we see that perfect curve running (with vanishing contact forces) is impossible, unless $C_o=0$. When $C_o=0$, a further condition is that the displacement $q_v^{\,a}=q_{v1}=q_{v2}$ is smaller than the lateral play of the wheelset with respect to the track and this yields a maximum value of Θ_o and a minimum value of r_o.

The wheelset forces are found from (4.2-6), (4.4-9) and (4.4-17):

$$F_i = \mp\tfrac{1}{2}K\,\frac{C_oC_o^{*}}{C_oC_o^{*}+\kappa}\,a r_o^{-1}\cos\phi_o,\quad M_i=\tfrac{1}{2}K\,\frac{C_o^{*}\pm\kappa\Gamma}{C_oC_o^{*}+\kappa}\,C_o a^2 r_o^{-1}\cos\phi_o \quad (i=1,2). \qquad (4.5\text{-}11)$$

4.6. The stability of the stationary motion in the case of vanishing speed

Before investigating the stability of the stationary motion for any arbitrary value of the speed V we first consider the case of vanishing speed, for which the homogenised differential equations (4.5-5) hold and in which the distance s is the independent variable.

For investigating the stability at vanishing speed, we substitute in (4.5-5)

$$\begin{pmatrix} q_v^{\,a} \\ q_\psi^{\,a} \end{pmatrix} = \begin{pmatrix} y_v^{\,a} \\ y_\psi^{\,a} \end{pmatrix} e^{p^{*}\underline{s}},\quad \begin{pmatrix} q_v^{\,b} \\ q_\psi^{\,b} \end{pmatrix} = \begin{pmatrix} y_v^{\,b} \\ y_\psi^{\,b} \end{pmatrix} e^{p^{*}\underline{s}}, \qquad (4.6\text{-}1)$$

Θ_o now being zero. Thus the eigenvalue problem

$$\begin{pmatrix} p^{*} & 0 & 0 & -\Gamma^{-1} \\ 0 & \kappa\beta^2p^{*}+C_o & \kappa\Gamma & 0 \\ 0 & -\Gamma^{-1} & p^{*}+C_o^{*} & -C_o^{*} \\ \kappa\Gamma & 0 & -C_o^{*} & \kappa\beta^2p^{*}+C_o^{*} \end{pmatrix}\begin{pmatrix} y_v^{\,a} \\ y_\psi^{\,a} \\ y_v^{\,b} \\ y_\psi^{\,b} \end{pmatrix} = \bar{0} \qquad (4.6\text{-}2)$$

comes about, with the characteristic equation

$$\sum_{k=0}^{4} A_k\, p^{*4-k} = 0, \tag{4.6-3}$$

where

$$\left.\begin{aligned} &A_o = (K\,\beta^2)^2,\ A_1 = \{C_o + C_o^*(1+\kappa\beta^2)\}\kappa\beta^2,\ A_2 = C_oC_o^*(1+\kappa\beta^2) \\ &\quad + 2\,\kappa^2\beta^2,\ A_3 = \kappa\{C_o + C_o^*(1+\kappa\beta^2)\},\ A_4 = \kappa(\kappa + C_oC_o^*). \end{aligned}\right\} \tag{4.6-4}$$

We can write (3) as

$$(\kappa\beta^2p^{*2} + C_op^* + \kappa)\{\kappa\beta^2p^{*2} + C_o^*(1+\kappa\beta^2)p^* + \kappa\} + \kappa\, C_oC_o^* = 0. \tag{4.6-5}$$

For $C_o = 0$, $C_o^* \neq 0$, this breaks up into the two second degree equations

$$\kappa\beta^2p^{*2} + \kappa = 0, \quad \kappa\beta^2p^{*2} + C_o^*(1+\kappa\beta^2)p^* + \kappa = 0. \tag{4.6-6^{a-b}}$$

In this case two roots are neutrally stable and the two other ones are asymptotically stable. This holds also for $C_o^* = 0$, $C_o = 0$. We shall show later on that for all other combinations all the four roots are asymptotically stable.

The case $C_o = 0$ is interesting because it involves perfect curve running. We determine the eigenvector from (2) and we find

$$y_v^{\,a} = 1,\ \ y_\psi^{\,a} = (\Gamma p^*)^2,\ \ y_v^{\,b} = y_\psi^{\,b} = \Gamma p^*. \tag{4.6-7}$$

When we put

$$p^* = j\,\frac{2\pi}{\underline{\lambda}} \tag{4.6-8}$$

we obtain from (6^a) exactly the Klingel wave length

$$\underline{\lambda} = \frac{2\pi}{\sqrt{-p^{*2}}} = 2\pi\beta, \quad \lambda = \underline{\lambda}\,\ell = 2\pi\sqrt{\frac{br}{\gamma_o\rho^*}}\ : \tag{4.6-9}$$

see (3.5-3) and (3.3-1).

For the displacements q_{vi} and $q_{\psi i}$ we can write

$$\left.\begin{aligned} q_{vi} &= \tfrac{1}{2}(q_v^{\,a} \pm q_v^{\,b}) = \tfrac{1}{2}(1 \pm \Gamma p^*)e^{p^*\underline{s}}, \\ q_{\psi i} &= \tfrac{1}{2}(q_\psi^{\,a} \mp q_\psi^{\,b}) = -\tfrac{1}{2}\Gamma p^*(1 \pm \Gamma p^*)e^{p^*\underline{s}}, \end{aligned}\right\} \tag{4.6-10}$$

so that

$$q_{vi} = \frac{\lambda^2}{4\pi^2\Gamma} q'_{\psi i}, \quad q_{\psi i} = -\Gamma\, q'_{vi}. \tag{4.6-11}$$

The wheelset forces of the stationary motion are found from (7) and (4.5-8):

$$f_v{}^a = f_\psi{}^a = f_v{}^b = f_\psi{}^b = 0: \tag{4.6-12}$$

both wheelsets are always purely rolling. This can be understood by looking after fig. 4.2-2. In the case in which we have only a shear spring with a rigidity C_s, we find from (4.4-21):

$$C_1 = C_2 = C_3 = 0, \quad C_4 = -C_5 = C_6 = C_s. \tag{4.6-13}$$

In the case of pure rolling, the shear spring is unloaded, so that

$$v_b - a\psi_b = 0, \quad q_v{}^b = q_\psi{}^b, \tag{4.6-14}$$

and this agrees with (7) and (1).

Returning to the general case we now investigate the equation (3) more in detail. Putting

$$x = C_o, \quad y = C_o{}^*, \quad a = \kappa\beta^2, \quad b = \kappa \tag{4.6-15}$$

we can reduce (4) to

$$\left.\begin{aligned} &A_o = a^2,\ A_1 = ax + a(1+a)y,\ A_2 = (1+a)xy + 2ab, \\ &A_3 = b\,x + b(1+a)y,\ A_4 = b\,xy + b^2. \end{aligned}\right\} \tag{4.6-16}$$

We calculate the Hurwitz determinants

$$D_2 = \begin{vmatrix} A_1 & A_3 \\ A_2 & A_o \end{vmatrix}, \quad D_3 = \begin{vmatrix} A_1 & A_3 & 0 \\ A_o & A_2 & A_4 \\ 0 & A_1 & A_3 \end{vmatrix} \tag{4.6-17}$$

and we find

$$\begin{aligned} D_2 &= a\{x + (1+a)y\}\{(1+a)xy + ab\}, \\ D_3 &= ab\,xy\{x + (1+a)y\}^2; \end{aligned} \tag{4.6-18}$$

thus we have

$$A_o > 0, \quad A_1 > 0, \quad D_2 > 0, \quad D_3 > 0, \quad A_4 > 0, \tag{4.6-19}$$

so that the stationary motion is always asymptotically stable when $a > 0$, $b > 0$, $x > 0$, $y > 0$.

In this way we may conclude that the motion of the vehicle is asymptotically stable for small velocities. In order to obtain as stable a vehicle as possible, it is desirable to choose such a combination of the four parameters β, κ, C_o and C_o^* that all the four roots of (3) are real and negative. Therefore, it is important to draw in an (x, y) plane the boundary of the region where the roots have this property.

On such a boundary the equation (3) has a double root, so that p^* must be a root of the two equations

$$f(p^*) = a^2p^{*4} + a\{x + (1 + a)y\}p^{*3} + \{(1 + a)xy + 2ab\}p^{*2} + b\{x + (1 + a)y\}p^* + b(xy + b) = 0, \tag{4.6-20^a}$$

and

$$f'(p^*) = 4a^2p^{*3} + 3a\{x + (1 + a)y\}p^{*2} + 2\{(1 + a)xy + 2ab\}p^* + b\{x + (1 + a)y\} = 0. \tag{4.6-20^b}$$

Now we write

$$\left.\begin{aligned} f(p^*) &= a_1\, xy + b_{11}\, x + b_{12}\, y + c_1, \\ f'(p^*) &= a_2\, xy + b_{21}\, x + b_{22}\, y + c_2 \end{aligned}\right\} \tag{4.6-21}$$

with

$$a_1 = (1+a)p^{*2} + b,\ a_2 = 2(1+a)p^*,\ b_{11} = (ap^{*2}+b)p^*,\ b_{12} = (1+a)(ap^{*2}+b)p^*,$$
$$b_{21} = 3ap^{*2} + b,\ b_{22} = (1+a)(3ap^{*2}+b),\ c_1 = (ap^{*2}+b)^2,\ c_2 = 4a(ap^{*2}+b)p^*. \tag{4.6-22}$$

Solving y both from (20^a) and from (20^b) yields

$$y = -\frac{b_{11}\, x + c_1}{a_1\, x + b_{12}} = -\frac{b_{21}\, x + c_2}{a_2\, x + b_{22}}, \tag{4.6-23}$$

so that x has to satisfy the quadratic equation

$$(a_1b_{21} - a_2b_{11})x^2 + (a_1c_2 - a_2c_1)x + b_{12}c_2 - b_{22}c_1 = 0. \qquad (4.6\text{-}24^a)$$

Likewise, y has to satisfy

$$(a_2b_{12} - a_1b_{22})y^2 - (a_1c_2 - a_2c_1)y + b_{21}c_1 - b_{11}c_2 = 0. \qquad (4.6\text{-}24^b)$$

Here we have used the property

$$b_{11}b_{22} - b_{12}b_{21} = 0 \qquad (4.6\text{-}25)$$

of the coefficients b_{ij}.

Now the relations (22) yield

$$a_1b_{21} - a_2b_{11} = a(1+a)p^{*4} - b(1-2a)p^{*2} + b^2, \qquad (4.6\text{-}26^a)$$

$$a_2b_{12} - a_1b_{22} = -(1+a)\{a(1+a)p^{*4} - b(1-2a)p^{*2} + b^2\}, \qquad (4.6\text{-}26^b)$$

$$a_1c_2 - a_2c_1 = 2(1+a)(ap^{*2} + b)(ap^{*2} + 2a - b)p^*, \qquad (4.6\text{-}27)$$

$$b_{12}c_2 - b_{22}c_1 = (1+a)(ap^{*2} + b)^2(ap^{*2} - b), \qquad (4.6\text{-}28^a)$$

$$b_{21}c_1 - b_{11}c_2 = -(ap^{*2} + b)^2(ap^{*2} - b). \qquad (4.6\text{-}28^b)$$

Multiplying (26^b) by (1+a) and adding up (26^a) we find that always one of the two relations

$$x = (1+a)y \qquad (4.6\text{-}29^a)$$

and

$$x + (1+a)y = -\frac{2(1+a)(ap^{*2}+b)(ap^{*2}+2a-b)p^*}{a(1+a)p^{*4}-b(1-2a)p^{*2}+b^2} \qquad (4.6\text{-}29^b)$$

holds.

The boundaries to be determined are found by calculating the quantities (26^a) etc. for various non-positive values of p^* and by solving x and y from (24^{a-b}). Special cases are:

$$p^* = 0,\ p^* = -\sqrt{b/a},\ p^* = -\infty,\ x = 0,\ y = \infty.$$

For $p^* = 0$ we find from (20^{a-b}) and also from (24^{a-b}) and (26^a)-(28^b):

$$x = \pm\sqrt{b(1+a)}, \quad y = \mp\sqrt{\frac{b}{1+a}}, \qquad (4.6\text{-}30)$$

in agreement with (29^b) but not with (29^a).

For $p^* = -\sqrt{b/a}$, (26^a)-(28^b) reduce to

$$\left.\begin{aligned} &a_1b_{21} - a_2b_{11} = 4b^2,\ a_2b_{12} - a_1b_{21} = -4(1+a)b^2, \\ &a_1c_2 - a_2c_1 = -8b(1+a)\sqrt{ab},\ b_{12}c_2 - b_{22}c_1 = b_{21}c_1 - b_{11}c_2 = 0, \end{aligned}\right\} \qquad (4.6\text{-}31)$$

so that (24^{a-b}) yield

$$x = 0 \quad \text{or} \quad x = 2(1+a)\sqrt{\frac{a}{b}},\quad y = 0 \quad \text{or} \quad 2\sqrt{\frac{a}{b}}\ . \qquad (4.6\text{-}32)$$

Here two combinations are possible:

$$x = 0,\quad y = 2\sqrt{a/b}, \text{ in agreement with } (29^b), \qquad (4.6\text{-}33^a)$$

$$x = 2(1+a)\sqrt{a/b},\quad y = 0, \text{ also in agreement with } (29^b). \qquad (4.6\text{-}33^b)$$

The third combination, $x = y = 0$, in agreement with (29^a), turns out to be in disagreement with $p^* = -\sqrt{b/a}$: as a matter of fact we find from (20^a) in that case: $p^* = \pm\, j\sqrt{b/a}$.

For $p^* \to -\infty$, (26^a)-(28^b) and (24^{a-b}) yield

$$x \to -ap^*,\ y \to -ap^*/(1+a), \qquad (4.6\text{-}34)$$

in agreement with both (29^a) and (29^b).

From (24^{a-b}) we see that x and y go to infinity when

$$a_1b_{21} - a_2b_{11} = 0,\quad a_2b_{12} - a_1b_{22} = 0, \qquad (4.6\text{-}35)$$

viz.

$$a(1+a)p^{*4} - b(1-2a)p^{*2} + b^2 = 0, \qquad (4.6\text{-}36)$$

so that

$$p^{*2} = b\,\frac{1 - 2a \pm \sqrt{1-8a}}{2a(1+a)} \qquad (4.6\text{-}37)$$

This can only occur when

$$a < \frac{1}{8},\quad \beta^2 < 1/8\kappa \qquad (4.6\text{-}38)$$

i.e., when the half wheelbase a and the half track gauge b satisfy the condition

$$a > 2b\sqrt{2\kappa}: \qquad (4.6\text{-}39)$$

for $b = 0{,}75$ m and $\kappa = 1$ the critical value of 2a is 4,2 m. Again the relation (29^b) holds and, when $x \to \infty$, y remains finite, vice versa.

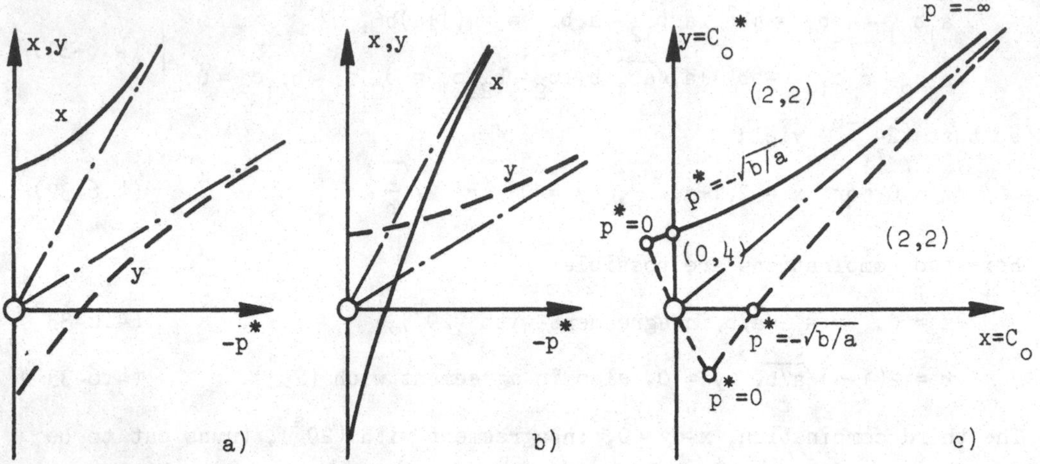

Fig. 4.6-1. The quantities x and y as functions of p^* for $\beta^2 > 1/8\kappa$.

We now can draw in an (x, y) plane the boundaries of the regions where the number of real roots is zero, two and four respectively.

In the case $\beta^2 > 1/8\kappa$ (fig. 4.6-1) there are three regions. The number of real roots in each region can be determined by investigating the point $x = y = 0$, in which two roots p^* are equal to $j\sqrt{b/a}$ and two to $-j\sqrt{b/a}$. Thus, the origin is in the region (0,4): zero real roots and four complex roots. The two adjacent regions must both of them be regions (2,2): two real and two complex roots.

In the case $\beta^2 < 1/8\kappa$ (fig. 4.6-2) there are regions where all the four roots are real and negative. This has a very much stabilizing effect. Unfortunately, the regions are very small; moreover, for bogies we always have $\beta^2 > 1/8\kappa$, so that these combinations of x and y never can be realized.

4.7. The stability of the fundamental motion for non-vanishing speed. The characteristic equation, the eigenvalues and the eigenvectors

We now determine the general solution of the homogeneous equations, related with the equations of motion (4.4-15). We use the same method as in section 3.4 and we put

$$^t\underline{\bar{f}}_a = (0,\ x_v{}^a,\ x_\psi{}^a)e^{\underline{pt}},\quad ^t\underline{\bar{f}}_b = (0,\ x_v{}^b,\ x_\psi{}^b)e^{\underline{pt}} \qquad (4.7\text{-}1)$$

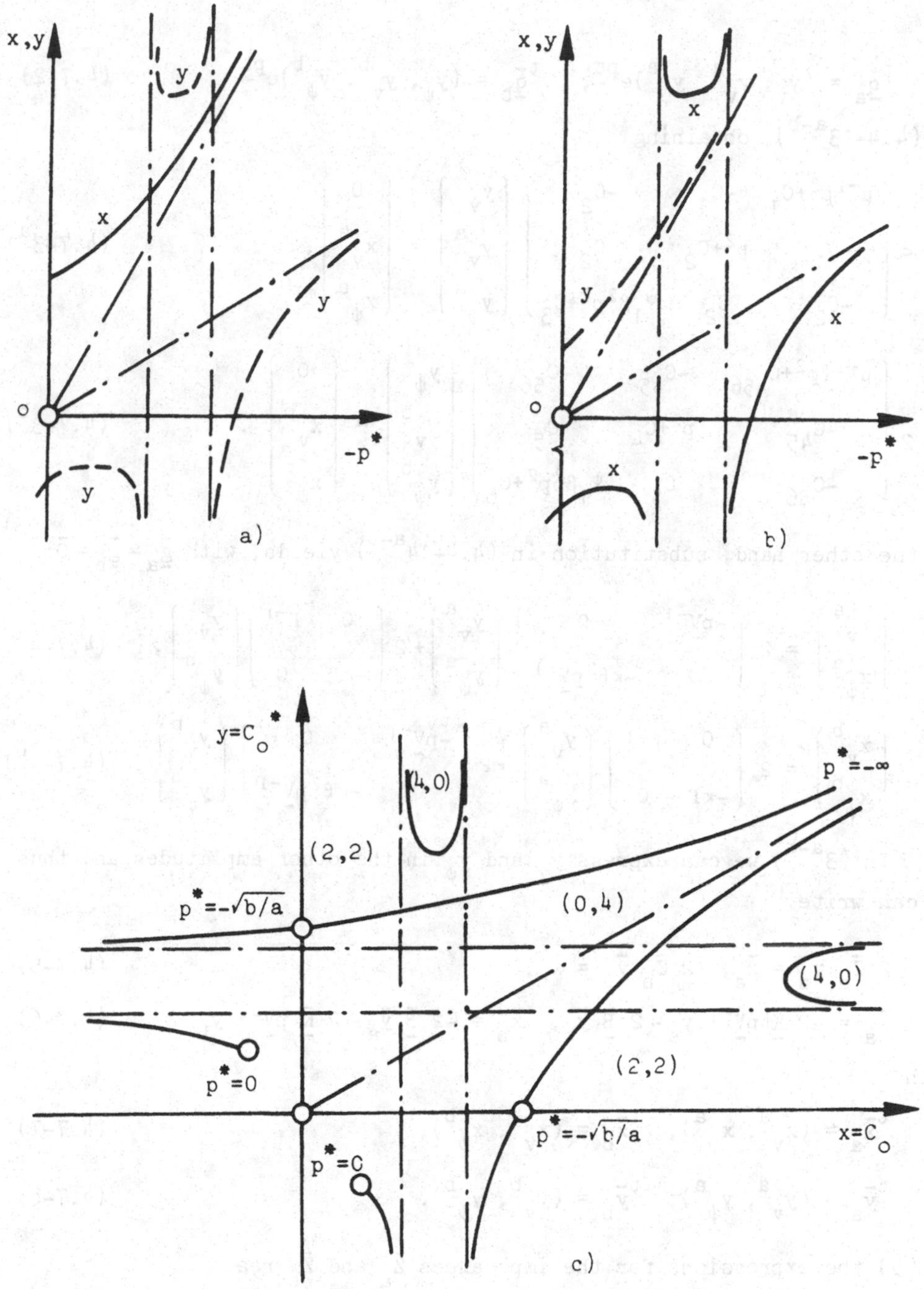

Fig. 4.6-2. The quantities x and y as functions of p^* for $\beta^2 < 1/8\kappa$.

and

$$ {}^{t}\underline{\bar{q}}_a = (y_v,\ y_v{}^a,\ y_\psi{}^a)e^{\underline{pt}},\qquad {}^{t}\underline{\bar{q}}_b = (y_\psi,\ y_v{}^b,\ y_\psi{}^b)e^{\underline{pt}} \tag{4.7-2} $$

in (4.4-13^{a-b}), obtaining

$$ 2\begin{pmatrix} \mu^{-1}p^2+C_1 & -C_1 & -C_2 \\ -C_1 & p^2+C_2 & C_2 \\ -C_2 & C_2 & \Psi_1\beta^2p^2+C_3 \end{pmatrix}\begin{pmatrix} y_v \\ y_v{}^a \\ y_\psi{}^a \end{pmatrix} = \begin{pmatrix} 0 \\ x_v{}^a \\ x_\psi{}^a \end{pmatrix}, \tag{4.7-3^a} $$

$$ 2\begin{pmatrix} \mu^{-1}\Psi p^2+C_{456} & -C_{45} & -C_{56} \\ -C_{45} & p^2+C_4 & C_5 \\ -C_{56} & C_5 & \Psi_1\beta^2p^2+C_6 \end{pmatrix}\begin{pmatrix} y_\psi \\ y_v{}^b \\ y_\psi{}^b \end{pmatrix} = \begin{pmatrix} 0 \\ x_v{}^b \\ x_\psi{}^b \end{pmatrix}. \tag{4.7-3^b} $$

On the other hand, substitution in (4.4-14^{a-b}) yields, with $\underline{\bar{g}}_a = \underline{\bar{g}}_b = \bar{0}$:

$$ \begin{pmatrix} x_v{}^a \\ x_\psi{}^a \end{pmatrix} = 2\begin{pmatrix} -p\underline{V}^{-1} & 0 \\ 0 & -\kappa\beta^2p\underline{V}^{-1} \end{pmatrix}\begin{pmatrix} y_v{}^a \\ y_\psi{}^a \end{pmatrix} + 2\begin{pmatrix} 0 & \Gamma^{-1} \\ -\kappa\Gamma & 0 \end{pmatrix}\begin{pmatrix} y_v{}^b \\ y_\psi{}^b \end{pmatrix}, \tag{4.7-4^a} $$

$$ \begin{pmatrix} x_v{}^b \\ x_\psi{}^b \end{pmatrix} = 2\begin{pmatrix} 0 & \Gamma^{-1} \\ -\kappa\Gamma & 0 \end{pmatrix}\begin{pmatrix} y_v{}^a \\ y_\psi{}^a \end{pmatrix} + 2\begin{pmatrix} -p\underline{V}^{-1} & 0 \\ 0 & -\kappa\beta^2p\underline{V}^{-1} \end{pmatrix}\begin{pmatrix} y_v{}^b \\ y_\psi{}^b \end{pmatrix}. \tag{4.7-4^b} $$

In (3^{a-b}) we can express y_v and y_ψ in the other amplitudes and thus we can write

$$ 2\,\bar{Z}_a\,\bar{y}_a = \bar{x}_a,\quad 2\,\bar{Z}_b\,\bar{y}_b = \bar{x}_b, \tag{4.7-5} $$

$$ \bar{x}_a = -2\,\underline{\bar{K}}\,p\underline{V}^{-1}\,\bar{y}_a - 2\,\underline{\bar{B}}\,\bar{y}_b,\quad \bar{x}_b = -2\,\underline{\bar{B}}\,\bar{y}_a - 2\,\underline{\bar{K}}\,p\underline{V}^{-1}\,\bar{y}_b. \tag{4.7-6} $$

with

$$ {}^{t}\bar{x}_a = (x_v{}^a,\ x_\psi{}^a),\quad {}^{t}\bar{x}_b = (x_v{}^b,\ x_\psi{}^b), \tag{4.7-7} $$

$$ {}^{t}\bar{y}_a = (y_v{}^a,\ y_\psi{}^a),\quad {}^{t}\bar{y}_b = (y_v{}^b,\ y_\psi{}^b). \tag{4.7-8} $$

In (5) the expressions for the impedances $\bar{Z}_a$ and $\bar{Z}_b$ read

$$\bar{Z}_a = \begin{pmatrix} Z_{vv}{}^a & Z_{v\psi}{}^a \\ Z_{\psi v}{}^a & Z_{\psi\psi}{}^a \end{pmatrix}, \quad \bar{Z}_b = \begin{pmatrix} Z_{vv}{}^b & Z_{v\psi}{}^b \\ Z_{\psi v}{}^b & Z_{\psi\psi}{}^b \end{pmatrix}, \tag{4.7-9}$$

$$Z_{vv}{}^a = p^2 + \frac{C_1 p^2}{p^2+\mu C_1}, \quad Z_{v\psi}{}^a = Z_{\psi v}{}^a = \frac{C_2 p^2}{p^2+\mu C_1}, \quad Z_{\psi\psi}{}^a = \Psi_1 \beta^2 p^2 + \frac{C_3 p^2 + \mu C_{13}}{p^2+\mu C_1}, \tag{4.7-10^a}$$

$$Z_{vv}{}^b = p^2 + \frac{\Psi p^2 C_4 + \mu C_{46}}{\Psi p^2 + \mu C_{456}}, \quad Z_{v\psi}{}^b = Z_{\psi v}{}^b = \frac{\Psi p^2 C_5 - \mu C_{46}}{\Psi p^2 + \mu C_{456}},$$

$$Z_{\psi\psi}{}^b = \Psi_1 \beta^2 p^2 + \frac{\Psi p^2 C_6 + \mu C_{46}}{\Psi p^2 + \mu C_{456}}. \tag{4.7-10^b}$$

This can be derived from (4.4-10) and (4.4-11); C_{46} and an analogous quantity, C_{13}, are defined by

$$C_{13} = C_1 C_3 - C_2{}^2, \quad C_{46} = C_4 C_6 - C_5{}^2. \tag{4.7-11}$$

This method admits an extension to the case in which the body rolling motion is taken into account: then it is only necessary to extend the impedance $\bar{Z}_a$ slightly.

The eigenvalue problem is defined by the two equations, obtained by eliminating $\bar{x}_a$ and $\bar{x}_b$ from (5) and (6).

The relations (5) and (6) can be represented by the flow diagram of fig. 4.7-1. Here we have used the admittances

$$\bar{Y}_a = \bar{Z}_a{}^{-1}, \quad \bar{Y}_b = \bar{Z}_b{}^{-1}. \tag{4.7-12}$$

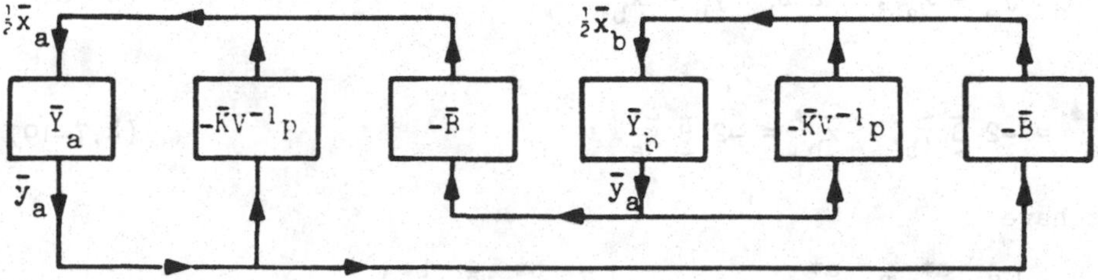

Fig. 4.7-1. The flow diagram for the two-axled vehicle.

The method can be improved by starting from the equations

$$2\,\underline{\bar{M}}\,\ddot{\underline{\bar{q}}}_a + 2\,\underline{\bar{K}}\,\underline{V}^{-1}\,\dot{\underline{\bar{q}}}_a + 2\,\underline{\bar{C}}_a\,\underline{\bar{q}}_a = \underline{\bar{f}}_a^{\,*},\quad 2\,\underline{\bar{M}}\,\ddot{\underline{\bar{q}}}_b + 2\,\underline{\bar{K}}\,\underline{V}^{-1}\,\dot{\underline{\bar{q}}}_b + 2\,\underline{\bar{C}}_b\,\underline{\bar{q}}_b = \underline{\bar{f}}_b^{\,*}, \tag{4.7-13}$$

$$\underline{\bar{f}}_a^{\,*} = -2\,\underline{\bar{B}}\,\underline{\bar{q}}_b,\quad \underline{\bar{f}}_b^{\,*} = -2\,\underline{\bar{B}}\,\underline{\bar{q}}_a \tag{4.7-14}$$

rather than from ($4.4\text{-}13^a$)-($4.4\text{-}14^b$), $\bar{g}_a$ and $\bar{g}_b$ being zero again. Then we replace (1) by

$$^t\underline{\bar{f}}_a^{\,*} = (x_v^{\,*},\, x_v^{\,a*},\, x_\psi^{\,a*})e^{pt},\quad ^t\underline{\bar{f}}_b^{\,*} = (x_\psi^{\,*},\, x_v^{\,b*},\, x_\psi^{\,b*})e^{pt}; \tag{4.7-15}$$

(3^{a-b}) has to be replaced by

$$2\begin{pmatrix} \mu^{-1}p^2+C_1 & -C_1 & -C_2 \\ -C_1 & p^2+\underline{V}^{-1}p+C_2 & C_2 \\ -C_2 & C_2 & \Psi_1\beta^2p^2+\kappa\beta^2\underline{V}^{-1}p+C_3 \end{pmatrix}\begin{pmatrix} y_v \\ y_v^{\,a} \\ y_\psi^{\,a} \end{pmatrix} = \begin{pmatrix} 0 \\ x_v^{\,a*} \\ x_\psi^{\,a*} \end{pmatrix}, \tag{4.7-16^a}$$

$$2\begin{pmatrix} \mu^{-1}\Psi p^2+C_{456} & -C_{45} & -C_{56} \\ -C_{45} & p^2+\underline{V}^{-1}p+C_4 & C_5 \\ -C_{56} & C_5 & \Psi_1\beta^2p^2+\kappa\beta^2\underline{V}^{-1}p+C_6 \end{pmatrix}\begin{pmatrix} y_\psi \\ y_v^{\,b} \\ y_\psi^{\,b} \end{pmatrix} = \begin{pmatrix} 0 \\ x_v^{\,b*} \\ x_\psi^{\,b*} \end{pmatrix} \tag{4.7-16^b}$$

and (4^{a-b}) by

$$x_v^{\,a*} = 2\Gamma^{-1}\,y_\psi^{\,b},\; x_\psi^{\,a*} = -2\kappa\Gamma\,y_v^{\,b};\; x_v^{\,b*} = 2\Gamma^{-1}\,y_\psi^{\,a},\; x_\psi^{\,b*} = -2\kappa\Gamma\,y_v^{\,a}. \tag{4.7-17}$$

Eliminating y_v and y_ψ again, we obtain

$$2\,\bar{Z}_a^{\,*}\,\bar{y}_a = \bar{x}_a^{\,*},\quad 2\,\bar{Z}_b^{\,*}\,\bar{y}_b = \bar{x}_b^{\,*} \tag{4.7-18}$$

and

$$\bar{x}_a^{\,*} = -2\,\underline{\bar{B}}\,\bar{y}_b,\quad \bar{x}_b^{\,*} = -2\,\underline{\bar{B}}\,\bar{y}_a. \tag{4.7-19}$$

Then we have

$$\bar{Z}_a^{\,*} = \begin{pmatrix} Z_{vv}^{\,a*} & Z_{v\psi}^{\,a*} \\ Z_{v\psi}^{\,a*} & Z_{\psi\psi}^{\,a*} \end{pmatrix},\quad \bar{Z}_b^{\,*} = \begin{pmatrix} \bar{Z}_{vv}^{\,b*} & \bar{Z}_{v\psi}^{\,b*} \\ \bar{Z}_{v\psi}^{\,b*} & \bar{Z}_{\psi\psi}^{\,b*} \end{pmatrix}, \tag{4.7-20}$$

$$Z_{vv}^{a*} = Z_{vv}^{a} + \underline{V}^{-1}p,\ Z_{v\psi}^{a*} = Z_{v\psi}^{a},\ Z_{\psi\psi}^{a*} = Z_{\psi\psi}^{a} + \kappa\beta^2\underline{V}^{-1}p, \qquad (4.7\text{-}21^a)$$

$$Z_{vv}^{b*} = Z_{vv}^{b} + \underline{V}^{-1}p,\ Z_{v\psi}^{b*} = Z_{v\psi}^{b},\ Z_{\psi\psi}^{b*} = Z_{\psi\psi}^{b} + \kappa\beta^2\underline{V}^{-1}p. \qquad (4.7\text{-}21^b)$$

From (18) and (19) we find the eigenvalue problem

$$\begin{pmatrix} \bar{Z}_a^* & \bar{B} \\ \bar{B} & \bar{Z}_b^* \end{pmatrix} \begin{pmatrix} \bar{y}_a \\ \bar{y}_b \end{pmatrix} = \bar{0}. \qquad (4.7\text{-}22)$$

The characteristic equation

$$\begin{vmatrix} \bar{Z}_a^* & \bar{B} \\ \bar{B} & \bar{Z}_b^* \end{vmatrix} = 0 \qquad (4.7\text{-}23)$$

becomes

$$\begin{vmatrix} Z_{vv}^{a*} & Z_{v\psi}^{a*} & 0 & -\Gamma^{-1} \\ Z_{v\psi}^{a*} & Z_{\psi\psi}^{a*} & \kappa\Gamma & 0 \\ 0 & -\Gamma^{-1} & Z_{vv}^{b*} & Z_{v\psi}^{b*} \\ \kappa\Gamma & 0 & Z_{v\psi}^{b*} & Z_{\psi\psi}^{b*} \end{vmatrix} = 0; \qquad (4.7\text{-}24)$$

writing this out yields

$$\begin{vmatrix} Z_{vv}^{a*} & Z_{v\psi}^{a*} \\ Z_{v\psi}^{a*} & Z_{\psi\psi}^{a*} \end{vmatrix} \cdot \begin{vmatrix} Z_{vv}^{b*} & Z_{v\psi}^{b*} \\ Z_{v\psi}^{b*} & Z_{\psi\psi}^{b*} \end{vmatrix} + \kappa\left(Z_{vv}^{a*} Z_{\psi\psi}^{b*} + Z_{\psi\psi}^{a*} Z_{vv}^{b*}\right)$$

$$- (\kappa^2\Gamma^2 + \Gamma^{-2}) Z_{v\psi}^{a*} Z_{v\psi}^{b*} + \kappa^2 = 0. \qquad (4.7\text{-}25)$$

Multiplying this equation with $(p^2+\mu C_1)(\Psi p^2+\mu C_{456})$ and using the relations (9)-(10^b) and (20)-(21^b) yields a 12th degree equation for p, as might be expected.

Lack of space prevents us to discuss the solution of the characteristic equation (25) and the root loci related to it. Suffice it to mention that approximate solutions can be found by means of perturbation techniques, using μ as the small parameter.

4.8. The fundamental problem for the two-axled vehicle

In designing a two-axled vehicle we fall into a conflicting situation because the requirement of perfect curving leads to another optimum combination of values of the elastic constants than the requirement of stability does.

In section 4.5 we found the values of the contact forces in curves, from which we concluded that a two-axled vehicle has a perfect curving behaviour, provided the coefficient C_o (4.5-6) is equal to zero, so that the coefficients C_1, C_2 and C_3 must satisfy the condition

$$C_1C_3 - C_2^2 = 0. \qquad (4.8\text{-}1)$$

The stability requirement has been fastened down in a quantitative way only for the case of small speeds. In section 4.6 we found the optimum values of the two combinations C_o and C_o^* of the six elastic constants. For $\beta^2 < 1/8\kappa$ they should be in the two regions (4,0) in fig. 4.6-2 and also for $\beta^2 > 1/8\kappa$ their values can be found such that stability is optimal.

A compromise solution can be found by taking for C_o the value of the (4,0) region a) in fig. 4.6-2, in which is much smaller than that of the (4,0) region b). The actual spring constants of two-axled cars have about these values. Note that they do not change when the rolling motion is taken into account.

However, some root loci may become very much unstable when the speed increases and it will be necessary in next future to investigate the behaviour of the root loci for higher speeds thoroughly. We have found already that in the case with only C_b and C_s differing from zero, it is possible to choose the constants in such a way that the vehicle is stable for all values of V.

A much better situation arises when the vehicle is designed with two bogies. Here the curving behaviour can be influenced by an appropriate design of steering rods and then we are completely free in the choice of the elastic constants: in that case they can be chosen in such a way that optimum stability even at high speeds is attained.

5. NONLINEARITIES; THE FIRST-ORDER THEORY FOR A SINGLE WHEELSET

For the single wheelset, described in chapter 3, general equations of motion have been derived [5,6]. For the motion on tangent track they can be considered as exact, whereas for the motion in a curve only the terms with r_o^{-1} have been taken into account.

From the general theory we can deduce a so-called theory of first order, in which only the linear restoring and inertia terms are retained, whereas the contact force terms remain non-linear: in the first order theory the products of forces and parasitic displacements have been neglected.

For this first-order theory the equations of motion read

$$m_i\ddot{u}_i + c_{xi}u_i = -\sum_{j=1}^{2} T_{xij}, \qquad (5\text{-}1^a)$$

$$m_i\ddot{v}_i + c_{yi}v_i = \sum_{j=1}^{2}\left\{(-1)^j N_{ij}\sin\gamma_{ij} - T_{yij}\cos\gamma_{ij}\right\}, \qquad (5\text{-}1^b)$$

$$m_i\ddot{w}_i + c_{zi}w_i = G_i - \sum_{j=1}^{2}\left\{N_{ij}\cos\gamma_{ij} + (-1)^j T_{yij}\sin\gamma_{ij}\right\}, \qquad (5\text{-}1^c)$$

$$J_i\ddot{\phi}_i + c_{zi}b_i^2\phi_i = \sum_{j=1}^{2}(-1)^j N_{ij}(b\cos\gamma_{ij} - r\sin\gamma_{ij}) + \sum_{j=1}^{2} T_{yij}(r\cos\gamma_{ij} + b\sin\gamma_{ij}), \qquad (5\text{-}2^a)$$

$$J_{yi}\ddot{\chi}_i = -r\sum_{j=1}^{2} T_{xij}, \qquad (5\text{-}2^b)$$

$$J_i\ddot{\psi}_i + c_{xi}b_i\,\psi_i = -b\sum_{j=1}^{2}(-1)^j T_{xij}. \qquad (5\text{-}2^c)$$

Here γ_{ij} is the real conicity. The tangential forces T_{xij} and T_{yij} are determined by linear formulae like $(2\text{-}4^{a-b})$ and (2-5) but for large displacements it is better to use non-linear expressions. The creep and spin quantities are given by

$$\upsilon_{xij} = V^{-1}(\dot{u}_i + r\,\dot{\chi}_i \mp b\,\dot{\psi}_i) - \zeta_{ij}^*/r, \qquad (5\text{-}3^a)$$

$$\upsilon_{yij} = V^{-1}\left\{\dot{v}_i \cos\gamma_{ij} \mp \dot{w}_i \sin\gamma_{ij} - \dot{\phi}_i (r \cos\gamma_{ij} + b \sin\gamma_{ij})\right\} - \psi_i \cos^{-1}\gamma_{ij}, \qquad (5\text{-}3^b)$$

$$\phi_{ij} = V^{-1}(\pm \dot{\chi}_i \sin\gamma_{ij} + \dot{\psi}_i \cos\gamma_{ij}) \mp r^{-1}\sin\gamma_{ij} - r\,\phi_i^{-1} \cos\gamma_{ij}, \qquad (5\text{-}4)$$

where ζ_{ij}^* is the momentaneous wheel radius.

The equations (1^a)-(4) should be completed by the constraint relations

$$w_i = w_i(v_i), \quad \phi_i = \phi_i(v_i) \qquad (5\text{-}5)$$

of the wheelsets. They can be determined from the rail and tyre profiles. Altogether we dispose on 12 equations: six equations of motion, four equations for the tangential forces and two constraint relations; on the other hand we have 12 unknowns: the six displacements, the two normal forces and the four tangential forces.

By means of the harmonic balance method [6,7] it is possible to obtain an approximate solution of the equations (1^a)-(5). The general result is shown in fig. 5-1. We usually find a limit-cycle and the amplitudes of the various displacements are functions of the speed V: this has been shown for the amplitude y_v of the lateral displacement. For all values of V the fundamental movement is possible but it becomes unstable above the speed V_c. For $V = V_c$ there is a branch point and often the speed decreases at increasing amplitude when we are on the limit-cycle characteristic. This means that when we are between V_c^* and V_c, a small perturba-

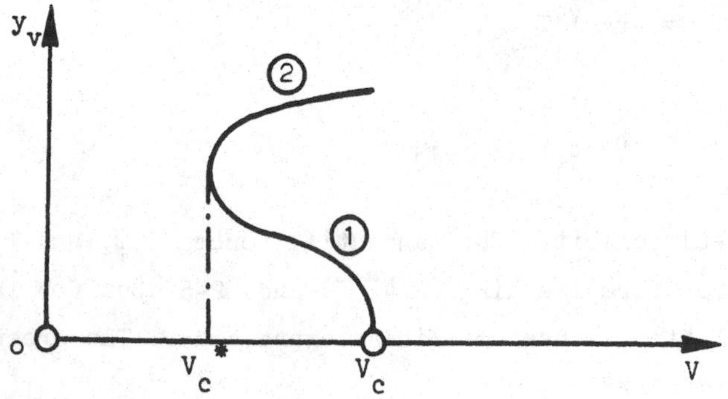

Fig. 5-1. The amplitude of the limit-cycle as a function of speed.

tion will only have the effect that the motion returns to the fundamental motion, whereas a large perturbation will cause the system to move according to the part (2) of the limit-cycle characteristic. So we see that the knowledge of the speed V_c is only of restricted importance.

However, it is expected that an increase of the speed V_c usually involves an increase of the speed V_c^*. Nevertheless, this aspect of the hunting problem should be investigated more completely. It is also desirable to extend the study to the case of a complete vehicle. In first approximation we can assume that the elastic interconnection between wheelset and body remains linear and then the impedance method, exposed in section 4.7, remains valid.

6. STOCHASTIC VIBRATIONS

Hitherto we assumed the track to be in perfect condition, that is, without deviations from the original straight or curved shape. In railway practice there are always small deviations. They can be taken into account in the following way. Each of the two rails can have a vertical displacement, a horizontal displacement and a rotation about the longitudinal axis. We can make two combinations of these six displacement quantities. The first combination is a shift of the track as a whole, in which its cross section does not deform, and it consists of a vertical displacement, a horizontal displacement and a rotation about a longitudinal axis. The second combination consists of a change of the track gauge, a rotation of the right-hand rail and a rotation of the left-hand rail.

The first combination can be taken into account easily: it only involves a shift of the originally straight line, described by the origin o_i, mentioned in section 3.1. But the second combination gives rise to much more complications, because it has an important influence on the constraint relations (5-5). Here there is still great scope for a profound investigation.

7. CONCLUSIONS

For obtaining a sufficient insight in the dynamical behaviour of a moving railway vehicle still much investigation work will have to be done. We conclude this paper with a small priority list.

a. First one should examine whether a vehicle with bogies can be designed in such a way that both perfect curve running and a stable behaviour even at high speeds is possible.

b. The influence of the six rigidities C_1, ... C_6 on the root loci and the eigenvectors should be fully investigated.

c. The influence of the most important non-linearities should be examined.

d. Also the behaviour of a vehicle on irregular track will have to be investigated.

Author likes to express his thanks to Mrs. E.F.C. Meijs-Burghgraef, Mrs. T.H. Scholte-Scheper and Mrs. L. de Vries, who have typed the manuscript of these lecture notes in a very accurate way; also to Mr. J.D. Diepstraten, who assisted him with numerical calculations, necessary for drawing up section 3.6.

REFERENCES

1. Carter, F.W., On the action of a locomotive driving wheel, *Proc. Roy. Soc.*, A 112, 151, 1926.
2. Kalker, J.J., *On the rolling contact of two elastic bodies in the presence of dry friction*, Thesis, Delft, 1967, 7+155 pp.
3. Kalker, J.J. Thesis, Delft, 90, 1967.
4. Kar, A.K., Wormely, D.N. and Hedrick, J.K., Generic rail truck characteristics, in: *Proc. 6th IAVSD-Symposium, Berlin, September 3-7, 1979*, Eds., Swets and Zeitlinger, 1980, 198.
5. Pater, A.D. de, *The exact theory of the motion of a single wheelset moving on a purely straight track*, Delft Un. of Techn., Dept. of Mech.

Engg., Lab. of Engg. Mechs. Report No. 648, Delft, 1979, 78 pp.

6. Pater, A.D. de, The general theory of the motion of a single wheelset moving through a curve with constant radius and cant, *Z. f. Angew. Math. u. Mech.*, 61, 277, 1981.
7. Pater, A.D. de, A non-linear model of a single wheelset moving with constant speed on a purely straight track, *Int. J. of Non-Linear Mechanics*, 15, 315, 1980.
8. Hauschild, W., The application of quasilinearization to the limit cycle behaviour of the nonlinear wheel-rail system, in: *Proc. 6th IAVSD-Symposium, Berlin, September 3-7, 1979*, Eds., Swets and Zeitlinger, 1980, 146.

DESIGN AND EVALUATION OF TRUCKS FOR

HIGH-SPEED WHEEL/RAIL APPLICATION

P. MEINKE, A. MIELCAREK
M.A.N.-Neue Technologie
Dachauer Str. 667
D-8000 München 50

1. INTRODUCTION

Increasing requirements of ride quality and speed demand an optimal design of high speed trucks.

Starting the description of a design strategy for trucks the main conditions of the vehicle operation, expecially the track performance, have to be analized. This is shown in the following chapter 2 for the example of the Federal Republic of Germany.

Another supposition for the design of "optimal" trucks is the existence of a requirement profile from the view of a service concept (specification), which defines on the one hand the running conditions to be evaluated, on the other hand the design criterias for weightening the achieved results (chapter 3).

For the possible development of a systematic design strategie first steps are available; as an example the design procedure may be described, which was worked out in the definition phase of a high speed train in Germany (chapter 4).

Finally possible improvements of the running behaviour of wheel/rail-vehicles due to the use of passive or active components are shown.

2. SPECIFICATION OF THE TRACK

2.1 Data on line and track

2.1.1 Definition of the layout of the line

The vehicle is to be designed for a track which has the following characteristics:

- Radius of curvature $R_r = 11.8\frac{v^2}{200}$ m
- Values for evaluation

Radius	Speed
120 m	5 km/h
190 m	40 km/h
500 m	90 km/h
800 m	120 km/h
2400 m	200 km/h
5300 m	300 km/h
7200 m	350 km/h

- Highest regular superelevation ü = 150 mm
  Superelevations are produced by raising the rail on the outside of the curve.
  Superelevation and gradient can be present simultaneously.
  Transition curves and superelevation ramps are s-shaped;

The beginnings and ends of ramps occur together.

- Length of transition curve

ü < 40 mm $\quad l_{ü} = 2{,}5\ V \dfrac{ü}{1000}$ m

ü > 45 mm $\quad l_{ü} = 12V \dfrac{ü}{1000}$ m

- Curves and straights are so that minimum travelling times of 2 s result in the straight or curve section

- Vertical curve radius R_a = 49000 m at 350 km/h (in accordance with relationship $R_a = 0.4\ V^2$)

- Maximum lingitudinal inclination of any section 12.5 ... 35 °/∘∘

With respect to further development of the wheel/rail system, the vehicle must also satisfy the following supplementary lay-out parameters:

- radius of curvature $R_{min} = 11.8 \dfrac{V^2}{300}$ m

- maximum permissible superelevation $ü_{max}$ = 200 mm

- vertical curve, summit $\quad R_{aK}$ = 27000 m (for V = 300 km/h)

- vertical curve, trough $\quad R_{aW}$ = 17000 m (for V = 300 km/h

The use on nodale points of the existing network, on repair tracks etc. demands compliance with the following additional requirements:

- Usability of standard points and crossing 1:9 when R = 190 m, V = 40 km/h and/or the points connection in the opposite direction 1:9 without a straight by the train

- Practicability of points with a rail cant of 1:∞ at 300 km/h in demonstration operation

- Shunting movements at low speed on repair tracks, when R = 100 m (single vehicle), with acceptance of constraints

Parameters for track dynamics and wheel/rail profiles

Rates of stiffness, damping and masses of the track are taken into account when the vehicle is evaluated.
The following basic parameters havė to used for evaluation.

Vertical stiffness of rail shackle

$c_v = (6)\ldots 9\cdot 10^7$ N/m

LEHR's damping, vertical

$D_v = (0,1)\ldots 0,3$

Covibrating mass of track, vertical

$m_v = 450$ kg

Transverse stiffness of rail shackle

$c_q = 2\cdot 10^7$ N/m

LEHR's damping, lateral

$D_q = 0,3$

Covibrating mass of track, lateral

$m_q = 450$ kg

Rotational stiffness of rail shackle

$c_\varphi = 3.3 \quad 10^5$ Nm/rad

LEHRS's damping with respect to rail rotation

$D_\varphi = 0.1$

Mass moment of inertia of rail

$\Theta_\varphi = 0.56$ kgm² (tracing point)

The profiles of wheel and rail is determined by the following data:

wheel profile	UIC/ORE S 1002
coefficient of friction	$\mu_o = 0.4$
rail profile	UIC 60
rail cant	1 : 40
track gauge 1	1432 mm
track gauge 2	1435 mm
track gauge 3	1438 mm
wheel gauge	1425 mm
wheel gauge measured from flange back to flange back	1360 mm
thickness of wheel flange	32.5 mm

The above data provide effective conicity values of $0.1 \leq \tan\gamma_E \leq 0.35$, which have to be used as a basis for the design speed of 350 km/h and for evaluation; consideration should be extended to value of $\tan\gamma_E \leq 0.45$ at 300 km/h (travelling over points).

Railway bridges

Evaluation of the design with respect to running dynamics for travelling over bridges has to be carried out on the basis of the following bridge data.

1. Bridge 1

   Type: steel truss bridge, open track
   Distance between pillars: 60 m
   Mass: 3500 kg/m
   Stiffness: $E\ I = 0.25 \cdot 10^{12}$ Nm²
   Single span bridge
   Single track

2. Bridge 2

Type: steel girder bridge, ballasted track

Distance between pillars: 49 m (length 50 m)

Mass: 2200 kg/m

Stiffness: 0 m to 10 m : $E\,I = 0.26\cdot10^{12}$ Nm²

10 m to 40 m : $E\,I = 0.32\cdot10^{12}$ Nm²

40 m to 50 m : $E\,I = 0.26\cdot10^{12}$ Nm²

Multispan bridge with 7 spans

Double track

It is recommended to have the rigid body eigenfrequencies of the cars $0.9 \leqq f \leqq 1.3$ Hz with respect to vertical transversal and rotational motion.

2.2 Track Irregularities

Description of irregularities

The theoretical analysis of the dynamic behaviour of wheel/rail-vehicles requires models of the track. Those exist for inertial consideration (resting observer), where the vehicles approach from the infinite, cross the considered part of the track, and remove to infinite other modells describe the track conditions in a coordinate system, which moves with the vehicles, and deliver permanently the track conditions relevant to the wheelsets.

Essential properties of the second modelling type are the disturbance functions describing surface and track irregularities. Based on the three position defects each rail

- vertical displacement $Z_{L,R}$
- lateral displacement $Y_{L,R}$
- Rotation angle $\varphi x_{L,R}$

track irregularities as

- vertical profile

$$\frac{Z_L + Z_R}{2}$$

- aligument

$$\frac{Y_L + Y_R}{2}$$

- cross level

$$\frac{Z_L - Z_R}{2a}$$

can be defined (fig. 1), whereas the alignment and the cross level influence the lateral dynamics and the vertical profile the vertical dynamics of the vehicle /3/.

Surface defects result from

- nonconstant rail profiles
- changes in rotational angle $\varphi_{x_{R,L}}$ of the rails and
- track gauge, variation $Y_L - Y_R$

which influence the nonlinear functions of contact geometry as

- difference of rollradii $\Delta r(u_y)$
- variation of contact angle $\Delta\delta\ (u_y)$
- running surface $\varphi x(u_y)$

and enter directly the system matrices.

So track irregularities are input for the investigation of the dynamic response behaviour of track guides vehicles, whereas surface defects influence already the stability of their motion. Therefore an evaluation of the running behaviour of such vehicles needs consequently the knowledge and the regard of track irregularities and surface defects /2/

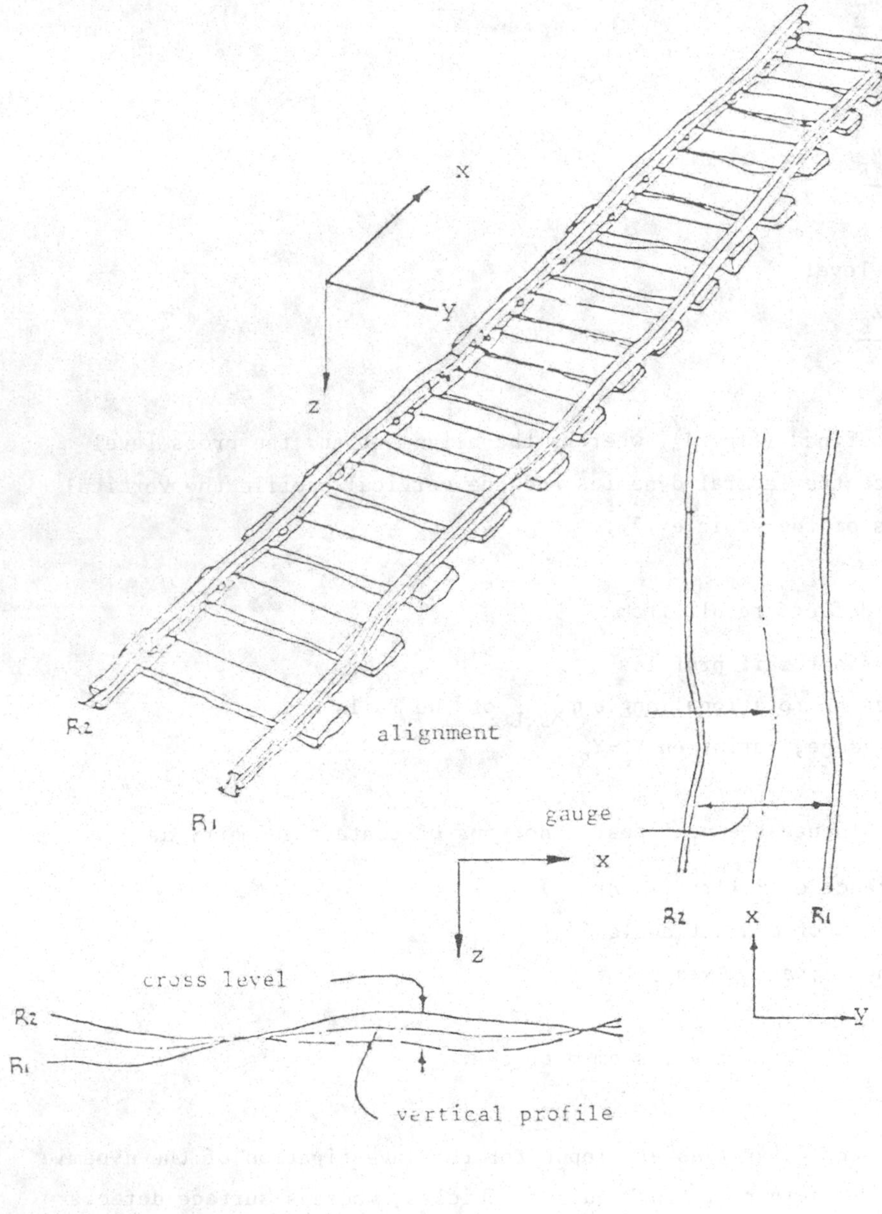

Fig. 1: Definition of track irregularities

Specification of positional tolerances of the track

A theoretical design of railway vehicles covering track irregularities needs measurements on tracks relative to an inertial basis. If the influence of surface defects is considered in a first assumption by sensitiveness analysis of contact geometry parameters, values of track irregularities have to be determined. Measured datas on an inertial basis of real tracks are quite rare at the time, but a lot of power density spectra is available for track irregularities in the literature. With in the theoretical investigations of the german high speed train the following analytical funktions have been used, for example /1/:

- Vertical profile

$$s_V\ (\Omega) = \frac{A_V \cdot \Omega_C^2}{(\Omega^2 + \Omega_r^2) \cdot (\Omega^2 + \Omega_c^2)}$$

- Alignment

$$S_A\ (\Omega) = \frac{A_A \cdot \Omega_C^2}{(\Omega^2 + \Omega_r^2) \cdot (\Omega^2 + \Omega_c^2)}$$

- Cross level

$$S_C\ (\Omega) = \frac{A_V/a^2\ \ \Omega_C^2 \cdot \Omega^2}{(\Omega^2 + \Omega_r^2) \cdot (\Omega^2 + \Omega_c^2) \cdot (\Omega^2 + \Omega_s^2)}$$

The following values are to be used in these equations

Ω_s = 0,4380 rad/m

Ω_c = 0,8246 rad/m

Ω_r = 0,0206 rad/m

and

$A_{V,A}$ = 5,9233 10^{-7} m rad
(lower level of interference)

$A_{V,A}$ = 1,5861 10^{-6} m rad
(higher level of interference)

a = 0,75 m

3. Evaluation procedure

For the design of running dynamics of wheel/rail-vehicles the calculation of relevant states of motion is necessary as an input into the evaluation procedure. In the case of the german high speed train with a design speed up to 350 km/h evaluation criterias have been chasen, which are described here as an example. Attaching evaluation criterias to the running states deliver an evaluation matrix for the considered vehicle (table 1) /1/. The criteria have still to be weighted, when the evaluation is in progress.

3.1 Eigenbehaviour

To determine travelling speeds which are relevant for the evaluation, even if they are below the design speed of 350 km/h, root locus plots of the vehicle are used.

Evaluation of speeds which are of interest is carried out using response criteria (particularly ride quality, acceleration and stroke).

The eigenbehaviour is covered for eigenfrequencies of 30 Hz max.

States of motion		Eigen-damping ν_{krit} Residual damping	Eigen-frequencies Eigen-vectors	Ride quality transv	Ride quality vert.	Force level trans.	Force level vert.	Stroke	Wear-Characteristics	Derailment safety
Motion on straights, eigenbehaviour	linear	X	X			X		X		
	non-linear	X	X	X		X		X		
Motion on straights, with track irregularities	linear			X	X	X		X		
	non-linear	X	X	X		X		X	X	
Motion in curves in same direction	quasistatic					X	X	X		
	dynamic without interferences	X	X							
Motion in opposite direction	quasistatic					X	X	X		
Motion on bridges	vertical dynamics				X		X	X		
Practicability of distored track										X
Summits trough	vertical dynamics							X		

Table 1: Matrix for evaluation of design with respect to running dynamics.

3.2 Evaluation of ride quality

Ride quality is evaluated by application of the "Wagenkastenlaufgüte" (wagon ride quality) in accordance with the W_Z-process. The point of reference for determination of the W_Z-values is on the floor of the wagon, in the centre above the bogie.

Lateral and vertical marginal values are fixed at

$$W_{z\ Grenz} = 3.0$$

Permissible maximum acceleration is fixed both laterally and vertically at 2.5 m/s.

The quality of emergency running conditions is not assessed.

The evaluation number W_Z is determined form the following equations:

- in the frequency domain

$$W_Z = \left(\sum_{i=1}^{n} (\bar{b}_i \cdot B_1\ (f_i))^2\right)^{0,15}$$

A PSD supplies that average acceleration value $\bar{b}_i$ which is associated with the supporting value of the frequency f_i and which is multiplied by the frequency-dependent evaluation factor $B_i(f_i)$.

- in the time domain

$$W_Z = \left(\frac{1}{T} \int_0^T (b(t) \cdot B)^2\ dt\right)^{0,15}$$

The frequency-dependent evaluation factor is taken into account by filtering the acceleration signal b (t).

3.3 Force level between wheel and rail

The lateral force level for running on irregulated track is evaluated by using ΣY of the individual wheels sts. The results are filtered with a break wave number of 64 Hz, for which the 24 m-value of the ΣY-forces is considered. The curve for this is assumed to have Gaussian distribution; this makes the average value equal to 0, the integral under the PSD is equal to the variance σ^2.

For the ΣY-forces, three times the standard deviation 3σ should amount to at most 2/3 of the Prud'homme marginal value:

$$3\sigma(\Sigma Y_{2m}) \leq 0.67\cdot 0{,}85\ (10 + 2Q_o/3)$$

This value should be taken as approximate.

For staggering and curving, the upper approximate values is to be used accordingly. To determine values for

- isolated gust
- 120 m curve
- counter curve

the full Prud'homme marginal value

$$\Sigma Y_{2m} \leq 0{,}85\ (10 + 2Q_o/3)$$

is used.

With respect to the permissible rail tension, the vertical wheel force (sum of static and dynamic wheel force) must not excees 160 kN.

3.4 Other criteria used in evaluation procedure

- Strokes

  Strokes are to be determined for the following states of motion:
  straight track with irregularities
  steady state in curving
  on bridge
  summits and troughs.

The vertical and lateral free play of the secondary suspension must not be used up.

The following permissible lateral amplitudes of wheelset motion in the track shall apply for linear calculations:

rail gauge 1: $\hat{y} \leqq \pm 3$ mm

rail gauge 2: $\hat{y} \leqq \pm 4$ mm

rail gauge 3: $\hat{y} \leqq \pm 5$ mm

- Wear characteristics

If necessary, a criterion for wear is used for a fine selection. This criterion represents the distribution of the friction power between wheel and rail over the section of the wheel profile, where y is the coordinate of the point of contact:

$$R(y) = \frac{v}{T} \int_{t=o}^{T} \sum_{i=1}^{3} (v_i \ T_i) \ (y(t)) \ dt$$

- Derailment safety

Derailment safety is to be proved to ORE B 55 (Report RP8).
This proof is also required for running on emergency suspension.

4. PRINCIPLES OF TRUCK DESIGN

4.1 Remarks on the Dynamics of Wheel/Rail-Vehicles

The track guiding concept of nowaday wheel/rail-technique is based mainly on rigid wheelsets, the profiles of which have variable conicity.

This basical concept of "cinematic guiding" described the first time 1984 by Klingel, also contains disadvantageous properties, which can result in severe problems: Besides specific resonance effects of carbody movement, already derivable from sole cinematics, the concept also includes a dynamic instability, the so called "huntig" behaviour. At operational conditions depending on vehicle speed and amplitude of wheelset motion unstable vibrations can appear. This phenomenon therefore has to be considered in more detail.

Within a linear investigation the critical speed is a characteristical criteria for evaluation of the running stability of a wheelset, a truck or a vehicle. The joined motion is a harmonic vibration. At low vehicle speeds the system is asymptotic stable, at higher speeds unstable, at least in a certain domain. At vehicle speeds below the critical speed motions, excited by external disturbances,are damped out whereas above critical speed the lowest extitation causes increasing vehicle vibration. In fact the amplitudes do not grow to invinite as linear theory supposes.

The nonlinearities of the system, as for instance the flange of the wheel are responsible that no derailment occurs even in a large range above linear stability limits. Now doubt that guiding changes under those conditions to weary and uncomfortable behaviour, but the derailment limit must not be reached either. A lot of questions concerning running behaviour therefore cannot be solved by linear theory, or its results may be taken only as a rough estimate, because it is valid only for small amplitudes. Especially the evaluation of the dynamic forces of vehicle and track needs the use of nonlinear theory.

The nonlinear stability behaviour can be evaluated by domains of limit cycles.

A stable limit cycle is present, if the simulation of time history results in a periodic motion. A stable limit cycle governs a domain. If in a simulation the initial conditions are within this domain the limit cycle does not depend from the initial conditions itself. The transient behaviour into a stable limit cycle therefore can be presented by increasing or decreasing amplitudes, dependent on the chosen initial conditions.

An unstable limit cycle may be determined within a simulation by studying neighboured transient solutions. The unstable limit cycle borders in these cases the domain of two stable limit cycles or of one stable limit cycle and a steady state solution with amplitude zero.

The equations of motion of wheelsets include the vehicle speed as a parameter. Numerical results of limit cycle calculations in wheel/rail-systems therefore often use the amplitude of the limit cycle upon vehicle speed. In such figures the linear stability border is given by a vertical line at the value of the critical speed.

If in the surrounding of an operational condition of the wheel/rail system (for instance the central position of the wheelset within the track) only a weak nonlinearity exists, then the dynamic behaviour gets closer to the linear solution with decreasing amplitude of the limit cycle. From this follows, that the linear stability border is the starting point of a stable or unstable section of limit cycle amplitudes.

Fig. 2 gives a general view on possible limit cycle amplitude dependent on vehicle speed of a rigig wheelset on a rigig track; the wheelset elastically coupled to the inertial system laterally and vertically. The lateral displacement of the wheelset up to flange contact is about ± 5 mm. The stable section of the limit cycle amplitude leads with increasing vehicle speed to the amplitude area of flange contact.

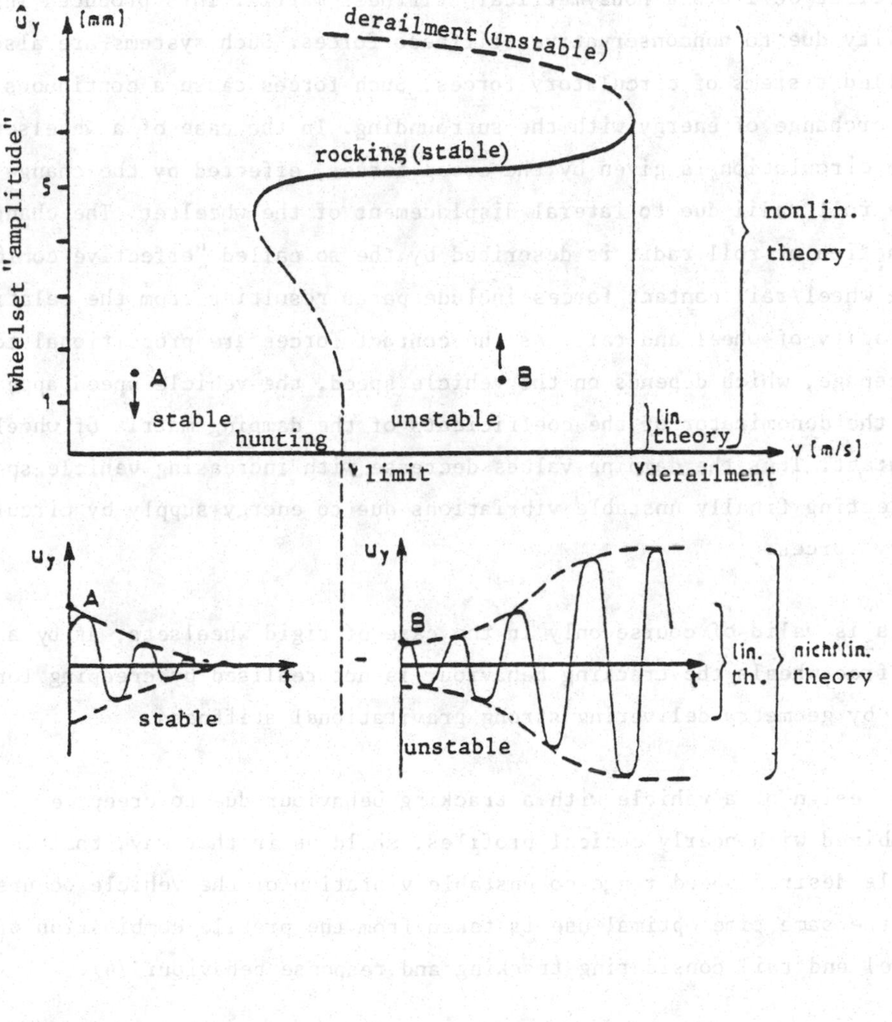

Fig. 2: General behaviour of wheelset amplitudes (limit cycles) up to derailment speed

The principle ability of instability is yet not based on the nonlinearity of the system, but remains also in linear modelling.

The linearized homogenious set of differential equations of motion of a wheelset delivers a nonsymetrical stiffness matrix. This produces instability due to nonconservative stiffness forces. Such systems are also called systems of circulatory forces. Such forces cause a continuous interchange of energy with the surrounding. In the case of a wheelset the circulation is given by the creep forces, effected by the change of the roll radii due to lateral displacement of the wheelset. The change function of roll radii is described by the so called "effective conicity" The wheel/rail contact forces include parts resulting from the relative velocity of wheel and rail. As the contact forces are proportional to creepage, which depends on the vehicle speed, the vehicle speed appears in the denominator of the coefficients of the damping matrix of wheel/rail contact. Thus the damping values decrease with increasing vehicle speed, effecting finally unstable vibrations due to energy supply by circulatory forces.

This is valid of course only in the case of rigid wheelsets, as by a set of free wheels the tracking behaviour is not realised by creeping forces, but by geometry delivering strong gravitational stiffness.

The design of a vehicle with a tracking behaviour due to creepage combined with nearly conical profiles, shold be in that way, that in the whole desired speed range no unstable vibration of the vehicle occurs and at the same time optimal use is taken from the profile combination of wheel and rail considering tracking and response behaviour /4/.

4.2 Linear Eigenbehaviour

The critical speed of a truck, coupled to an inertial system by secundary suspension, forms a stability mountain system, if it is plotted as a function of the primary suspension (fig. 3).

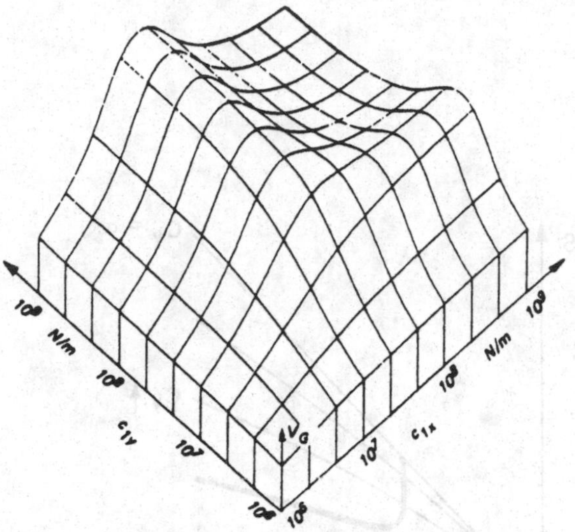

Fig. 3: Stability mountain system

A transformation of the top values, that is to say of the optimal critical speed stiffness values, which are connected as following

$$c_S = \frac{d^2/_4\, c_x\, c_y}{a^2/_4 c_y + d^2/_4 c_x}$$

d = supporting basis of the axlebearing connection

a = distance of the axles

$$c_B = \frac{d^2}{4} \qquad c_x$$

delivers the relation as shown in fig. 4

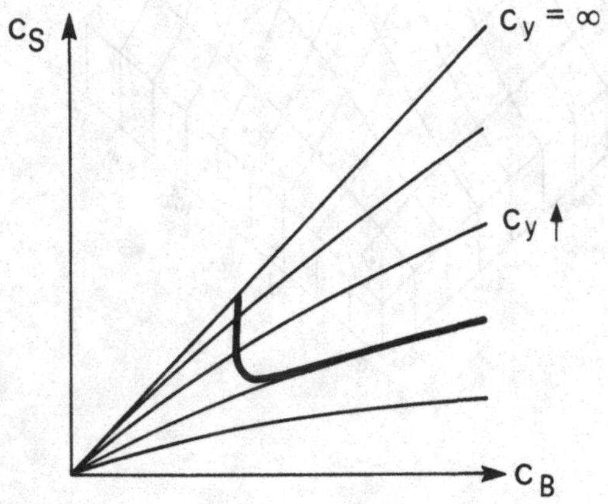

Fig. 4: Top line of a conventional truck

It is evident, that not all the value couples of c_S, c_B can be reached, having a conventional suspension. Therefore it is necessary to use additional coupling between the wheelsets, by which the optimal stiffness values of the whole can be designed as shown in fig. 5.

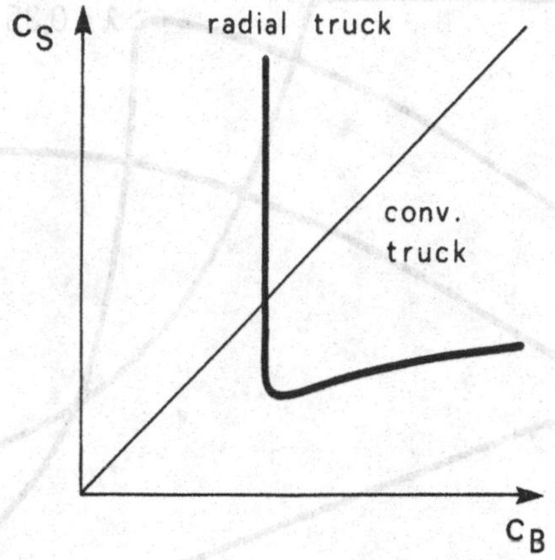

Fig. 5: Top line of radial truck

This top line, that is to say the "crown" of the stability mountain system mentioned before, moves depending on the basical geometry of the wheel/rail profiles.

If the critical for different gauges are calculated and for each of these examples a projection of the stability mountain system is shown, then the optimal combinations of sheering stiffnesses c_S and of bending stiffnesses c_B can be found out for all the conicity values /5/.

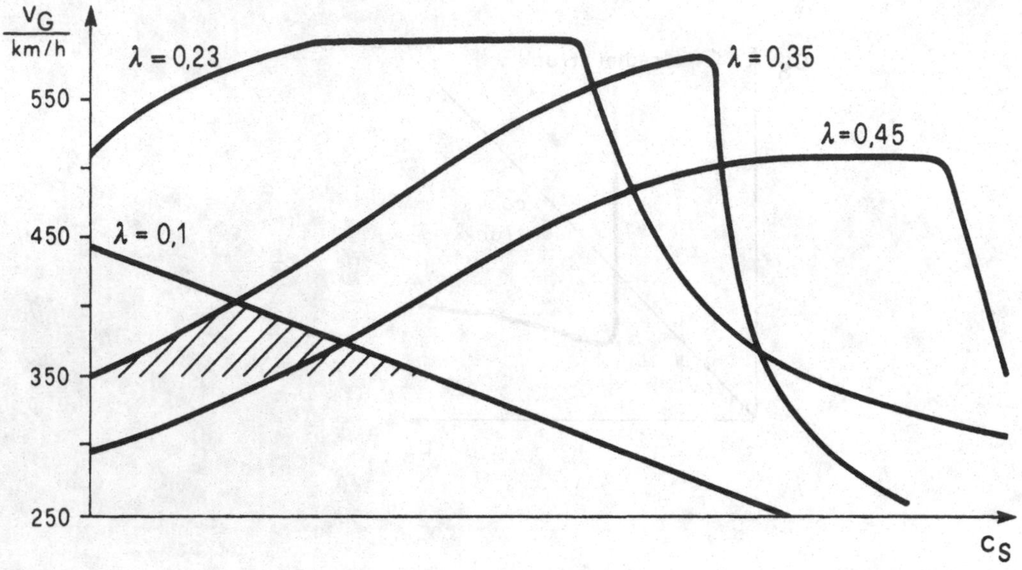

Fig. 6: Critical speed dependent on coupling stiffnesses with different conicities at c_S/c_B = const.

4.3 Ride quality

The essential result of linear ride quality calculation (horizontal and vertical) is: The lower the secondary suspension is tuned, the better is the ride quality, reasonable damping values supposed. The primary suspension is of lower influence. In reality it is not always possible to tune as low as one desires. There exist amplitude limitations vertical as well as horizontal: Nonlinear influences therefore become effectful /6/.

The best possible ride quality will be reached by the optimisation of the vibration isolation of the car body in connection to the trucks. An optimal approach of the secondary suspension, regarding the marginal design conditions, implicates renouncing at running stability-increasing measures on secondary level. An optimisation of interconnections in the truck design concerning the running behaviour should therefore be limited on primary level /5/.

Consequently the problems of curving dynamics and of the actual design as well as ceeping constant the defined parameters of hunting dampers at operational conditions is avcided.

4.4 Steady State Curving

Analysing the top line points of equal values concerning running stability in the c_S, c_B-diagramme, a different curving behaviour depending on the curve radius can be seen. The steady state curving can be evaluated on following items

- sum of lateral forces Σ Y acting between wheel and rail
- the actual hunting angle φ_Z of the forward wheelset
- the resulting maximal longitudial wheel force

The maximum value of Σ Y at the top line is shown in fig. 7 taken into consideration the curve radii between 190 m and 7200 m.

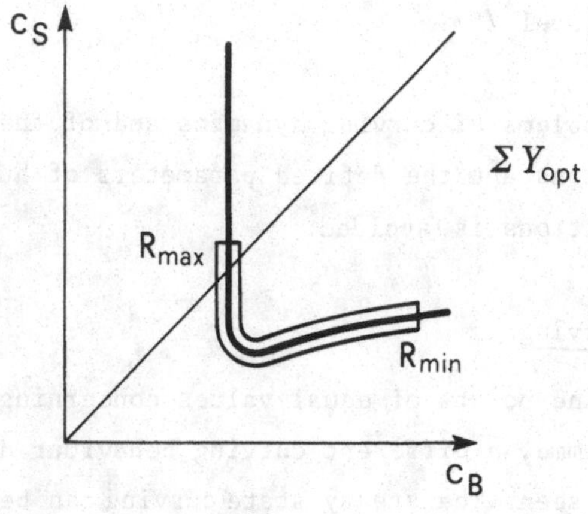

Fig. 7: Minimum of lateral forces as a function of the top line values and the curve radius

Above a radius of 3000 m the lateral forces change into a minimum in the transition area between conventional and radial truck design.
The hunting angle φ_Z of the forward wheelset compared with the ideal radial position depends also on the curve radius for the different c_S, c_B-combinations and turns also into a minimum (fig. 8).

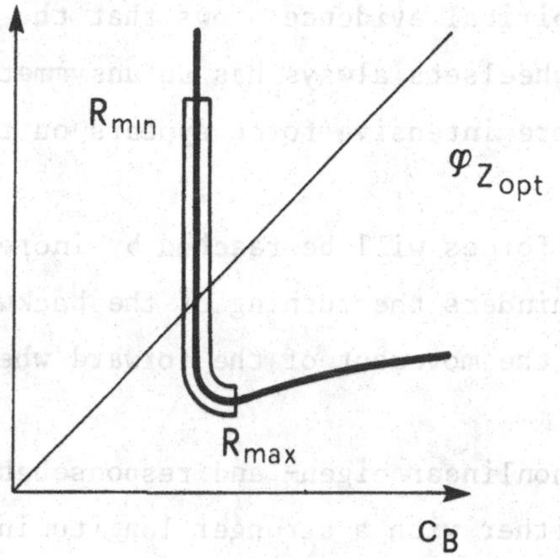

Fig. 8: Minimum of hunting angle φ_Z as a function of the top line values and the curve radii

Considering these evaluation criterias a truck for high speeds should therefore be designed with the stiffnesses c_S and c_B at the border between conventional and radial coupling /5/.

4.5 Nonlinear Eigen- and Responsebehaviour

The nonlinear stability behaviour is to be evaluated according to limit cycles.

A calculation of the limit cycles for a truck with a design of the sheer- and bending stiffness in the transition area between conventional and radial coupling shows periodical movements with small amplitudes $u_y < 1$ mm.

This stable limit cycle does not appear in a design with higher bending stiffness any more.

To evaluate the responsebehaviour due to track irregularities the level of the lateral forces of the forward and the backward wheelset is taken into account. The empirical evidence shows that the radial truck due to the coupling of the wheelsets always has an unsymmetrical force distribution, by which the more intensive force appears on the backward wheelsets.

A reduction of these forces will be reached by increasing the bending stiffness c_B, which hinders the turning of the backward wheelset due to stronger coupling to the movement of the forward wheelset.

The analysis of the nonlinear eigen- and responsebehaviour recommends therefore a design either with a stronger longitudinal coupling of the wheelsets to the truck frame or to increase the bending stiffness of the wheelset coupling.

4.6 Determination of the design values

As conclusions can be stated

- the application of coupling between wheelsets permits a truck design which does not require a hunting damper to reach high speed level
- concerning the attainable critical speed optimal combinations of sheer- and bending stiffness can be yield
- the analysis of curving behaviour recommends a design with the lowest possible values of sheer- and bending stiffnesses
- to reach the optimal design according to nonlinear eigen- and response behaviour mainly high bending stiffness values are suggested.

So a compromise between optimal curving behaviour and nonlinear response behaviour is required.

5. CONVENTIONAL TRUCKS

The use of conventional trucks in high speed vehicles is only possible with the aid of measures to increase the running stability, such as hunting dampers between truck and carbody.

Hunting stabilisation consists of a moment coupling between the truck and the carbody with reference to the vertical axis. Various rotary stabilizing concepts are known in wheel-rail technology.

- rotational restraint

  The basic idea lying behind a restriction on rotation is to restrain the position of the truck in relation to the carbody as far as its tendency to turn out round the vertical axis is concerned. To ensure curve negotiation, this turning moment restriction is subject to an upper limit. In other words, when the uppermoment limit is exceeded, the truck can be turned. The most familiar hardware using this rotational restriction principle comprises the bolster beam type of truck design with bolster guides free from longitudinal play and the carbody supported on the bolster by means of sliding pads. The maximum turning moment is thus governed by the weight of the carbody, the friction coefficient and the support base width. On non-bolster truck designs an equivalent degree of rotational restraint can be achieved by locating hydraulic cylinders on either side of the truck to stiffen up as the truck attempts to turn. In such cases the turning moment is limited by pressure control valves.

On both of the design principles stated here, material elasticity prevents the truck from being held absolutely rigidly.

From the operating point of view, therefore, restrictions on rotation are to be regarded as a resilient restrain on the truck. Below the maximum turning moments very high stiffness values are encountered.

- Damping rotary movement

  Rotary damping, designed to eliminate hunting action, aims to damp out the turning movements of the truck. This is achieved, for example, in wheel-rail technology by positioning two seperate hydraulic dampers between the truck and the carbody. In practice, the efficacy of rotatory-movement damping begins to be demonstrated only at very high damping constants, so that special dampers designs are needed.

If we examine actual truck rotary damping system techniques, we find that the damper pivot mounts and the damper itself are never free from elasticity. In real terms, the damper must always be interpreted as a series circuit comprising spring and damper. The elasticity allowance is calculated from the individual elasticities arranged in series. The following factors contribute individual elasticities: compressiblity of the damping medium, elasticity of rubber seals, damper housing and piston rod, connecting lugs, rubber bushings, connecting pins, pivot brackets and mounts /9/.

It could be shown that it is worth striving to achieve a degree of pivot stiffness. However, the decisive gain is in the ability to achieve lower damping constants. These enable the vehicle's curve negotiating abilities to be improved, and the moment caused by the truck turning is dissipated more rapidly.

In fig. 9 as an example a truck is shown including this design concept of a rotational restraint.

Besides an improved tuning of primary and secondary suspension in the cased of powered trucks also the motor suspension can be taken into account. Based on a modern high speed train of the Deutsche Bundesbahn, in /8/ are presented the possibilities to improve the running behaviour of powered trucks by the use of laterally elastically suspended motors.

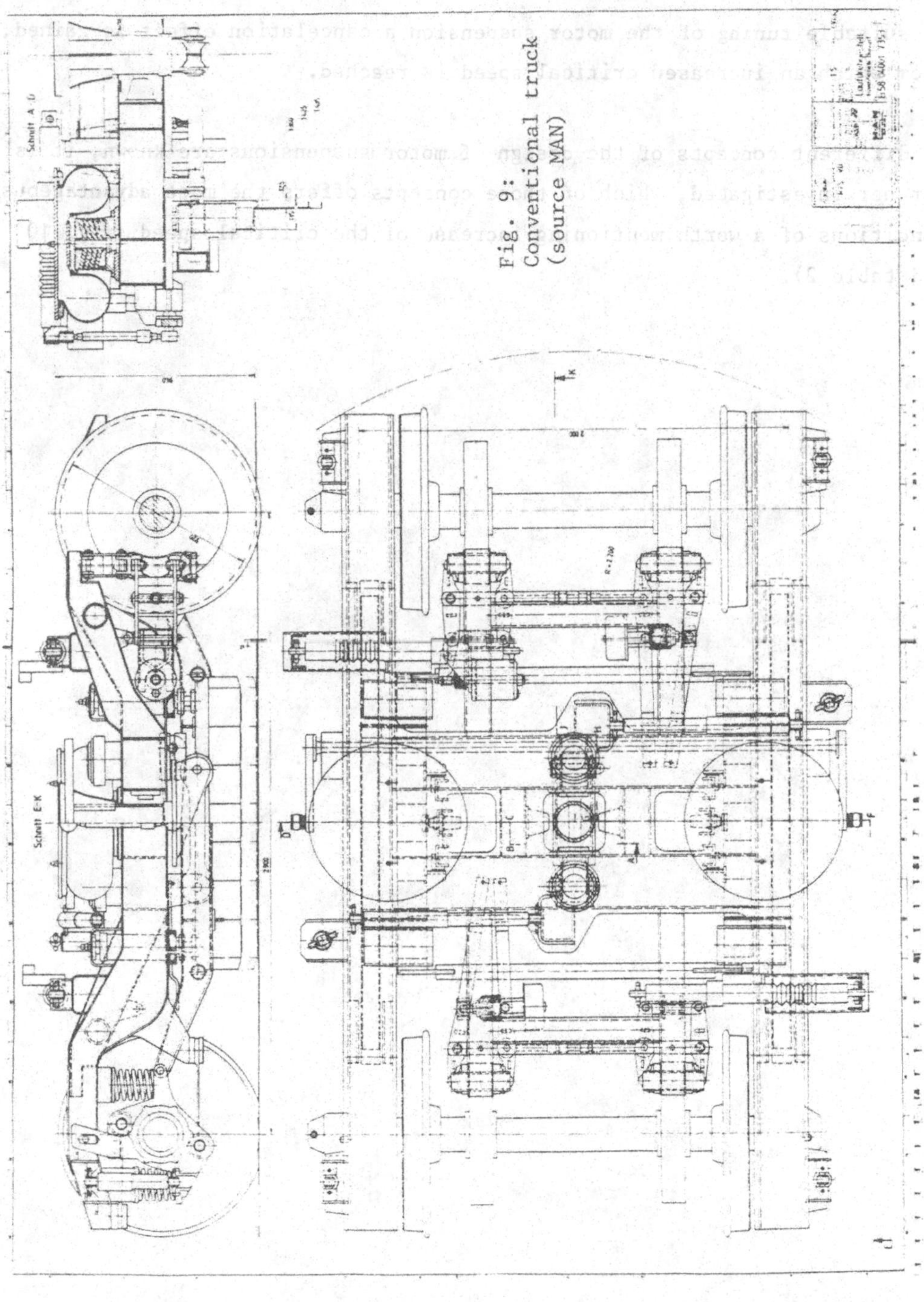

Fig. 9: Conventional truck (source MAN)

By suitable tuning of the motor suspension a cancelation effect is gained, from which an increased critical speed is reached.

As different concepts of the design of motor suspensions are known, it is further investigated, which of those concepts offers the most advantageous conditions of a worth mentioning increase of the critical speed (fig. 10 and table 2).

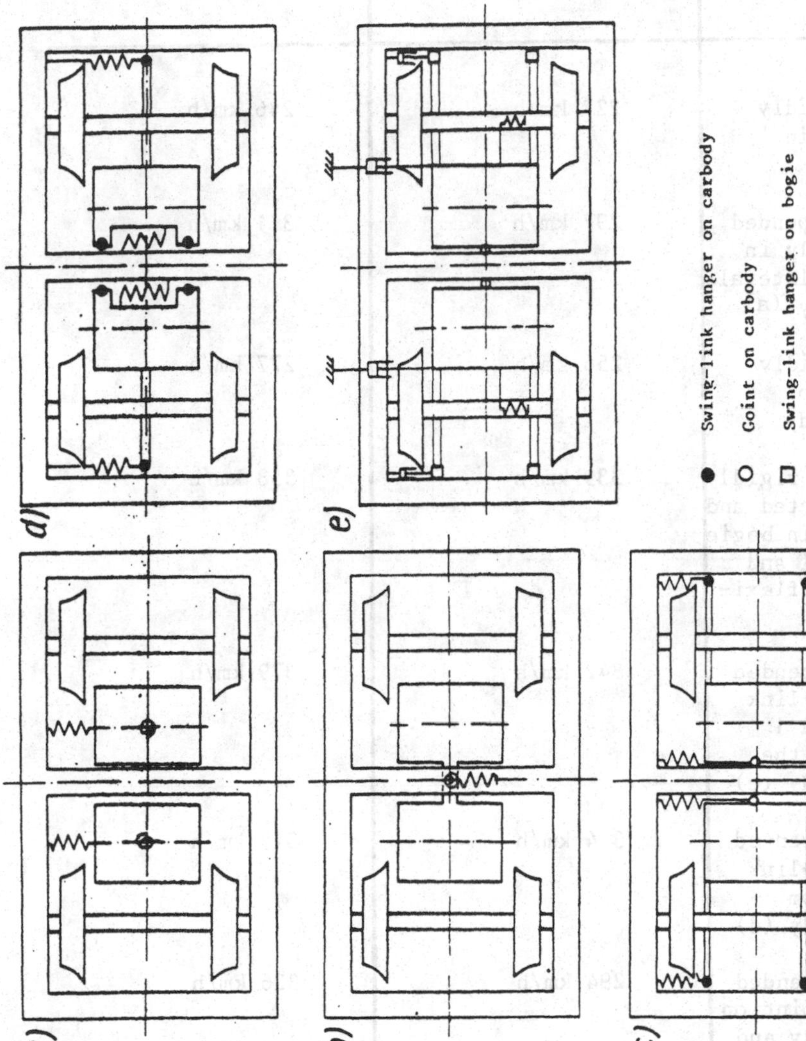

Fig. 10: Concepts of motor suspensions in powered trucks /8/

Concept of motor suspension	primary stiffnesses $c_{1x} = 2.154*10^7$ N/m, $c_{1y} = 1.0*10^7$ N/m	$c_{1x} = 2.154*10^7$ N/m, $c_{1y} = 2.154*10^7$ N/m
Motors rigidly installed in bogie	227 km/h	246 km/h
Motors suspended individually in bogie for lateral flexibility (a)	297 km/h	311 km/h
Motors rigidly attached to vehicle body	256 km/h	277 km/h
Motor pair rigidly interconnected and suspended in bogie for lateral and rotational flexibility (b)	339 km/h	358 km/h
Motors suspended by 2 swing-link hangers and 1 joint from the vehicle body (c)	342 km/h	379 km/h
Motors suspended by 3 swing-link hangers from vehicle body (d)	314 km/h	343 km/h
Motors suspended from one joint on vehicle body and two swing-link hangers on bogie (e)	294 km/h	326 km/h

Table 2: Reachable critical speeds of different motor suspension concepts

6. RADIAL TRUCK

In a radial truck additional couplings are introduced between the wheelsets realizing certain bending and sheer stiffnesses c_B and c_S for stable running behaviour to an upper speed range. In fig. 11 the principle difference in the design of conventional and radial trucks is shown as the influence of additional wheelset couplings on critical speed, the speed at the stability border.

On the basis of the design recommendations given in chapter 2 the actual design (fig. 12) is chosen quite near to the top line presenting the optimal critical speed behaviour, close to the border of conventional and radial coupling, with slight increase of the bending stiffness c_B. The primary suspensions are designed as low as possible. The additional sheer- and bending stiffness is realized by using an elastic supplementary frame coupled to the axle bearing as a part of the primary suspension /6/.

In fig. 13 a survey (sensibility analysis) is given of the critical speed dependent on the primary suspension of the wheelsets within different values of sheer- and bending stiffnesses as stability mountain systems. For $c_S = c_B = 0$ the well-known stability behaviour of a conventional truck is presented. The "stability mountain" at optimal combinations of primary stiffnesses reaches in this case about 400 km/h.

Due to the supplementary frame of the radial truck the critical speeds are increased, so that they can be found all over 400 km/h independent of the actual combination of the primary stiffness.

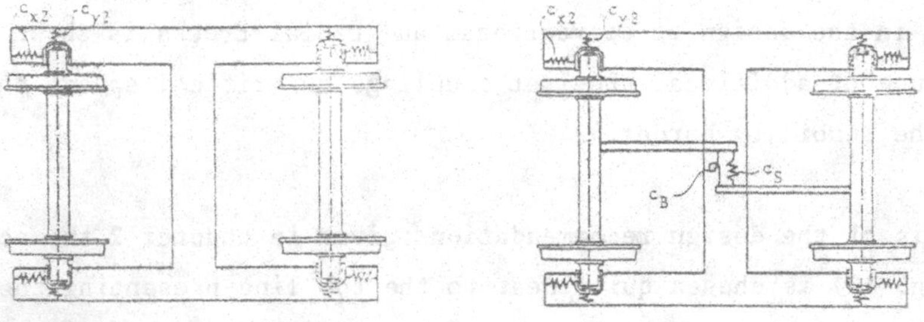

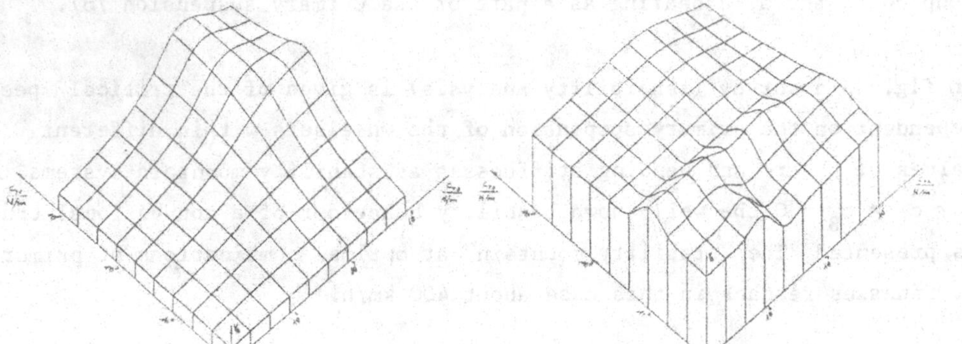

Fig. 11 A) Conventional truck B) Radial truck

Critical speed as a function of stiffnesses longitudinal (c_{x2}) and lateral (c_{ys}) between wheelsets and bogie frame

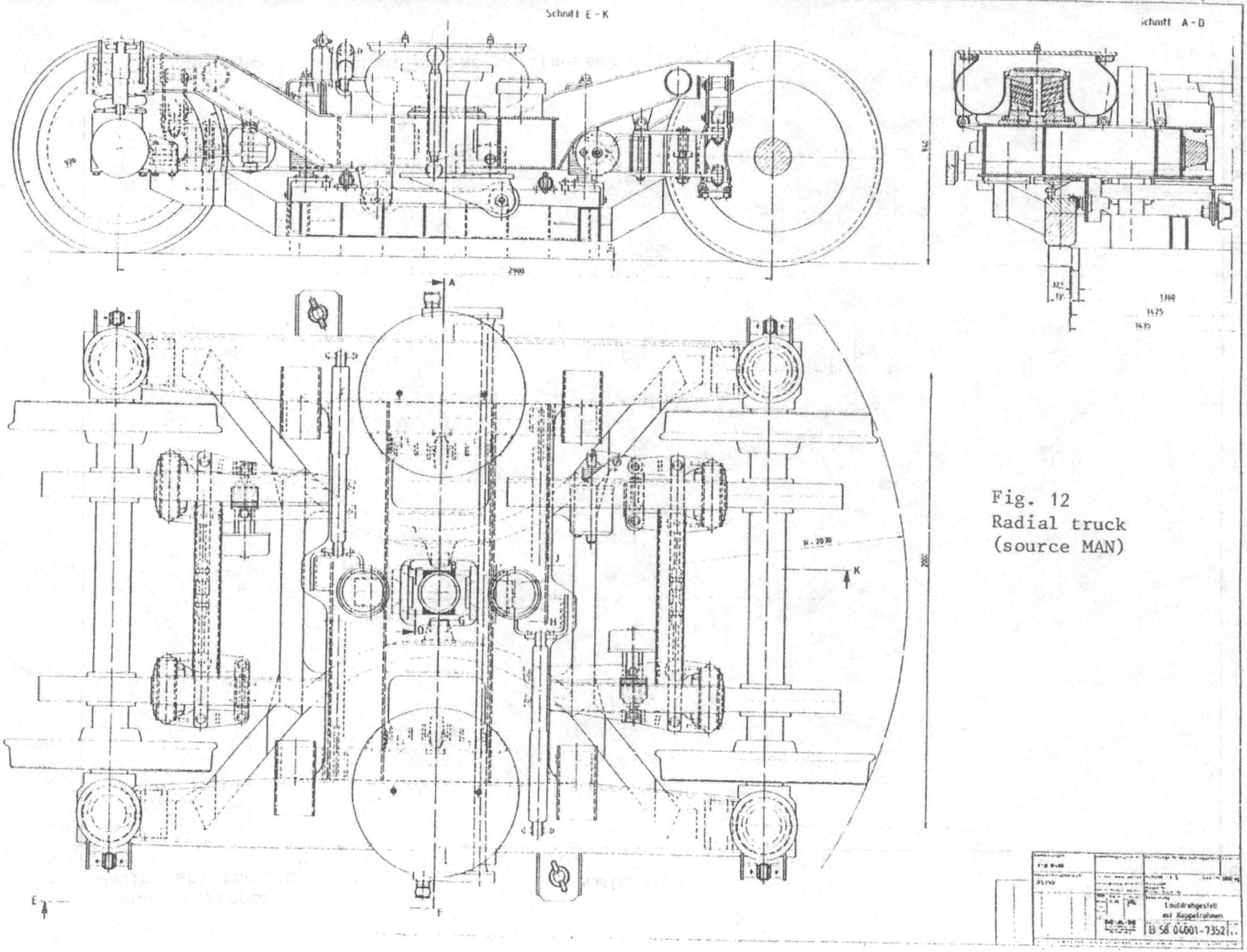

Fig. 12
Radial truck
(source MAN)

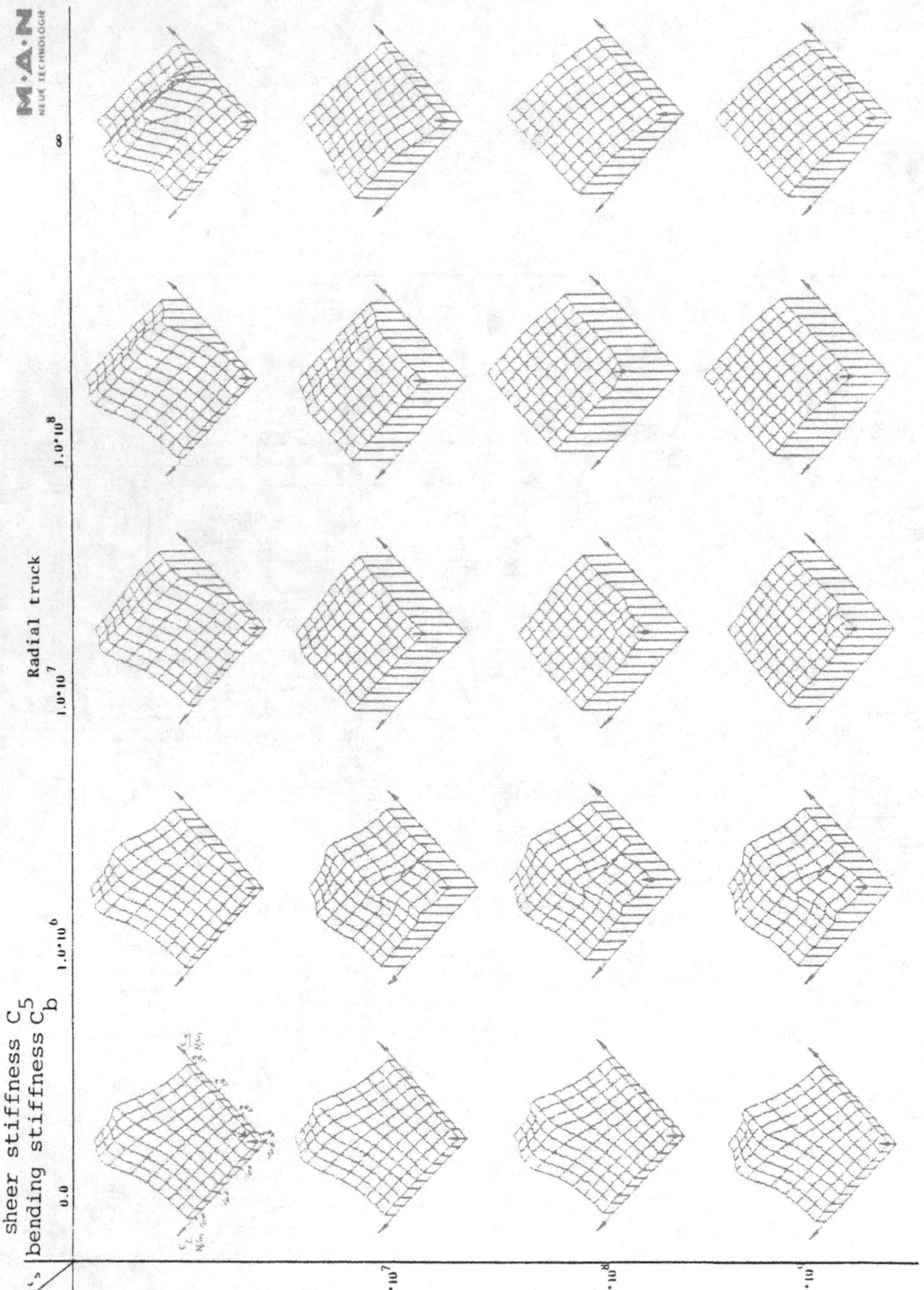

Fig. 13: Critical speed dependent on the stiffnesses c_x, c_y, c_s, c_b

7. ACTIVE COMPONENTS FOR WHEEL/RAIL-SYSTEMS

The application of active components for wheel/rail-systems (especially for future systems of high-speed trains) should be seen under the aspects

- improvement of ride quality
- stabil running behaviour at high speeds
- good curving performance
- reduction of wheel/rail tread wear

One of the possible applications is the active rotational restraint, developed by MAN. It operates between carbody and the bogies of wheel/rail-vehicles and provides, among other advantages, a better stability in retention with a good curving performance.

7.1 Active rotational restraint

The system of the active rotational restraint is consisting of three main components:

- sensors
- electronic regulator
- hydraulic actuator

Basing on the measurement of the sensors the regulator computes a signal, which the electro-hydraulic servo valve in connection with the actuator converts into a force. This force is introduced into the bogies on the one hand and the car body on the other hand and stabilizes the running behaviour of the vehicle (fig. 14).
Redundant systems and electronic safty-circuits guarantee the safe function of the active rotational restraint in case of failure or bad function of components /9/.

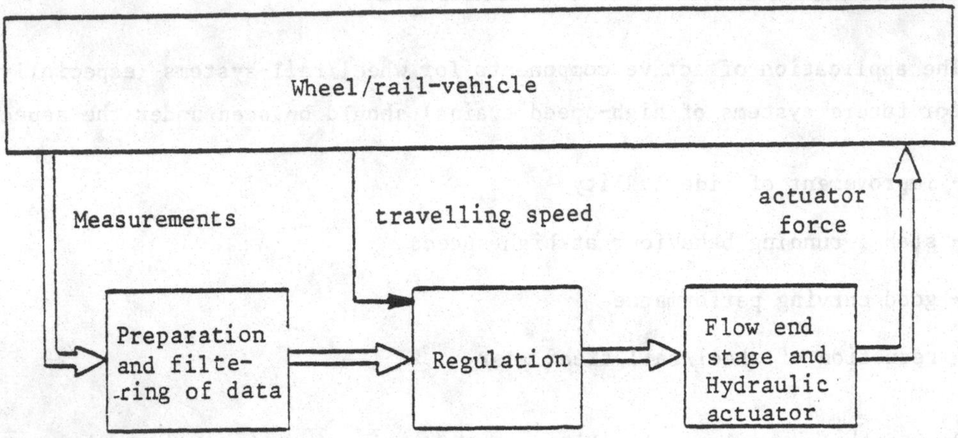

Fig. 14: Structure of the wheel/rail-system with active rotational restraint

Design of Regulator

The regulator is the main component of the active system. Therefore the design of the structure and calculation of parameters of the regulator are very important, as it has a great influence on the dynamic behaviour of the vehicle.

One way to evaluate the effect of the active rotational restraint is by looking at the root loci of the wheel/rail-system. Fig. 15 illustrates the improvement which can be achieved by using the active system (Curve 1: Wheel/rail-system without active control; Curve 2: System with active control); the unstable eigenvalues of the system move to the left side of the y-axis (region of stable dynamic behaviour) by implementing the active rotational restraint. Furthermore, it is possible to draw all eigenvalues to the left, until they show a damping more than 10 % (over the entire speed range).

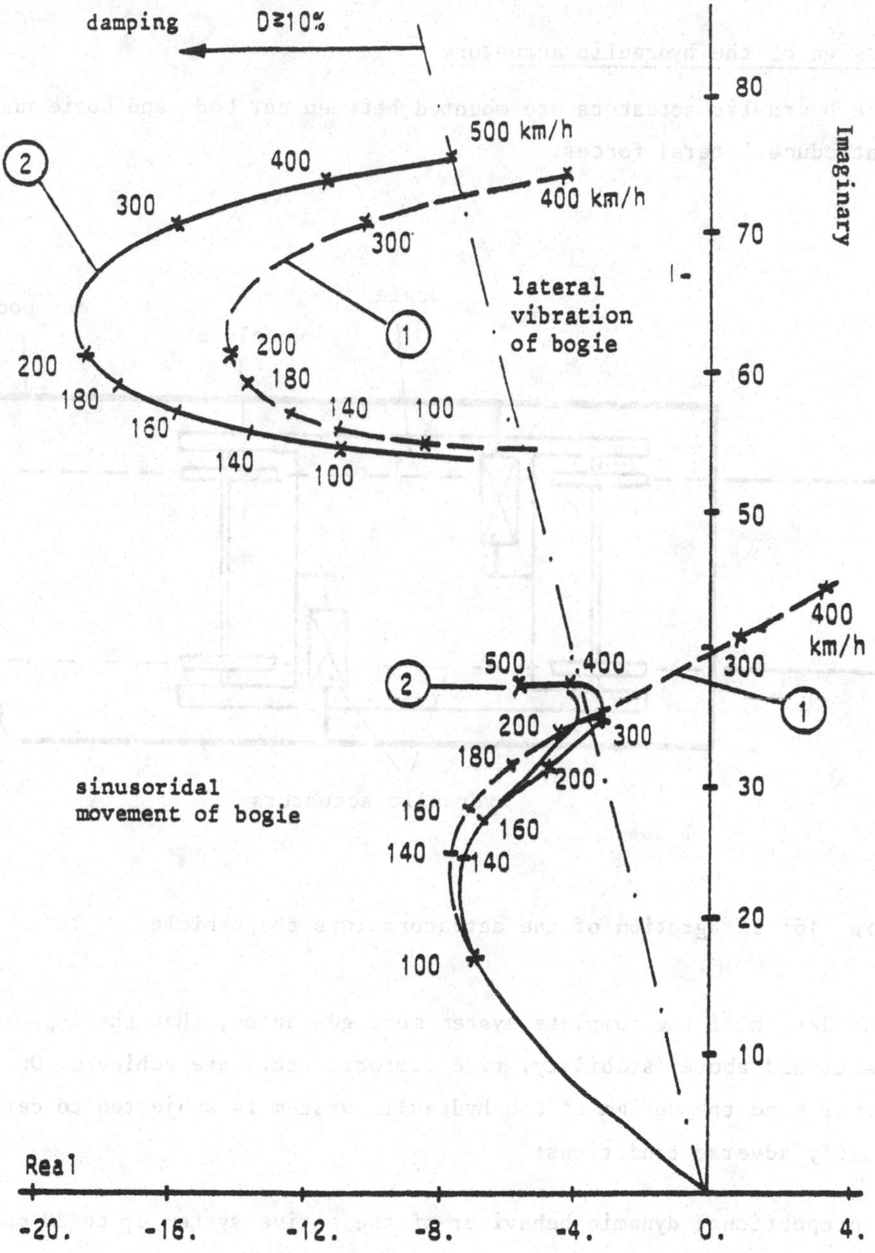

Fig. 15: Root loci for a wheel/rail-system

Curve 1: without active control

Curve 2: with active control

Design of the hydraulic actuators

The hydraulic actuators are mounted between car body and bogie and introduce lateral forces.

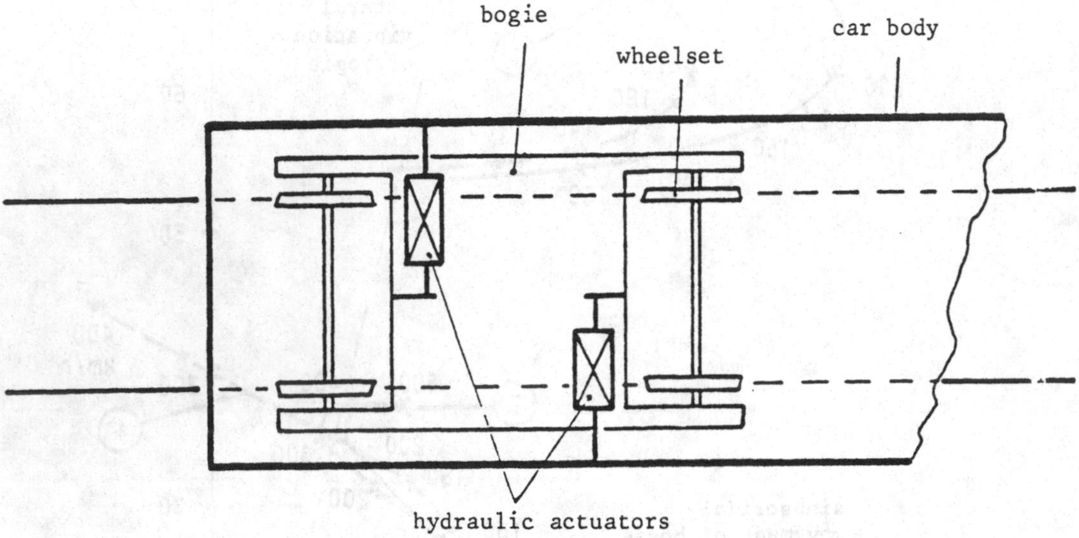

Fig. 16: Integration of the actuators into the vehicle

The design of the complete system must guarantee, that the improvements mentioned above (stability, ride comfort, etc.) are achieved. On the other hand the design of the hydraulic system is subjected to certain, partly adverse conditions:

- proportional dynamic behaviour of the active system up to 30 cps
- force of the actuators up to 50 kN
- solid construction of all components in regard to environment in the vehicle (water, dust, heat, etc.)

- high reliability in connection with a fail-safe behaviour, which presents from hazardous events in case of failure or bad function of components
- restricted space for fitting in the actuators into the vehicle

Fig. 17 shows the hydraulic actuator (source of force) designed for integration in wheel/rail-vehicles.

Safety engineering

Failures in operation of the active rotational restraint can be hazardous to persons and property, expecially when occuring at high travelling speed. Therefore the design of the system has to stick to the following safety criterias drawn up for track-guided traffic systems:

- safety assurance in case of single failures
- independence of various, partly redundant components in order to prevent simultanous malfunction
- quick failure detection for minimizing risk probability in case of multiple failures

In order to enable continuous consideration of safety and availability apsects in the design phase as well as in the implementation phase the system is checked by qualitative and quantitative analyses (eg. failure mode and effects analysis, computer-aided evaluation of BOOLE and MARKOV probability models).

The result is a safety concept based on redundancy and fail-safe strategies, which satisfies the safety requirements. The sensors are implemented twice (2-out-of-2 redundancy), whereas the regulator and the flow end stage are arranged in three channels with two analog and one digital computers (hybrid 2-out-of-3 redundancy).

Thy hydraulic actuator itself is fail-safe, as it works as a passive damper by opening a bypass. The coefficient of damping can be varied in order to guarantee a stable dynamic behaviour at maximum speed. However, in this case the performance in the other criterias (ride comfort, wheel/rail tread wear, curving) will be reduced.

Furthermore, several measurements are drawn out of the hydraulic and electronic system and put to a digital computer for checking the system.

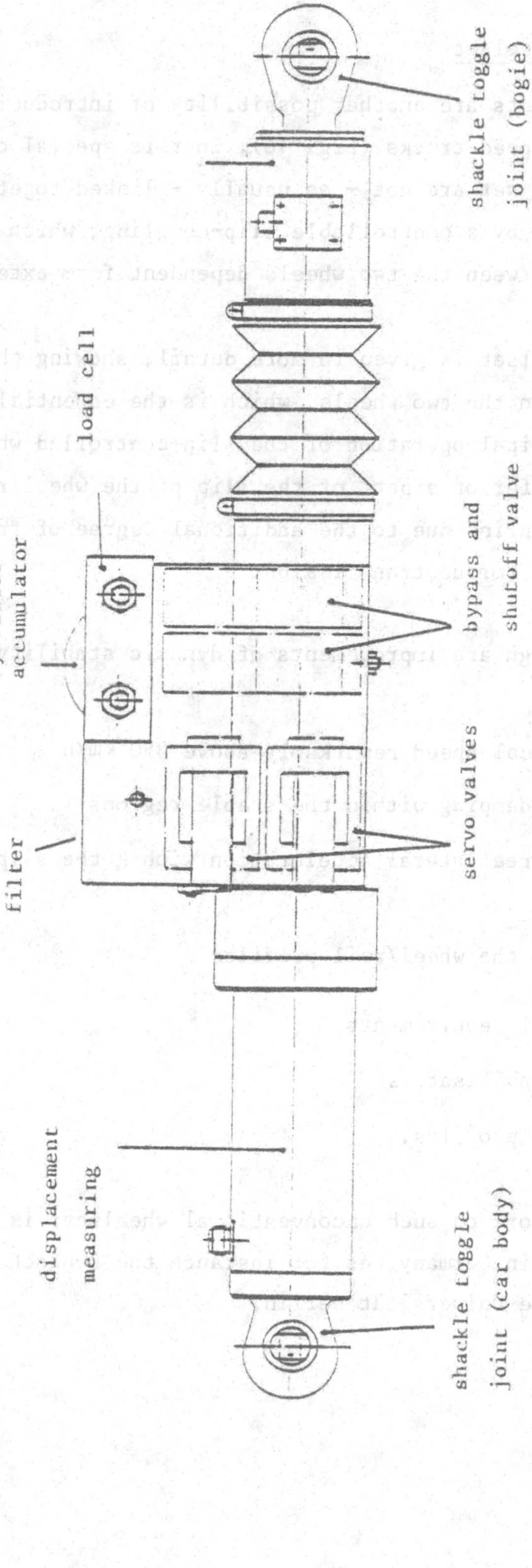

Fig. 17: Hydraulic actuator of the active rotational restraint (source MAN)

7.2 Slip controlled wheelset

Slip controlled wheelsets are another possibility of introducing active components into high speed trucks (fig. 18). In this special configuration the two wheels of each set are not - as usually - linked together by a nearly rigid axle, but by a controllable slip-coupling, which permits a torque transmission between the two wheels dependent form external signals.

In fig. 19 such a wheelset is given in more detail, showing the magnetic powder coupling between the two wheels, which is the essential feature of this design. The principal operation of the slip-controlled wheelset depends upon the transfer of a part of the slip of the wheel/rail contact area into the slip-coupling due to the additional degree of freedom of rotation by controlled torque transmission.

The goals of this design are improvements of dynamic stability and guidance behaviour by

- increasing the critical speed remarkably above 350 km/h
- improvement of eigendamping within the stable regions
- acceptance of more free lateral acceleration within the slip boundaries
- reduction of wear of the wheel/rail profiles

including as additional requirements

- omission of other stabilisators
- use of standard wear profiles.

At least theoretical work on such unconventional wheelsets is done also at other institutions in Germany, as for instance the Deutsche Bundesbahn and the Technische Universität Berlin.

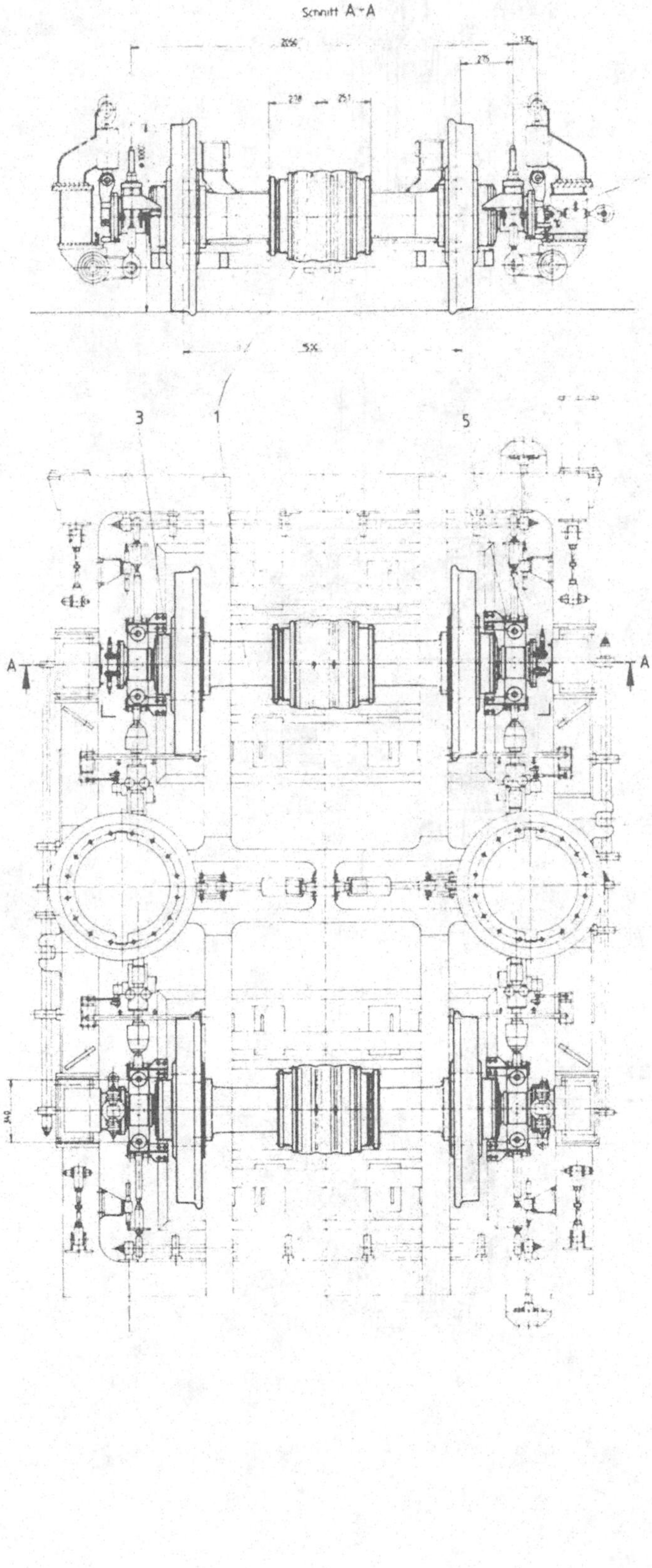

Fig. 18: Truck design including slip-controlled wheelsets (source MBB)

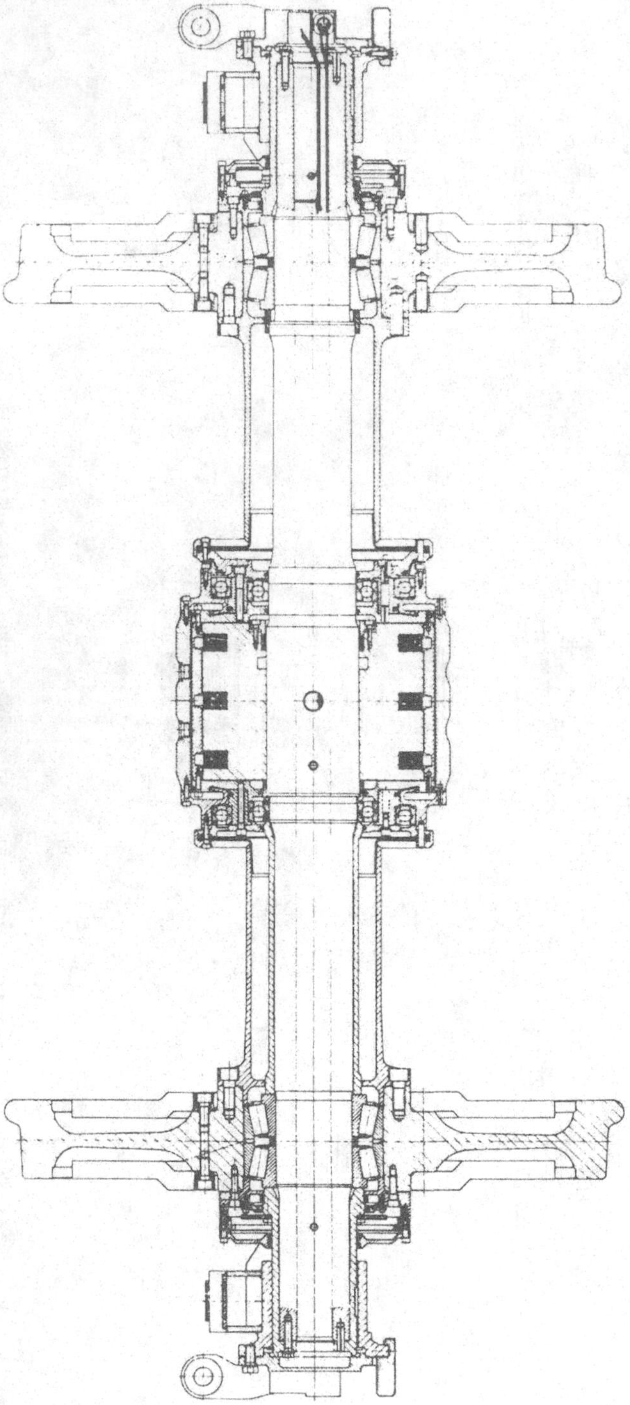

Fig. 19: Slip-controlled wheelset (source MBB)

7.3 ROLLAB-Unversal Railway Bogie

Another application of active components in wheel/rail-systems is the Universal Railway Bogie, developed by ROLLAB. The main design features are shown in fig. 20.

Liquid-Gas-Suspenion

With this type of suspension a considerable improvement of the suspension properties can be achieved as compared to conventional spring-damping system. It is thus possible to change during operation, the characteristics of the system with regard to damping and stiffness and accommodate it to values expendient not only to the rail conditions but also to the car load. The control of these stiffness-damping properties is made by means of valves (throttling) and change of the gas volume in the system.

Automatic System for Powered Banking

Specical Control Systems are cooperating with the lateral control system (see before) in order to accomodate in each situation the attitute (tilting and position) of the carriage to correct values.

Automatic Control of Height

The carriages can be kept at a constant predetermined height over the rails (independent of the loading).

Axles line up with the curve radius

As a consequence of an active and automatic adjustment of the axles to focus at the centre of the curve the wear on wheels and rail will be reduced and the running characteristics of the train will be improved.

Lateral Control of Carriages

In curves and in combination with tilting of carriages it is also an advantage to be able to control the carriages laterally. This control system can be applied in other situations as well.

Active Rail Brake

During braking the cinematic energy can be converted and fed back to the mains. It can also be converted into heat or mechanical rotary energy. As an alternative railbrakes can be used as active brakes during normal as well as under emergency operations. A third braking system could be a disc brake. The bogie could be driven by hydraulic motors, which could be operated as pumps for braking purposes.

In principle the technology for a system of this kind has already been solved and part solutions have already been verified by experiments.

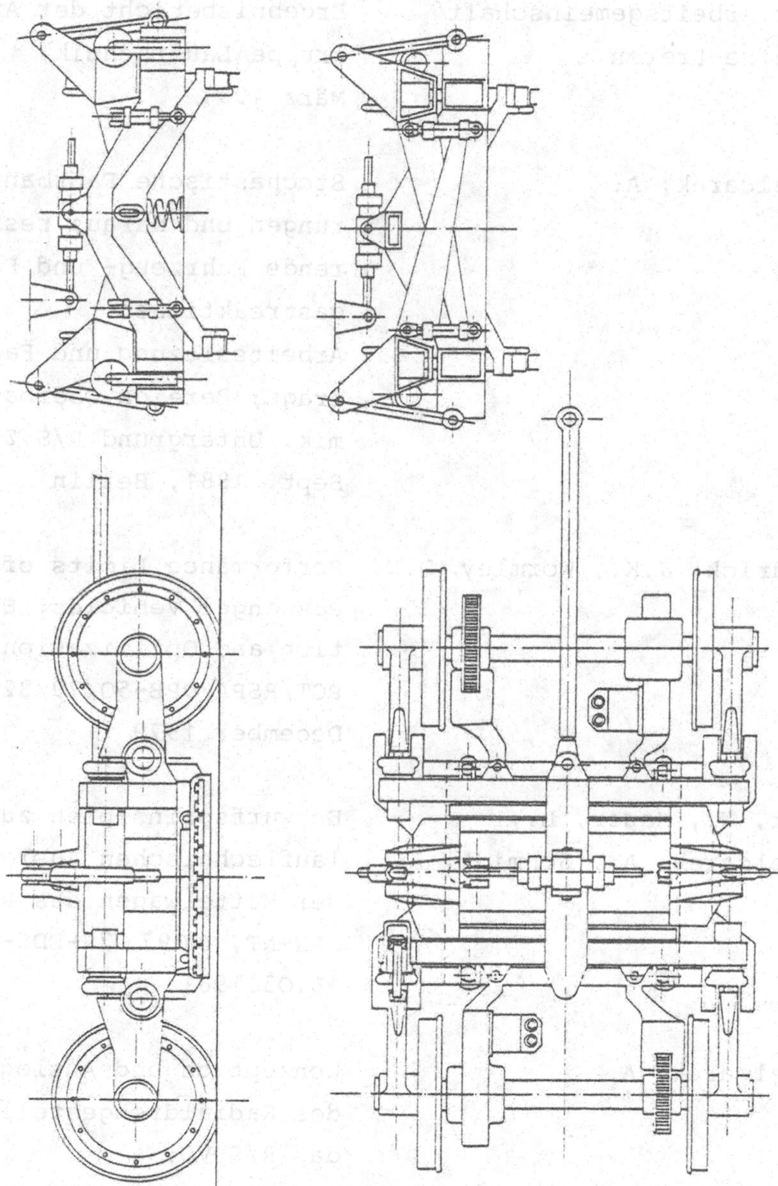

Fig. 20: Truck design of ROLLAB

LIST OF REFERENCES

/1/ Arbeitsgruppe Lauftechnik der Arbeitsgemeinschaft Rheine-Freren — Definitionsphase R/S-VD, Ergebnisbericht der Arbeitsgruppe Lauftechnik, März 1981

/2/ Mielcarek, A. — Stochastische Fahrbahnstörungen und daraus resultierende Fahrzeug- und Fahrgastreaktionen Arbeitssitzung und Fachvorträge; Bereich Oberbaudynamik, Untergrund R/S,ZFF, Sept. 1981, Berlin

/3/ Hedrick, J.K., Wormley, D.N. — Performance Limits of Rail Passenger Vehicles: Evaluation and Optimization POT/RSPA/DPB-50/79/32, December 1979

/4/ Kik, W., Mauer, L., Mielcarek, A., Schmidt, A. — Entwurfsprinzipien zur lauftechnischen Auslegung der Mittelwagen des R/S-VD MAN-NT, B 097 079-EDS-001, 18.05.1981

/5/ Mielcarek, A. — Konzeption und Auslegung des Radialdrehgestells für das R/S-VD, MAN-NT, 01.12.1981

/6/	Gasch, R., Knothe, K.	Rechnergestützte Laufwerksentwicklung - Stand und Perspektiven - Forschungsprogramm Bahntechnik (R/S) Statusseminar 1982
/7/	Mauer, L.	MIT/MAN Zusammenarbeit, Verbesserung der Auslegung von Schienenfahrzeugen, Technischer Wissensstand, MAN-NT, EDS-022, 16.10.1979
/8/	Mielcarek, A., Meinke, P.	Querelastisch aufgehängte Fahrmotoren zur Erhöhung der Grenzgeschwindigkeit von Rad/Schiene-Fahrzeugen VDI-Berichte Nr. 381, 1980
/9/	Örley, H.	Drehgestell hydraulisch gedämpft fluid, Dezember 1981
/10/	Günther, C.R.	Schlupfgeregelte Radsätze für ein Hochgeschwindigkeits Drehgestell Statusseminar V, Spurgeführter Fernverkehr-Rad/Schiene-Technik April 1978

MATHEMATICAL MODELING AND CONTROL SYSTEM DESIGN
OF MAGLEV VEHICLES

K. Popp
Universität Hannover, FRG

1. INTRODUCTION

Magnetically levitated vehicles are under development for applications in rapid transit systems in highly populated areas as well as for high speed transportation over large distances. The feasibility of electromagnetic guidance and control, particularly for high speed operations in connection with the use of linear induction motors for propulsion, has been shown by various test-vehicle runs, cf. Table 1.

VEHICLE	YEAR	VEHICLE LENGHT	VEHICLE MASS	REACHED SPEED	GUIDEWAY LENGHT
MAGNETMOBIL	1971	7.6 m	5.8 t	90 km/h	660 m
TRANSRAPID 02	1971	11.7 m	11.3 t	164 km/h	930 m
TRANSRAPID 04	1975	15.0 m	20.0 t	253 km/h	2 400 m
KOMET	1975	8.5 m	8.8 t	401 km/h	1 300 m
KOMET M	1977	8.5 m	11.0 t	400 km/h	1 300 m
TRANSRAPID 05	1979	26.0 m	36.0 t	75 km/h	908 m
TRANSRAPID 06	PLANNED FOR 1982	54.0 m	120.0 t	300-400 km/h	31 400 m

Table 1: Maglev Vehicles in Germany.

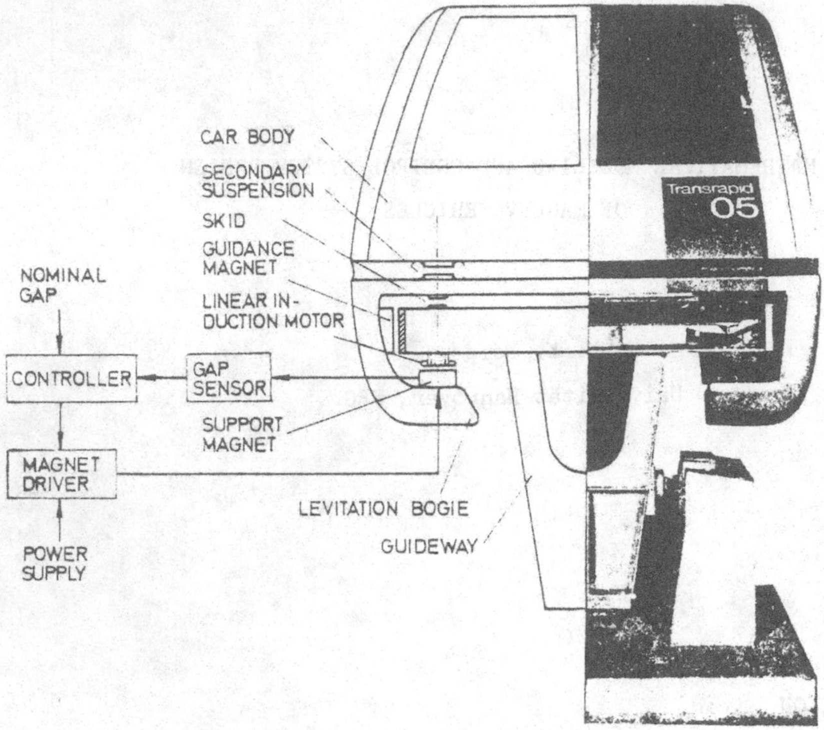

Fig.1: Front view at TRANSRAPID 05 and electromagnetic levitation principle.

Test speeds of more than 400 km/h already have been reached. Recently the first public Maglev vehicle TRANSRAPID 05, see Fig. 1, has been presented at the International Transportation Exhibition in Hamburg. In order to reduce guideway construction costs, Maglev vehicles are going to be operated on flexible elevated guideways. Thus, there is a strong dynamic coupling between vehicle motion and guideway vibration and also an interaction with the suspension control system. For the evaluation of motion stability, ride comfort and safety as well as for overall system optimization a dynamic analysis of the entire system consisting of vehicle, guideway and suspension control is required.

A short literature review indicates the number of different fields which are involved in the present problem area. The basic concept for Maglev vehicles was described by Kemper[21-23] in his patent forty years ago. The present state of the development in Germany is summarized by Reister, Zurek[43] and Zurek[52], while Bahke[1], Muckli[29] give a comparison of different high speed ground transportation systems. The future role of tracked levitated transportation systems is critically evaluated by Ward[50] The research on vibration of structures under moving loads has a long tradition going back to the mid 19th century. A summary of the classical cases is given by Frýba[5]. Recent results paricularly on research in air cushion vehicle-guideway systems are reviewed by Richardson, Wormley[44]. Vehicle models are usually achieved by the multibody approach. The dynamics of multibody systems is treated in Magnus[26], Müller, Schiehlen[32]. Other important subjects are guideway irregularity models and ride comfort evaluation, cf. Hedrick et.al.[13-14], Müller, Popp[33], Hullender[16-17], Snyder, Wormley[48], ISO 2631[18], Smith et. al.[46]. Guideway roughness measurements have been performed by Sussman[49] and Caywood, Rubinstein[3]. An integrated analysis of irregularities and comfort may be found in Müller et.al.[34, 55].

Control concepts for Maglev vehicles are reported in Gottzein[6-11], Rothmayer[45], a number of experimental results being included in the first cited papers. Müller et.al.[30-31] used disturbance accomodation techniques to regard deterministic disturbances, while Breinl[53-54] designed control systems with low disturbance sensitivity. In all cases a rigid guideway and/or a vehicle in standstill is assumed for the control system design and linear time invariant control laws are obtained. On the other hand, Meisinger[27-28] considers the control system of a moving flexible vehicle on a flexible guideway which results in linear periodic time variable feedback laws. Günther[12] treats the secondary suspension control. Other related contributions are due to Katz et.al.[20], Pollard, Williams[35], Jayawant[19], and Yamamura[51]. The numerical analysis of the closed-loop dynamics of Maglev vehicles can effectively be performed using the simulation programs reported in Caywood et.al.[2], Kortüm et.al.[24-25], and Duffek et. al.[4]. The author's work on research concerning the Maglev

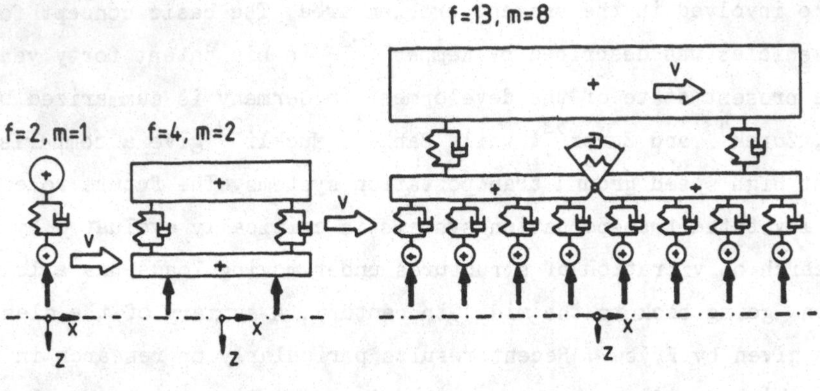

Fig.2: Vehicle models of different complexity.

vehicle-guideway system ist given in Popp[36-42, 56]

2. MODELING OF THE OPEN-LOOP SYSTEM

The vehicle-guideway system under consideration is a very complex system with many degress of freedom. However, the motions in a vertical plane, i.e. heave, pitch and vertical bending are decoupled from the other motions in the usual case of a symmetric construction. Here, only these dominant motions are considered. Futhermore, we assume small displacements except the forward motion of the vehicle which may take place on a straight track with constant speed v . The dynamical equations are obtained by separating the vehicle from the guideway and introducing magnetic suspension forces according to the principle of interaction, cf. Fig. 2. The subsystems are mathematically described and also the disturbances and put together, resulting in the open-loop description. Thus, the entire open-loop model contains:

vehicle model, magnetic suspension model, guideway model and guideway roughness model.

The vehicles are usually modeled as mass point systems or multibody systems, cf. Fig. 2. The complexity of the models depends on the purpose of the analysis. The mathematical description ist given by a system of ordinary differential equations of second order. The action of the suspension magnets is replaced by single forces. Although eletromagnets are highly nonlinear elements, a linearized model described by an ordinary first order differential equation has proven itself well, cf. Gottzein, Lange[9].

The guideway model consists of an infinite sequence of identical and uncoupled Bernoulli-Euler beam-structure elements of lenght L , see Chapt. 1, Fig. 6 . The total guideway deflection can be split into a deterministic part caused by the loading of the moving magnetic forces, and a stochastic part produced by random guideway irregularities. The deterministic deflection of a guideway element is governed by the wellknown fourth order partial differential equation, which can be replaced by an infinite set of ordinary differential equations using the modal expansion. For our purposes an approximate solution comprising a finite number of eigenmodes is sufficient. In order to keep the mathematical description of the entire guideway as simple as possible, only the minimum number of elements is regarded, which results in a periodic shifting of elements under the moving vehicle. The random guideway irregularities are caused by vertical offset, random walk, camber, and surface roughness of its elements, cf. Chapt. 1. Measurements have proven that the total guideway roughness can be modeled as a stationary, Gaussian, ergodic random variable with zero mean value. A very simple but useful model is due to Sussman[49]. The corresponding vehicle excitation can be described by a colored noise process, where the time delays due to the distances between the suspension forces, see Fig. 3, have to be regarded, cf. Müller et.al.[33,34]. The colored noise excitation process can be generated from a white noise process by means of a shape filter, i.e. a system of linear differential equations with white noise input, cf. Chapt. 1.

Although the details of the mathematical description of the entire open-loop system depend on the chosen subsystem models, the resulting state equation Σ always has the same structure, cf. Popp[42,56].

$$\dot{\underline{x}}(t) = \underline{A}(t)\underline{x}(t) + \underline{B}\,\underline{u}(t) + \underline{b}_d(t) + \underline{b}_s(t) \quad , \tag{1}$$

$$\underline{x}(t=\nu T+o)=\underline{U}\,\underline{x}(t=\nu T-o),\ \nu=1,2,\ldots,\ \underline{x}(+o) = \underline{x}_o, \quad (\Sigma) \tag{2}$$

where $\underline{x}(t)$ is the nx1-state vector and $\underline{u}(t)$ is the mx1-control vector; $\underline{A}(t)$ denotes the nxn-system matrix, $\underline{b}_d(t)$ and $\underline{b}_s(t)$ are the nx1-vectors of deterministic disturbances (e.g. moving static loads) and stochastic disturbances, respectively. The system Σ is periodic time dependent, i.e. $\underline{A}(t+T)=\underline{A}(t)$, $\underline{b}_d(t+T)=\underline{b}_d(t)$, $T = L/v$, because a periodically bilt guideway (element lenght L) and a constant speed v have been assumed. Furthermore, it shows periodically jumping states, eq. (2), due to the fact that the vehicle front reaches a resting guideway element at the end of each period. Thus some guideway states become suddenly zero. Eq. (2) describes also the shifting of the guideway elements. The dimension n of the state vector which corresponds to the order of the system matrix is given by

$$n = 2f + m + 2\bar{n}\bar{f} + s \ , \ 1\leq s \leq m \ , \tag{3}$$

where f and $\bar{f}$ represent the degress of freedom of the vehicle and of the guideway element, respectively, $\bar{n}$ is the minimum number of guideway elements within the system bounds, m is the number of magnet forces and s is the order of the shape filter. In a realistic system description the total system order n is quite large. Thus, the open-loop system can mathematically be described by a

- linear, periodic high order state equation with periodically jumping states.

The state equation (3) has to be completed by the measurement equation

$$\underline{y}(t) = \underline{C}\,\underline{x}(t) + \underline{q}(t) \quad , \tag{4}$$

where the measurment vector $\underline{y}(t)$ represents the real output of the system and provides information about the states, $\underline{C}$ is the measurement matrix and $\underline{q}(t)$ the sensor noise. In technical applications, one usually measures the vertical acceleration, air gap, magnetic flux and current of the suspension magnets.

3. CONTROL SYSTEM DESIGN

The most important technical design criteria for the control system are

- asymptotic stability of all motions, best possible safety and reliability, sufficient ride comfort, low power consumption.

Since the Maglev vehicle is unstable without active control of the primary suspension, the first criterion is essential to operate the system. But the other criteria are important as well. The vehicle must not touch the guideway, not even under unusual dynamical conditions. The comfort specifications based e.g. on ISO 2631, have to be met, along with redundancy specifications. On the other hand the costs have to be as small as possible, calling for low power consumption.

There are different feasible control concepts for Maglev vehicles which should be touched upon:

i) Centralized mode control or decentralized single magnet control (magnetic wheel).
ii) Level control or air gap control.

i) In a centralized mode control all data processing is done in an onboard computer so that information about distinct modes, e.g. heave or pitch is available. The centrally computed mode signals are fed back in mode controllers. The suspenion magnets are usually fixed on the car body. In a decentralized control regime the magnets are decoupled from each other and flexibly mounted on the car body. Each magnet has an individual power supply, sensors and controller. These components result in a single unit which is sometimes called a magnetic wheel. Here redundancy is realized by configuration. Single magnet control also offers benefits in the control system design.

ii) Level or z-control provides an active decoupling from the real track, i.e. each magnet and thus the car body are kept at a constant level above an ideal track. Level control results in good ride comfort, since the vertical accelerations are very small. However, the nominal air gap has to be large (≈ 15 mm). In contrast, the gap or s-control keeps the nominal gap widths constant and results in best safety conditions. The nominal air gap can be small (≈ 6-8 mm). In order, to meet the comfort spezifications, an active secondary suspension might be necessary.

Naturally, centralized control and level control are often combined as well as decentralized control and gap control. In early test vehicles, cf. Table 1, centralized mode control was implemented, while in recent vehicles decentralized gap control has been used. For any control concept the control system design must start with lower oder models. The control design philosophy characterized by a flow chart of the analysis procedure is outlined in Fig. 3. As a first step, the sophisticated high order model Σ, cf. (1), (2), has to be simplified to a lower order model $\tilde{\Sigma}$. The problem of order reduction can be solved either mathematically by mode truncation (condensation) or, as in the present case, by physical assumptions such as zero speed and/or a rigid guideway. In addition, the disturbances are usually neglected since they are assumed to be small. Thus, the simplified low order model can be stated as

$$\begin{aligned} &\dot{\underline{\tilde{x}}}(t) = \underline{\tilde{A}}(t)\,\underline{\tilde{x}}(t) + \underline{\tilde{B}}\,\underline{u}(t), \quad \underline{\tilde{A}}(t+T) = \underline{\tilde{A}}(t), \\ &\underline{\tilde{x}}_{\nu+} = \underline{\tilde{U}}\,\underline{\tilde{x}}_{\nu-}, \quad \underline{\tilde{x}}_{o+} = \underline{\tilde{x}}_{o}, \quad \nu = 1,2,\ldots, \qquad \textcircled{\tilde{\Sigma}} \\ &\underline{\tilde{y}}(t) = \underline{\tilde{C}}(t)\,\underline{\tilde{x}}(t)\ . \end{aligned} \tag{5}$$

The system order now is

$$\tilde{n} = 2\tilde{f} + \tilde{m} + 2\tilde{n}\tilde{\tilde{f}}\ , \quad 1 \le \tilde{m} \le m\ . \tag{6}$$

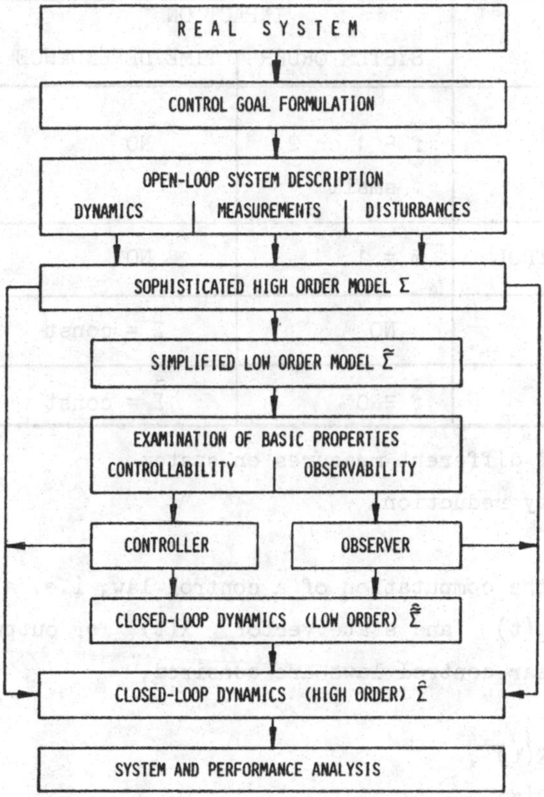

Fig.3: Flow chart of system analysis.

The impact of different measures on system simplification is shown in detail in Table 2. As can be seen, best results are gained by assuming a rigid guideway, particularly in connection with single magnet control. Then, the simplified system can mathematically be described by a

- linear, constant, reduced order state equation without jumping states.

The next step after system reduction, cf. Fig. 4, is the actual control synthesis. Only a few remarks concerning this topic shall be made here. More details about these methods may be found in Popp[42], and practical solutions are given in Gottzein et.al.[6-11], and Breinl[53, 54]. The aim of

MEASURE	INPACT ON		
	SYSTEM ORDER	TIME DEPENDENCE	JUMPS
MODE TRUNCATION	$\tilde{\tilde{f}} \approx 1$ or 2 $\tilde{f}$ small	NO	NO
SINGLE MAGNET CONTROL	$\tilde{m} = 1$	NO	NO
SPEED v = 0	NO	$\tilde{\Sigma}$ = const	$\tilde{\underline{U}} = \underline{E}$
RIGID GUIDEWAY	$\tilde{\tilde{f}} \equiv 0$	$\tilde{\Sigma}$ = const	$\tilde{\underline{U}} = \underline{E}$

Table 2: Impact of different measures on system complexity reduction.

the synthesis in the computation of a control law, i.e. a relation between control vector $\underline{u}(t)$ and state vector $\tilde{\underline{x}}(t)$ or output vector $\tilde{\underline{y}}(t)$. Only linear control laws are consired,

$$\underline{u}(t) = -\underline{K}_x(t)\, \tilde{\underline{x}}(t)\ , \tag{7}$$

$$\underline{u}(t) = -\underline{K}_y(t)\, \tilde{\underline{y}}(t)\ , \tag{8}$$

where the feedback matrices $\underline{K}_x(t)$, $\underline{K}_y(t)$ comprise the control gains. Although it is possible to solve the optimal state feedback problem for linear periodic time-varying systems $\tilde{\Sigma}(t)$ (5) with jumping states, resulting in a periodic gain matrix $\underline{K}_x(t+T)=\underline{K}_x(t)$, cf. Meisinger[27,28], one is more interested in constant gain matrices $\underline{K}_x$ = const, $\underline{K}_y$ = const, since they are much easier to implement. Starting from $\tilde{\Sigma}$ (5), where $\tilde{\underline{A}}$ = const, $\tilde{\underline{U}} = \underline{E}$, the constant gains can be obtained using pole assignment, Theorem 1, or optimal control with respect to quadratic cost functionals, Theorem 2:

<u>Theorem 1</u> (pole assignment):

If and only if the linear $\tilde{n}$th-order system $\tilde{\Sigma}$,

$$\dot{\tilde{\underline{x}}}(t) = \tilde{\underline{A}}\, \tilde{\underline{x}}(t) + \tilde{\underline{B}}\, \underline{u}(t)\ ,\quad \tilde{\underline{x}}(\tau_o) = \underline{x}_o$$

is completely controllable (i.e. Rank $\underline{Q}_c$ = = Rank $[\underline{\tilde{B}} \,|\, \underline{\tilde{A}}\,\underline{\tilde{B}} \,|\, \underline{\tilde{A}}^2\,\underline{\tilde{B}} \,|\, \ldots\ldots \,|\, \underline{\tilde{A}}^{n-1}\,\underline{\tilde{B}}] = \tilde{n}$), then a suitable state feedback

$$\underline{u}(t) = -\,\underline{K}_x\,\underline{\tilde{x}}(t)$$

can be found such that the eigenvalues $\hat{\lambda}_i$, i = 1,..., $\tilde{n}$, of the closed - loop system

$$\dot{\underline{\tilde{x}}}(t) = \underline{\hat{A}}\,\underline{\tilde{x}}(t),\; \underline{\hat{A}} = \underline{\tilde{A}} - \underline{\tilde{B}}\,\underline{K}_x,$$

can arbitrarily be located in the complex plane (with the restriction that complex eigenvalues occure in complex conjugate pairs).

In case of a completely controllable $\tilde{n}$th-order single input system,

$$\dot{\underline{\tilde{x}}}(t) = \underline{\tilde{A}}\,\underline{\tilde{x}}(t) + \underline{b}\,u(t),$$

and for a set of arbitrarily chosen eigenvalues $\hat{\lambda}_i$, i = 1,..., $\tilde{n}$, which result in the characteristic polynomial

$$\hat{p}(\lambda) = (\lambda-\hat{\lambda}_1)(\lambda-\hat{\lambda}_2)\ldots.(\lambda-\hat{\lambda}_{\tilde{n}}) = \lambda^{\tilde{n}}+\hat{a}_1\lambda^{\tilde{n}-1}+\ldots.+\hat{a}_{\tilde{n}},$$

the gain vector $\underline{k}_x^T$ in the feedback-law $u(t) = -\,\underline{k}_x^T\,\underline{\tilde{x}}(t)$ can uniquely be calculated by

$$\underline{k}_x^T = \underline{e}_{\tilde{n}}^T\,\underline{Q}_c^{-1}\,\hat{p}\,(\underline{\tilde{A}}),$$

where

$$\underline{e}_{\tilde{n}}^T = [0,\,0,\ldots.,\,0,1],$$

$$\underline{Q}_c = [\underline{b} \,|\, \underline{\tilde{A}}\,\underline{b} \,|\, \underline{\tilde{A}}^2\underline{b} \,|\, \ldots. \,|\, \underline{\tilde{A}}^{\tilde{n}-1}\underline{b}],$$

$$\hat{p}(\underline{\tilde{A}}) = \underline{\tilde{A}}^{\tilde{n}} + \hat{a}_1\,\underline{\tilde{A}}^{\tilde{n}-1} + \ldots + \hat{a}_{\tilde{n}-1}\,\underline{\tilde{A}} + \hat{a}_{\tilde{n}}\,E\,.$$

Theorem 2 (linear optimal regulator):
Consider the linear ñth-order system $\tilde{\Sigma}$,

$$\dot{\underline{\tilde{x}}}(t) = \underline{\tilde{A}}\, \underline{\tilde{x}}(t) + \underline{\tilde{B}}\, \underline{u}\,(t) \ , \ \underline{\tilde{x}}(t_o) = \underline{x}_o \ ,$$

and the functional

$$J(\underline{\tilde{x}}(t), \underline{u}(t)) = \frac{1}{2} \int_{t_o}^{\infty} [\underline{\tilde{x}}^T(t) \underline{Q}\, \underline{\tilde{x}}\,(t) + \underline{u}^T(t)\, \underline{R}\, \underline{u}(t)\,]dt \ ,$$

where $\underline{Q} = \underline{Q}^T \geq \underline{0}$, $\underline{R} = \underline{R}^T > \underline{0}$, $(\underline{\tilde{A}}, \underline{Q})$ completely controllable.

If and only if $\tilde{\Sigma}$ is completely controllable then exists a unique optimal state feedback

$$\underline{u}^*(t) = - \underline{R}^{-1}\, \underline{\tilde{B}}^T\, \underline{P}\, \underline{\tilde{x}}(t) \equiv -\underline{K}_x\, \underline{\tilde{x}}(t) \ ,$$

where $\underline{P} = \underline{P}^T > \underline{0}$ is the unique solution of the Riccati equation

$$\underline{\tilde{A}}^T\, \underline{P} + \underline{P}\, \underline{\tilde{A}} - \underline{P}\, \underline{\tilde{B}}\, \underline{R}^{-1}\, \underline{\tilde{B}}^T\, \underline{P} + \underline{Q} = \underline{0}$$

such that

$$J \rightarrow \min.$$

The minimum criterion value J^* is given by $J^* = \underline{x}_o^T\, \underline{P}\, \underline{x}_o$ and the closed-loop system

$$\dot{\underline{\tilde{x}}}(t) = \underline{\hat{A}}\, \underline{\tilde{x}}(t) \ , \ \underline{\hat{A}} = \underline{\tilde{A}} - \underline{\tilde{B}}\, \underline{K}_x = \underline{\tilde{A}} - \underline{\tilde{B}}\, \underline{R}^{-1}\, \underline{\tilde{B}}^T\, \underline{P}$$

is asymptotically stable.

Both methods are suitable for multivariable systems and allow the design of state controllers as well as state observers. In the case of output feedback design, cf. (8), parameter optimization methods may be used.

As the final step in getting the closed-loop system description, the control law (7) or (8) based on the reduced order system Σ is introduced in the original higher order system Σ . This results in

$$\dot{\underline{x}}(t) = \underline{\hat{A}}(t)\,\underline{x}(t) + \underline{b}_d(t) + \underline{b}_s(t)$$
$$\hat{\Sigma}(t) \qquad (9)$$
$$\underline{x}_{\nu+} = \underline{U}\,\underline{x}_{\nu-}\ ,\ \underline{x}_{o+} = \underline{x}_o\ ,\ \nu = 1,2,\ldots,$$

where the periodicity condition $\underline{\hat{A}}(t+T) = \underline{\hat{A}}(t)$ holds. However, even if an optimal state feedback law (7) with respect to $\tilde{\Sigma}(t)$ is computed and implemented in the original higher order system $\Sigma(t)$, then this results in an incomplete state feedback with respect to $\Sigma(t)$ and nothing can be said about the stability of $\hat{\Sigma}(t)$ or other properties. Thus, a performance analysis of the closed-loop system $\hat{\Sigma}(t)$ is required.

4. SYSTEM ANALYSIS

The dynamic analysis of the closed-loop system $\hat{\Sigma}(t)$ is carried out starting with the homogeneous solution followed by a stability investigation. The performance of the Maglev vehicle-guideway system depends essentially on the system responses of the present disturbances. Since linearity is given, the responses due to deterministic and stochastic disturbances can be calculated separately and then be superposed. Generally, exact methods are preferred, but some approximate results are also dealt with. Usually, for steady-state response calculations, numerical simulation methods are employed. Here, the wellknown analytical methods for periodic system based on Floquet theory are applied and extended to jumping states. But only the main results are shwon; derivations and proofs may be found in Popp[42].

4.1. Homogeneous solution and stability analysis

Consider the homogeneous part of the closed-loop system $\hat{\Sigma}(t)$ (9).

$$\underline{\dot{x}}(t)=\underline{\hat{A}}(t)\ \underline{x}(t)\ ,\ \underline{\hat{A}}(t+T)=\underline{\hat{A}}(t),\ \underline{x}_{\nu+}=\underline{U}\ \underline{x}_{\nu-},$$

$$\underline{x}_{o+}=\underline{x}_o\ ,\ \nu=1,2,\dots\ . \tag{10}$$

The solution of (10) can be obtained by piecewise calculation using the time representation $t = \tau + \nu T$, $0 \le \tau \le T = L/v$ and Floquet theory,

$$\underline{x}(t=\tau+\nu T)=\underline{\Phi}(t)\ [\underline{U}\ \underline{\Phi}(T)]^{\nu}\ \underline{x}_o\ ,\ 0 \le \tau \le T,$$

$$\nu = 1,2,\dots, \tag{11}$$

where the transition matrix $\underline{\Phi}(t)$ follows from

$$\underline{\dot{\Phi}}(\tau) = \underline{A}(\tau)\ \underline{\Phi}(\tau),\ \underline{\Phi}(0) = \underline{E}\ . \tag{12}$$

For $\underline{U} = \underline{E}$ eq. (11) represents the wellknown result for a periodic system without jumping states. The stability of system (10) can be determined by investigating the solution (11) for $t \to \infty$ or equivalently for $\nu \to \infty$. It is obvious that the stability behavior depends uniquely on the eigenvalues σ of the growth matrix $\underline{U}\ \underline{\Phi}(T)$ which is abbreviated by $\underline{\Psi}$. The characteristic equation thus becomes

$$\det(\sigma\underline{E} - \underline{\Psi}) = 0\ , \tag{13}$$

and the following stability theorem applies, cf. Hsu[15]:

Theorem 3 (stability behavior):
The homogeneous system (10) is

i) asymptotically stable if and only if all eigenvalues of the growth matrix $\underline{\Psi}$ have absolute values less than one,

$$|\sigma_i| < 1,\ i = 1,\dots,n,$$

ii) stable if and only if all eigenvalues have absolute values not greater than one, $|\sigma_i| \leq 1$, $i = 1,\ldots,n$, and for all eigenvalues λ_r with absolute values equal to one the defect d_r of the characteristic matrix $(\sigma_r \underline{E} - \underline{\Psi})$ is equal to the multiplicity m_r of these eigenvalues, $d_r = m_r$,

iii) unstable if and only if there is at least one eigenvalue which has absolute value greater than one, $|\sigma_i| > 1$, or there are multiple eigenvalues σ_s which have absolute value equal to one $|\sigma_s| = 1$, and the corresponding defect d_s of the characteristic matrix $(\sigma_s \underline{E} - \underline{\Psi})$ is less than the multiplicity m_s of these eigenvalues, $d_s < m_s$.

We are interested only in asymptotic stability. For this purpose, it is convenient to introduce the spectral radius $spr(\underline{\Psi})$ of the growth matrix.

$$spr(\underline{\Psi}) = \max_i |\sigma_i| \quad , \quad i = 1,\ldots, n . \tag{14}$$

Then, asymptotic stability is given if and only if

$$spr(\underline{\Psi}) < 1 . \tag{15}$$

4.2 Steady-state response to deterministic disturbances

The inhomogeneous deterministic part of the closed-loop system $\hat{\Sigma}(t)$ (9) can be written as

$$\begin{aligned} &\dot{\underline{x}}(t)=\hat{\underline{A}}(t)\,\underline{x}(t)+\underline{b}_d(t),\ \hat{\underline{A}}(t+T)=\hat{\underline{A}}(t)\ , \\ &\underline{b}_d(t+T) = \underline{b}_d(t)\ ,\ \underline{x}_{\nu+} = \underline{U}\,\underline{x}_\nu\ , \\ &\underline{x}_{o+} = \underline{x}_o\ ,\ \nu = 1,2,\ldots\ . \end{aligned} \tag{16}$$

The application of the usual simulation methods requires that eq. (16) be numerically integrated over a sufficiently long time. However, the steady-state response of an asymptotically stable system can also be found from the general solution,

$$\underline{x}(t=\tau+\nu T)=\underline{\Phi}(\tau)[\underline{\Psi}^{\nu}\underline{x}_o + \sum_{k=1}^{\nu} \underline{\Psi}^k \underline{c}_d(T) + \underline{c}_d(\tau)], \quad \nu=0,1,\ldots$$

$$\underline{c}_d(\tau) = \int_0^{\tau} \underline{\Phi}^{-1}(\overset{*}{\tau})\, \underline{b}_d(\overset{*}{\tau})\, d\overset{*}{\tau} , \quad 0 \le \tau \le T . \tag{17}$$

The steady-state solution $\underline{x}_{\infty}(t)$ is obtained for $t \to \infty$ or equivalently $\nu \to \infty$. Then, since asymptotic stability is assumed, the first term in brackets vanishes and the remaining infinite geometrical matrix series converges to $\underline{S}_d$,

$$\underline{S}_d = \sum_{k=1}^{\infty} \underline{\Psi}^k = (\underline{E}-\underline{\Psi})^{-1}-\underline{E} = (\underline{E}-\underline{\Psi})^{-1}\underline{\Psi} . \tag{18}$$

Thus, the steady-state response corresponding to deterministic disturbances is given by

$$\underline{x}_{\infty}(t)=\lim_{\nu\to\infty} \underline{x}(t=\tau+\nu T) = \underline{\Phi}(\tau)[\underline{S}_d\underline{c}_d(T)+\underline{c}_d(\tau)] =$$

$$= \underline{x}_{\infty}(t+T) , \quad 0 \le \tau\le T , \tag{19}$$

which is a periodic function of time. Here, numerical integrations have to be carried out only for one single period. The simulation method can be recommended only for highly damped systems, while the computation of the formal solution (19) yields good results in all other cases.

4.3 Steady-state response to stochastic disturbances

Analogous to (16) the stochastic part of the closed-loop system $\hat{\Sigma}(t)$ (9) is

$$\dot{\underline{x}}(t)=\hat{\underline{A}}(t)\, \underline{x}(t)+\underline{b}_s(t) , \quad \hat{\underline{A}}(t+T)=\hat{\underline{A}}(t) , \quad \underline{b}_s(t)\sim(\underline{0},\underline{Q}_b) ,$$

$$\underline{x}_{\nu+} = \underline{U}\, \underline{x}_{\nu-} , \quad \underline{x}_{o+} = \underline{x}_o \sim (\underline{0}, \underline{P}_{xo}) , \quad \nu = 1,2, \ldots . \tag{20}$$

Here, $\underline{b}_s(t)$ is assumed to be a Gaussian white noise vector process with zero mean, characterized by the constant intensity matrix Q_b The initial condition $\underline{x}_o$ is now a random vector, where a vanishing mean value is assumed and where $\underline{P}_{xo}$ denotes the covariance matrix. Since the system is periodic and linear, the response $\underline{x}(t)$ is generally a nonstationary Gaussian vector process which can be characterized by the mean vector $\underline{m}_x(t)$ and the covariance matrix $\underline{P}_x(t)$

$$\underline{m}_x(t) = E\{\underline{x}(t)\},$$
$$\underline{P}_x(t) = E\{[\underline{x}(t)-\underline{m}_x(t)]\,[\underline{x}(t)-\underline{m}_x(t)]^T\} = \underline{P}_x^T(t). \tag{21}$$

However, due to the vanishing initial mean vector, $\underline{m}_x(0) = \underline{0}$, the mean vector $\underline{m}_x(t)$ vanishes identically. The covariance matrix $\underline{P}_x(t)$ is the solution of the Liapunov differential equation

$$\dot{\underline{P}}_x(t) = \hat{\underline{A}}(t)\,\underline{P}_x(t)+\underline{P}_x(t)\,\hat{\underline{A}}^T(t) + \underline{Q}_b,$$
$$\underline{P}_{x\nu+} = \underline{U}\,\underline{P}_{x\nu-}\,\underline{U}^T, \quad \underline{P}_{xo+} = \underline{P}_{xo}, \quad \nu = 1,2,\ldots, \tag{22}$$

where the jump condition for the covariance matrix follows upon introduction of the state jumps in the definition (21). Eq. (22) is the starting point for the numerical calculation of $\underline{P}_x(t)$. However, the analytical solution is preferable here, cf. Popp[42]; it is given by

$$\underline{P}_x(t=\tau+\nu T)=\underline{\Phi}(t)\Big[\sum_{k=1}^{\nu}\underline{\Psi}^k\underline{C}_s(T)\,(\underline{\Psi}^k)^T+\underline{C}_s(t)\Big]\,\underline{\Phi}^T(t),$$
$$\nu = 0,1,\ldots, \tag{23}$$

$$\underline{C}_s(\tau) = \int_0^{\tau}\underline{\Phi}^{-1}(\overset{*}{\tau})\,\underline{Q}_b\,(\underline{\Phi}^{-1}(\overset{*}{\tau}))^T\,d\overset{*}{\tau}, \quad 0 \le \tau \le T.$$

The steady-state solution $\underline{P}_{x\infty}(t)$ follows from (23) for

$$t \to \infty \quad \text{or} \quad \nu \to \infty.$$

$$\underline{P}_{x\infty}(t)=\lim_{\nu\to\infty}\underline{P}_x(t=\tau+\nu T) = \underline{\Phi}(\tau)[\underline{S}_s + \underline{C}_s(\tau)]\,\underline{\Phi}^T(\tau) = \underline{P}_{x\infty}(t + T), \tag{24}$$

where asymptotic stability has again been assumed. Here, $\underline{S}_s$ is solution of the algebraic Stein equation,

$$\underline{S}_s - \underline{\Psi}\, \underline{S}_s\, \underline{\Psi}^T = \underline{\Psi}\, \underline{C}_s(T)\, \underline{\Psi}^T . \tag{25}$$

The steady-state solution $\underline{P}_{x\infty}(t)$ is periodic and can be obtained by numerical integration over only one single period along with some algebraic operations. The simulation method can be recommended only for highly damped systems of low order, while in all other cases the solution (24) is more advantageous. In any case, the time dependent solution is laborious. Thus, a simple time invariant approximation shall briefly be mentioned. It starts with the reduced-order system $\bar{\Sigma}$ (5), where a rigid guideway is assumed but the stochastic disturbances due to guideway irregularities are not neglected. The state equation of the closed-loop system then has the form

$$\begin{aligned} \dot{\tilde{\underline{x}}}(t) &= \hat{\tilde{\underline{A}}}\, \tilde{\underline{x}}(t) + \tilde{\underline{b}}_s(t), \quad \tilde{\underline{x}}_o \sim (\underline{0}, \tilde{\underline{P}}_{xo}) , \\ \tilde{\underline{b}}_s(t) &\sim (\underline{0}, \tilde{\underline{Q}}_b), \end{aligned} \tag{26}$$

which is time invariant and exhibite no jumps in the state variables. The steady-state covariance matrix $\tilde{\underline{P}}_{x\infty}$ now is constant and follows from the algebraic Liapunov equation

$$\hat{\tilde{\underline{A}}}\, \tilde{\underline{P}}_{x\infty} + \tilde{\underline{P}}_{x\infty}\, \hat{\tilde{\underline{A}}}^T + \tilde{\underline{Q}}_b = \underline{0} , \tag{27}$$

which can easily be solved using Smith's[47], method, for example. Comparisons by means of examples show, that $\tilde{\underline{P}}_{x\infty}$ is a good approximation of $\underline{P}_{x\infty}(t)$

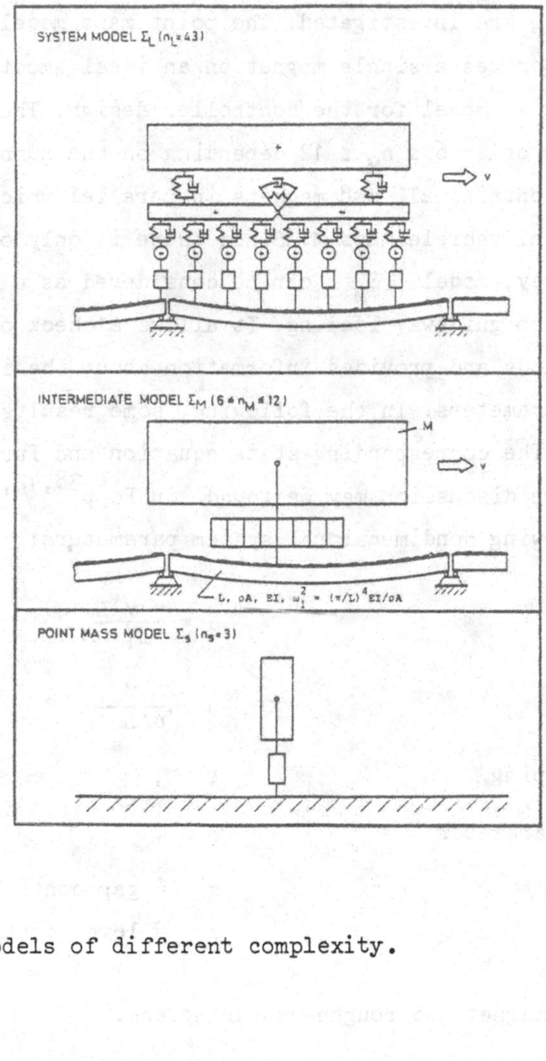

Fig.4: System models of different complexity.

5. EXMAPLE

As an example a vehicle model Σ_L with f=13 degrees of freedom and m=8 suspension magnets is considered, cf. Fig. 4, and single magnet control is chosen. The vehicle may be operated on a single span guideway, where $\bar{n}$=2 spans are coupled during the period T . If we only regard $\bar{f}$=2 modes to describe the guideway deflection and use the simple shape filter of order s=1, then the total system order is n_L=43. Prior to analyzing the model Σ_L, as is done in Popp[42], two simplified models, Σ_M and Σ_s, are investigated. The point mass model Σ_s (system order $n_s=3$) characterizes a single magnet on an ideal smooth and rigid guideway and serves as model for the controller design. The intermediate model Σ_M (system order $6 \le n_M \le 12$ depending on the number of included guideway modes $\bar{f}$) contains all m=8 magnets in parallel which is sufficient to levitate the total vehicle mass M. Since there is only one contact point on the guideway, model Σ_M can be considered as a worst case model with respect to guideway loading. It allows a check on the developed analysis methods and provides information about the influence of different design parameters. In the following, some results for the model Σ_M are shown. The corresponding state equation and further results as well as an extensive discussion may be found in Popp[38,42,56]. The results depend on the following nondimensional system parameters:

Speed ratio	$\alpha = \dfrac{\pi\, v/L}{\omega_1}$	,
Mass ratio	$\mu \quad \dfrac{M}{\rho A L}$	,
Modal beam damping	ζ	,
Control gain parameter	β	,
Control typ	s	$\hat{=}$ gap control,
	z	$\hat{=}$ level control,

as well as on some magnet and roughness parameters.

5.1 Stability analysis

The stability analysis is performed as outlined in section 4.1. It turns out that gap or s-control is more critical than level or z-control, due to the strong dynamic interaction between vehicle and guideway. Some stability results are shown in Fig. 5, where the spectral radius spr $(\underline{U}\ \underline{\Phi}\ (T))$ of the growth matrix is plotted versus the speed ratio α. In Figs. 5a)- 5b) only the frist beam mode is included, $\bar{f}=1$, while Figs. 5 c), 5 d) show the results when the frist two modes, $\bar{f}=2$, are taken into account. In either case the modal beam damping is assumed to be $\zeta_1 = \zeta_2 = 0.05$. It can be seen that the spectral radius spr $(\underline{U}\ \underline{\Phi}(T))$ increases with higher speeds α , larger mass ratio μ and smaller control gain parameter β . For $\bar{f}=1$ asymptotic stability is given for all considered parameters, while for $\bar{f}=2$ the parameter combinations $\alpha = 1$, $\mu = 1$, $\beta = 1.42$ leads to instability.

5.2. Responses to deterministic disturbances

In order to observe the transient behavior, the time histories of the vertical vehicle motion and the guideway motion under the vehicle are plotted until steady state is reached. The results are shown in Fig. 6 for level or z-control and in Fig. 7 for gap or s-control, where $\bar{f}=2$ guideway modes are regarded and the speed ratio α is varied. In either case steady-state is reached after two or three periods and the control goals are almost completely realized. However, for s-control, Fig. 7, the car body acceleration increases considerably with the speed α which may restrict the validity of the model.

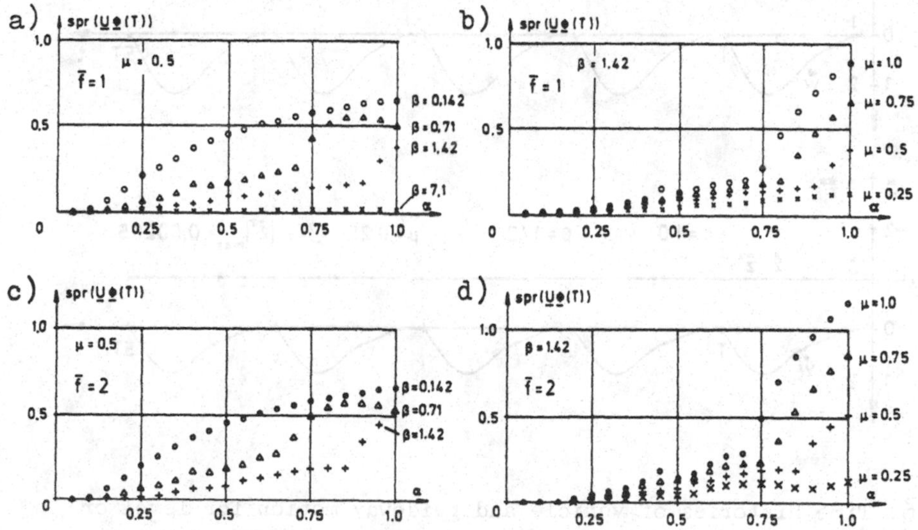

Fig.5: Results of the stability analysis.

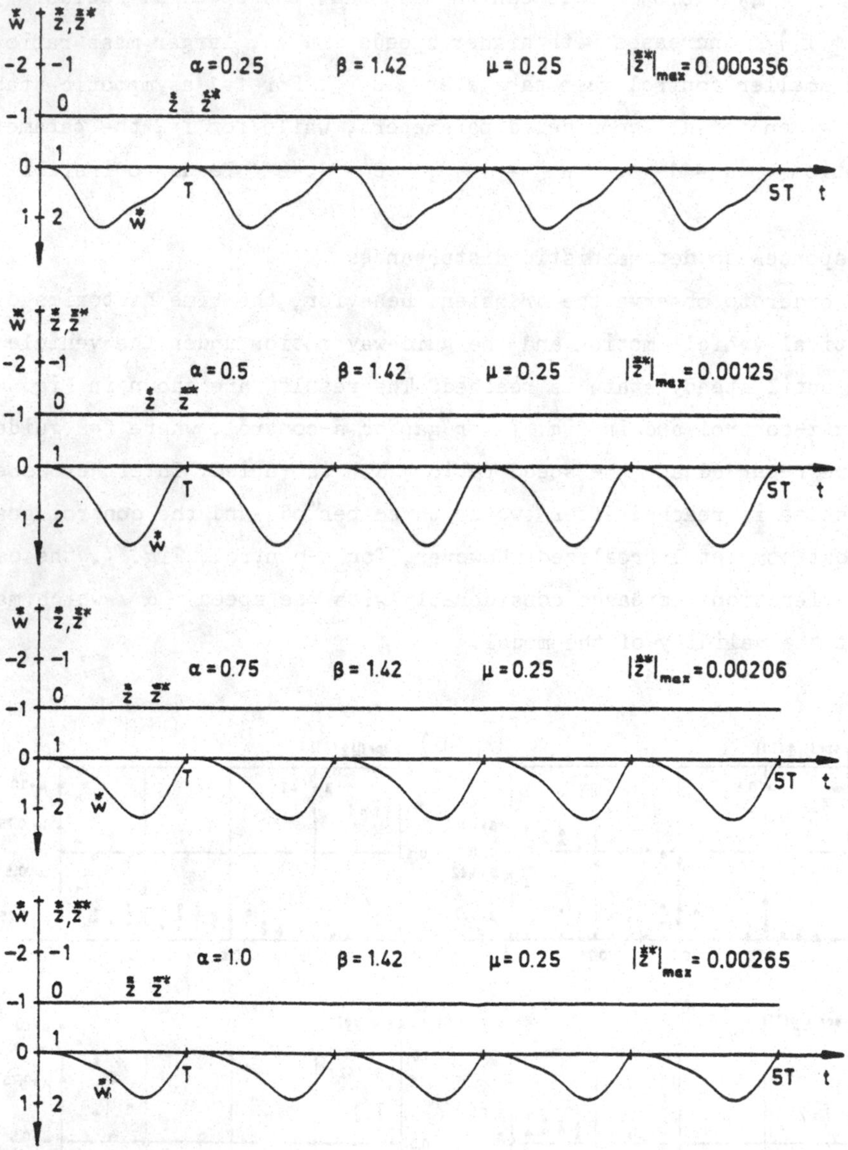

Fig. 6: Time histories of vehicle and guideway motion for level or z-control ($\overset{*}{w}$, $\overset{*}{z}$ and $\overset{**}{z}$ denote nondimensional guideway deflection, vehicle deflection and vehicle acceleration, respectively).

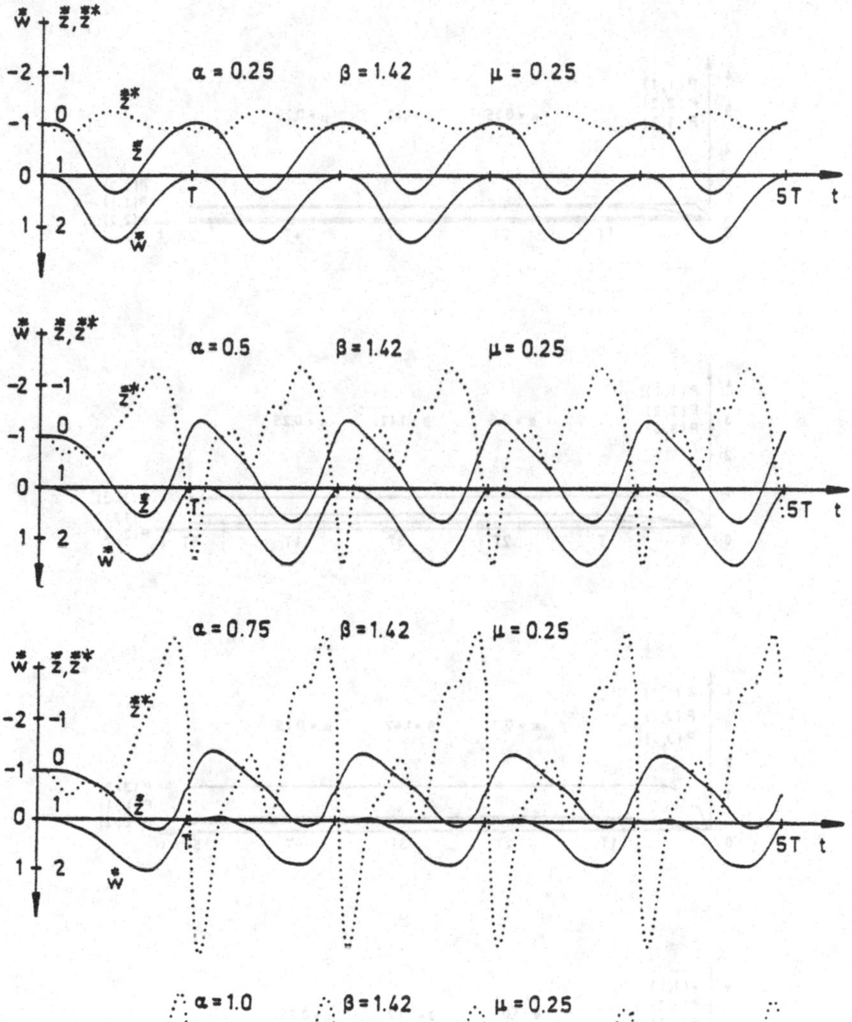

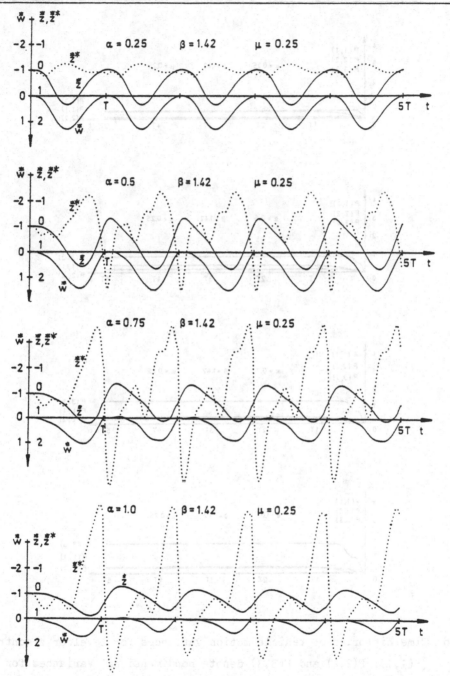

Fig. 7: Time histories of vehicle and guideway motion for gap or s-control ($\overset{*}{w}$, $\overset{*}{z}$ and $\overset{**}{z}$ denote nondimensional guideway deflection, vehicle deflection and vehicle acceleration, respectively).

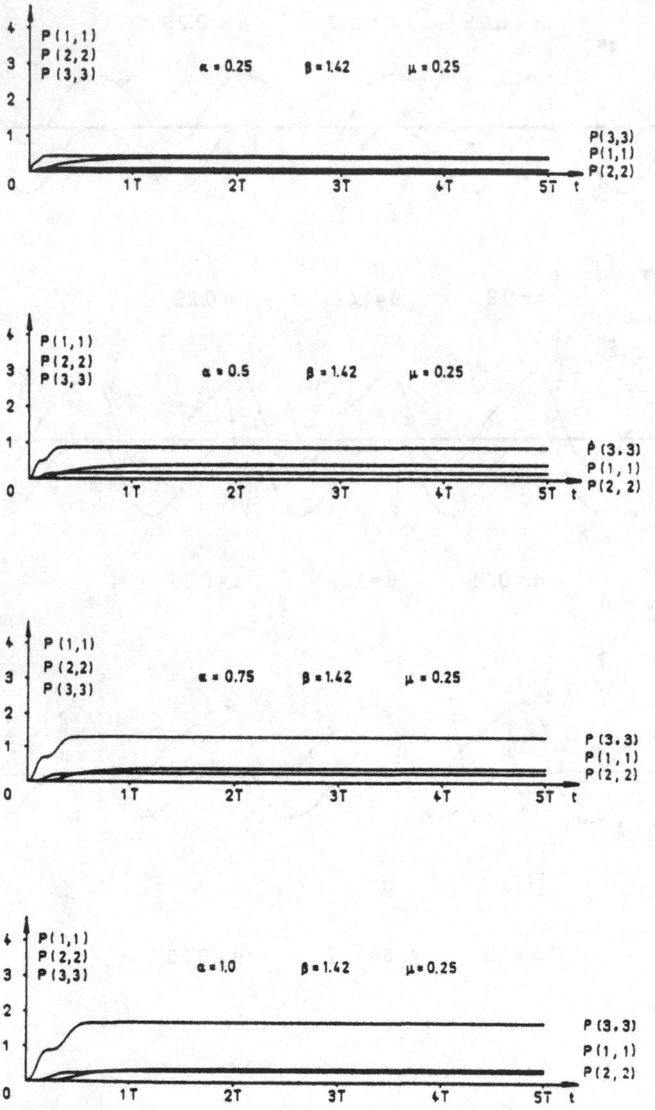

Fig. 8: Time histories of vehicle motion variances for level or z-control (P(1,1), P(2,2) and P(3,3) denote nondimensional variances for vehicle deflection, velocity and acceleration, respectively, where the roughness model /49/ has been used).

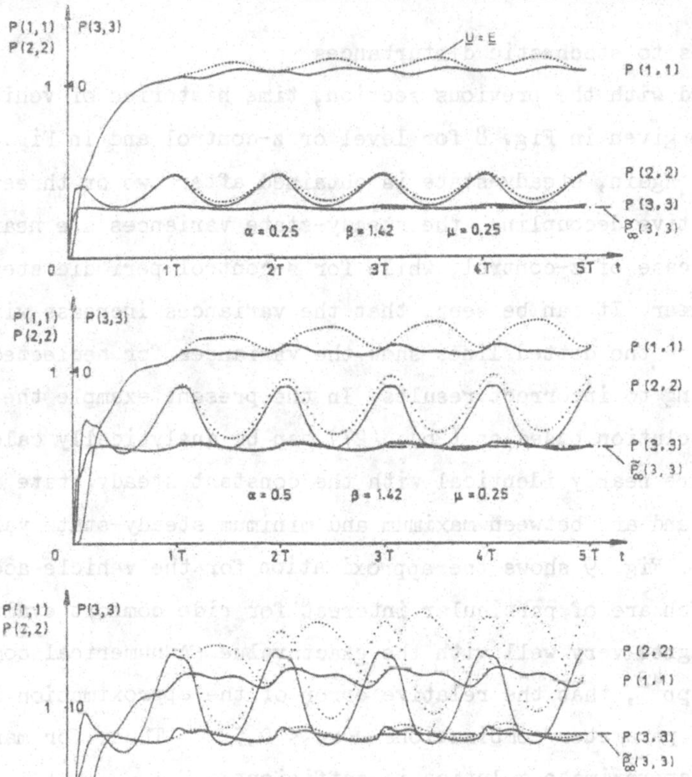

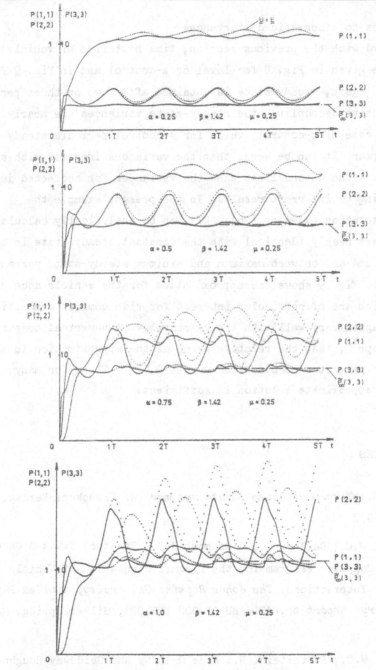

Fig. 9: Time histories of vehicle motion variances for gap or s-control (P(1,1), P(2,2), P(3,3) denote nondimensional variances for vehicle deflection, velocity and acceleration, respectively, where the roughness model /49/ has been used).

5.3. Responses to stochastic disturbances

In accord with the previous section, time histories of vehicle motion variances are given in Fig. 8 for level or z-control and in Fig. 9 for gap or s-control. Again, steady-state is obtained after two or three periods. Due to the active decoupling, the steady-state variances are nearly constant in the case of z-control, while for s-control periodic steady-state variances appear. It can be seen, that the variances increase with speed α . In Fig. 9 the dotted lines show the variances for neclected jumps, clearly leading to incorrect results. In the present example the approximate solution based on (26), (27) can be analytically calculated. The results are nearly identical with the constant steady state in case of z-control and are between maximum and minimum steady-state variances for s-control. Fig. 9 shows the approximation for the vehicle acceleration variances which are of particular interest for ride comfort evaluations. The results agree very well with the exact value. A numerical comparison shows, cf. Popp[42], that the relative error of the approximation is smaller than 10 % for parameter combinations $\alpha\mu < 0,5$ Thus, for many purposes the approximate solution is sufficient.

REFERENCES

1. Bahke, E.: *Transportsysteme heute und morgen.* Krauskopf-Verlag, Mainz, 1973.

2. Caywood, W.C., Dailey, G., O'Connor, J.S., Stadter, J.T.: A General Purpose Computer Program for the Dynamic Simulation of Vehicle-Guideway Interactions. *The Johns Hopkins University, Applied Physics Laboratory, Report* No. APL/JHU CP 008 TPR 021, Silver Spring, Md., 1972.

3. Caywood, W.C., Rubinstein, N.: Ride Quality and Guideway Roughness Measurements of the Transpo'72 PRT System. *High Speed Ground Transp. J.* 8, No. 3, 1974, pp. 214-225.

4. Duffek, W., Kortüm, W., Wallrapp, O.: A General Purpose Programm for the Simulation of Vehicle-Guideway Interaction Dynamics. *5th VDS - 2nd IUTAM Symposium on Dynamics of Vehicles on Roads and Tracks,* Wien, 1977.

5. Frýba, L.: *Vibration of Solids and Structures Under Moving Loads.* Noordhoff Int. Publ., Groningen, 1972.

6. Gottzein, E., Brock, K.-H., Schneider, E., Pfefferl, J.: Control Aspects of a Magnetic Levitation High Speed Test Vehicle. *Automatica* 13, 1977, pp. 201-223.

7. Gottzein, E., Crämer, W., Ossenberg, F.W., Roche, Ch.: Optimal Control of a Maglev Vehicle. *Proc. of the IUTAM Symposium on the Dynamics of Vehicles on Roads and Railway Tracks,* Delft, 1975, pp. 504-530.

8. Gottzein, E., Crämer, W.: Critical Evaluation of Multivariable Control Techniques based on Maglev Vehicle Design. *4th IFAC Symp. Multivariable Technological System,* Fredericton, N.B., Canada, July 4-7, 1977.

9. Gottzein, E., Lange, B.: Magnetic Suspension Control System for the German High Speed Train. 5th IFAC Symposium Automatic Control in Space, Genua, Juni 1973, in: *Automatica* 11, No. 5, 1975.

10. Gottzein, E., Lange, B., Ossenberg-Franzes, F.: Control System Concept for a Passenger Carrying Maglev Vehicle. *High Speed Ground Transp. J.* 9, No. 1, 1975, pp. 435-447.

11. Gottzein, E., Miller, L., Meisinger, R.: Magnetic Suspension Control System for High Speed Ground Transportation Vehicles. *World Electrotechnical Congress.* Section 7, Paper 07, Moscow, 1977.

12. Guenther, Chr.: A New Approach to High-Speed Tracked Vehicle Suspension Synthesis. In: Leondes, C.T. (ed.), *Control and Dynamic Systems,* Vol. 13, New York, San Francisco, London, 1977, pp. 71-133.

13. Hedrick, J.K. Billington, G.F., Dreesbach, D.A.: Analysis, Design and Optimization of High Speed Vehicle Suspension Using State Variable Techniques. *J. Dyn. Syst. Meas. Control,* Trans. ASME 96, Ser. G, 1974,

pp. 193-203.

14. Hedrick, J.K., Ravera, R.J., Anders, J.R.: The Effect of Elevated Guideway Construction Tolerances on Vehicle Ride Quality. *J. Dyn. Syst. Meas. Control,* Trans. ASME 97, Ser. G, 1975, pp. 408-416.

15. Hsu, C.S.: Impulsive Parametric Excitation: Theory. *J. Appl. Mechanics,* Trans. ASME, Juni 1972, pp. 551-558.

16. Hullender, D.A.: Analytical Models for Certain Guideway Irregularties. *J. Dyn. Syst. Meas. Control,* Trans. ASME 97, Ser. G, 1975, pp. 417-423.

17. Hullender, D.A., Bartley, T.M.: Defining Guideway Irregularity Power Spectra in Terms of Construction Tolerances and Constraints. *High Speed Ground Transp. J.* 9, No. 1, 1975, pp. 356-368.

18. *ISO* 2631: Guide for the Evaluation of Human Exposure to Whole-body Vibration, 1st Ed. 1974.

19. Jayawant, B.V.: Dynamical Aspects of Passenger Carrying Vehicles using Controlled D.C. Electromagnets. *5th VSD - 2nd IUTAM Symposium on Dynamics of Vehicles on Roads and Tracks.* Wien 1977.

20. Katz, R.M., Nene, V.D., Ravera, R.J., Skalski, C.A.: Performance of Magnetic Suspensions for High Speed Vehicles Operating over Flexible Guideways. *J. Dyn. Syst. Meas. Control,* Trans. ASME 96, Ser. G, 1974, pp. 204-212.

21. Kemper, H.: Schwebebahn mit räderlosen Fahrzeugen, die an eisernen Fahrschienen mittels magnetischer Felder entlang geführt wird. *Patentschrift* Nr. 643316, 1937 (Patenterteilung 1934).

22. Kemper, H.: Schwebende Aufhängung durch elektromagnetische Kräfte: Eine Möglichkeit für eine grundsätzlich neue Fortbewegungsart. *ETZ,* April 1938, pp. 391-395.

23. Kemper, H.: Elektrisch angetriebene Eisenbahnfahrzeuge mit elektromagnetischer Schwebeführung *ETZ-A,* Januar 1953, pp. 11-14.

24. Kortüm, W., Lehner, M., Richter, R.: Multibody Systems Containing Active Elements: Algorithmic Generation of Linearized System Equations,

System Analysis and Order-Reduction. *IUTAM Symposium Dynamics of Multibody Systems*, Munich, Aug. 1977.

25. Kortüm, W., Richter, R.: Simulation of Multibody Vehicles Moving over Elastic Guideways. *Vehicle System Dynamics* 6, 1977, pp. 21-35.

26. Magnus, K. (Ed.): Dynamics of Multibody Systems. *Proc. IUTAM Symp.* Munich 1977. Springer, Berlin, Heidelberg, New York, 1978.

27. Meisinger, R.: Optimale Regelung periodischer Systeme mit sprungförmiger Zustandsänderung. *ZAMM* 57, 1977, pp. 79-81.

28. Meisinger, R.: Beiträge zur Regelung einer Magnetschwebebahn auf elastischem Fahrweg. *Dissertation*, TU München, 1977.

29. Muckli, W.: Bahnsysteme mit berührungsfreier Fahrtechnik. *ZEV-Glasers Annalen* 100, No. 1, 1976, pp. 16-19.

30. Müller, P.C., Bremer, H., Breinl, W.: Tragregelsysteme mit Störgrößen-Kompensation für Magnetschwebefahrzeuge. *Regelungstechnik* 24, 1976, pp. 257-265.

31. Müller, P.C.: Design of Optimal State-Observers and its Application to Maglev Vehicle Suspension Control. *4th IFAC Symp. Multivariable Technological Systems*, Fredericton, N.B., Canada, July 4-7, 1977.

32. Müller, P.C., Schiehlen, W.: *Lineare Schwingungen*. Akademische Verlagsgesellschaft, Wiesbaden, 1976.

33. Müller, P.C., Popp, K.: Kovarianzanalyse linearer Zufallsschwingungen mit zeitlich verschobenen Erregerprozessen. *ZAMM* 59, 1979, pp. T 144-T 146.

34. Müller, P.C., Popp, K., Schiehlen, W.: Covariance Analysis of Nonlinear Stochastic Guideway-Vehicle-Systems. *6th IAVSD Symp. on Dynamics of Vehicles on Roads and Tracks*, Berlin 1979.

35. Pollard, M.G., Williams, R.A.: The Dynamic Behaviour of a Low Speed Electro-Magnetic Suspension. *Proc. IUTAM Symp. on the Dynamics of Vehicles on Roads and Railway Tracks*, Delft, 1975, pp. 445-478.

36. Popp, K.: Näherungslösung für die Durchsenkungen eines Balkens unter

einer Folge von wandernden Lasten. *Ing.-Arch.* 46, 1977, pp. 85-95.

37. Popp, K.: Stabilitätsuntersuchung für das System Magnetschwebefahrzeug-Fahrweg. *ZAMM* 58, 1978, pp. T 165 - T 168.

38. Popp, K., Habeck, R., Breinl, W.: Untersuchungen zur Dynamik von Magnetschwebefahrzeugen auf elastischen Fahrwegen. *Ing.-Arch.* 46, 1977, pp. 1-19.

39. Popp, K., Müller, P.C.: On the Stability of Interactive Multibody Systems with an Application to Maglev-Vehicle-Guideway Control System. *IUTAM Symposium Dynamics of Multibody Systems,* München, 1977.

40. Popp, K., Schiehlen, W.: Dynamics of Magnetically Levitated Vehicles on Flexible Guideways. *Proc. IUTAM Symp. on the Dynamics of Vehicles on Roads and Railway Tracks,* Delft, Aug. 1975, pp. 479-503.

41. Popp, K.: Zufallsschwingungen von Fahrzeugen auf elastischem Fahrweg am Beispiel einer Magnetschwebebahn. *ZAMM* 60, 1980, pp. 70-73.

42. Popp, K.: Beiträge zur Dynamik von Magnetschwebebahnen auf geständerten Fahrwegen. Habilitationsschrift, TU München 1978. *Fortschr.-Ber. VDI-Z.* Series 12, No. 35, Düsseldorf 1978.

43. Reister, D., Zurek, R.: Entwicklungsstand der elektro-magnetischen Schwebetechnik für eine Hochleistungsschnellbahn. *ETR* 25, No. 3, 1976, pp. 155-160.

44. Richardson, H.H., Wormley, D.M.: Transportation Vehicle/Beam-Elevated Guideway Dynamic Interactions: A State-of-the-Art-Review. *J. Dyn. Syst. Meas. Control,* Trans. ASME 96, Ser. G, 1974, pp. 169-179.

45. Rothmayer, W.: Elektromagnetische Trag- und Führungssysteme (EMS). In: *Lehrgang der Carl-Cranz-Gesellschaft* 0 R2.2 Simulationsmodell für Spurgebundene Fahrzeuge, Oberpfaffenhofen, 1977.

46. Smith, C.C., McGehee, D.Y., Healey, A.J.: The Prediction of Passenger Riding Comfort from Acceleration Data. *J. Dyn. Syst. Meas. Control,* Trans. ASME 100, Ser. G, 1978, pp. 34-41.

47. Smith, R.A.: Matrix Equation XA + BX = C . *SIAM J. Appl. Math.* 16, 1968, pp. 198-201.

48. Snyder III, J.E., Wormley, D.N.: Dynamic Interactions Between Vehicles and Elevated, Flexible Randomly Irregular Guideways. *J. Dyn. Syst. Meas. Control,* Trans. ASME 99, Ser. G, 1977, pp. 23-33.

49. Sussmann, N.E.: Statistical Ground Excitation Models for High Speed Vehicle Dynamic Analysis. *High Speed Ground Transp. J.* 8, No. 3, 1974, oo, 145-154.

50. Ward, J.D.: The Furure Roles for Tracked Levitated Systems, *J. Dyn. Syst. Meas. Control,* Trans. ASME 96, Ser. G, 1974, pp. 1-11.

51. Yamamura, S.: Perfomance Analysis of Electromagnetically Levitated Vehicle. *World Electrotechnical Congress,* Section 7, Paper 09, Moscow, 1977.

52. Zurek, R.: Method of Levitation for Tracked High-Speed Traffic. *Endeavour,* Vol. 2, No. 3, 1978.

53. Breinl, W.: Entwurf eines parameterunempfindlichen Reglers am Beispiel einer Magnetschwebebahn. *Regelungstechnik* 28, 1980, pp. 87-92.

54. Breinl, W.: Entwurf eines unempfindlichen Tragregelsystems für ein Magnetschwebefahrzeug. *Dissertation,* TU München, 1980.

55. Müller, P.C., Popp, K., Schiehlen, W.: Berechnungsverfahren für stochastische Fahrzeugschwingungen. *Ing.-Arch.* 49, 1980.

56. Popp, K.: Contributions to the Dynamic Analysis of Maglev Vehicles on Elevated Guideways. *Shock and Vibration Bulletin,* 1980.

SOME DESIGN CRITERIA FOR THE LAYOUT OF MAGLEV-VEHICLE-SYSTEMS

Wolfgang Crämer
Simulation and Control Department,
Spaceflight Division, Messerschmitt-Bölkow-Blohm GmbH, Munich, FRG

1. INTRODUCTION

Control system synthesis for MAGLEV-vehicles was developed consequently parallel with the structure of the vehicle.
Centralized control concepts, applied to rigid body test vehicles can easily be adapted to changes in the control algorithms, just by changing the onboard-computer-program. For application vehicles, however, a centralized control concept leads to high system complexity, since quadruple redundancy is necessary to meet the required safety-standards. The large necessary magnet gap for rigid body vehicles moreover results in high power-consumption. So parallel to the development of modular structure vehicles, decentralized control was applied, where controller and observer-design can be done for the "magnet-wheel".
A decentralized control concept makes use of the redundant number of magnets in a MAGLEV-system and thus guarantees reliability by configuration. The weak coupling of magnet and vehicle additionally allows smaller magnet gaps and lower power consumption.

2. VEHICLE CONCEPTS

2.1 Rigid body vehicles

Rigid body vehicles, i.e. vehicles without any suspension system, have been used at the first trials with MAGLEV-vehicles.

A rigid body vehicle will not be able to adapt to guideway disturbances with wavelength , shorter than vehicle length which leads to the necessity of larger nominal gaps to absorb such disturbances. And even wavelengths greater than vehicle length may not be followed by the vehicle for high speed, because of ridecomfort requirements.

A large nominal gap causes heavy magnets, high nominal magnet voltage and magnet current and for given maximum values of voltage and current results in a smaller dynamic range.

2.2 Modular structure vehicles

As a better approximation to an operational vehicle (Fig.2) therefore modular structure vehicles have been built, where the magnets are attached via a primary suspension to magnet frames, which are connected with the cabin by a secondary suspension (Fig.1, Fig.2).

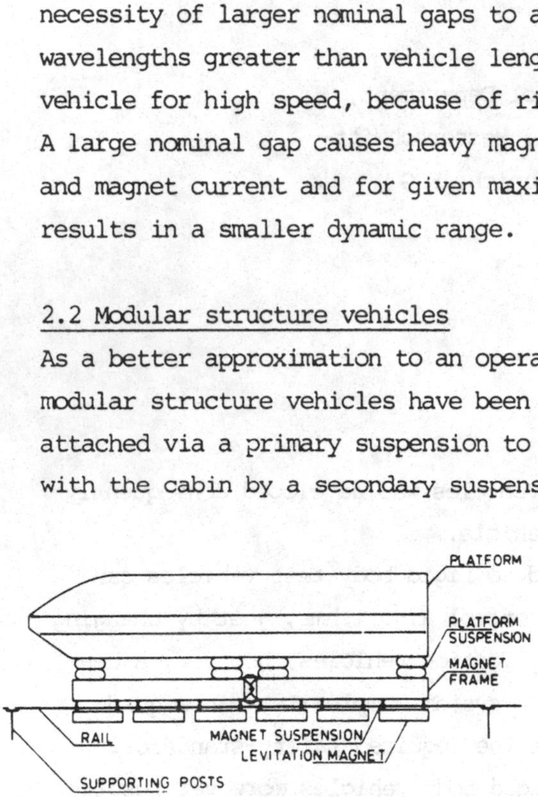

Fig.1: Modular structure concept of KOMET II and TRO5

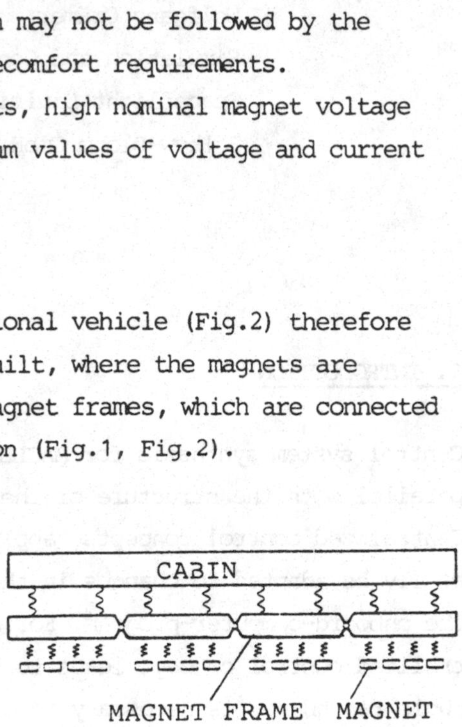

Fig.2: Mechanical structure of operational systems (TRO6)

As this structure, for suitable choice of the suspension elements, allows to adapt to guideway irregularities even at high speed, a smaller nominal gap may be realized, leading to a considerable reduction in magnet weight and power, required in the levitation system. Additionally ride comfort requirements can be fulfilled with comparatively simple passive suspension elements for realistic guideways. By comparison of the behaviour of KOMET I and KOMET II (Fig.3) the advantages of the modular structure were proven.

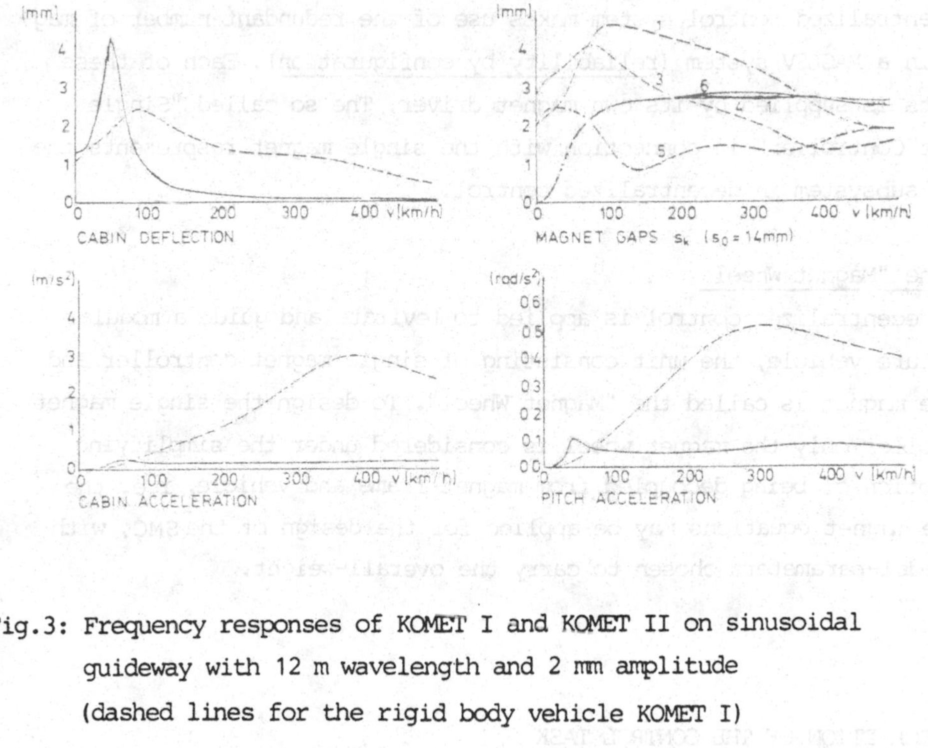

Fig.3: Frequency responses of KOMET I and KOMET II on sinusoidal guideway with 12 m wavelength and 2 mm amplitude (dashed lines for the rigid body vehicle KOMET I)

3. CONTROL CONCEPTS

3.1 Centralized control concept

In a centralized control concept all the data processing required by the control laws is done in the onboard computer. The advantage of such a configuration in test vehicles is obvious: Changes in control parameters or even control algorithms can be realized without any modification of hardware by only changing the onboard software. In case of an application vehicle, such a centralized control leads to high system complexity, since quadruple redundancy is necessary to meet the required safety standards.

3.2 Decentralized control concept

A decentralized control system makes use of the redundant number of magnets in a MAGLEV system (reliability by configuration). Each of these magnets is supplied by its own magnet driver. The so called "Single Magnet Controller" in connection with the single magnet respresents the basic subsystem in decentralized control.

3.3 The "Magnet Wheel"

When decentralized control is applied to levitate and guide a modular structure vehicle, the unit consisting of single magnet controller and single magnet is called the "Magnet Wheel". To design the single magnet controller only the magnet wheel is considered under the simplifying assumption of being decoupled from magnet frame and vehicle, i.e. the single magnet equations may be applied for the design of the SMC, with the model-parameters chosen to carry the overall-weight.

4. DESCRIPTION OF THE CONTROL TASK

4.1 Single magnet model

Under the simplifying assumption of a quadratic relation between magnet force and magnet current the nonlinear magnet equations are

$$P_M = \frac{\mu_o AN^2}{4} \frac{I_M^2}{(s_M+\varepsilon)^2} \quad (1); \quad U_M = \frac{\mu_o AN^2}{2} \frac{\dot{I}_M}{(s_M+\varepsilon)} - \frac{\mu_o AN^2}{2} \frac{I_M \dot{s}_M}{(s_M+\varepsilon)^2} + I_M R + \dot{I}_M L_S \quad (2)$$

with ε replacing the iron path and L_s the stray-inductivity.

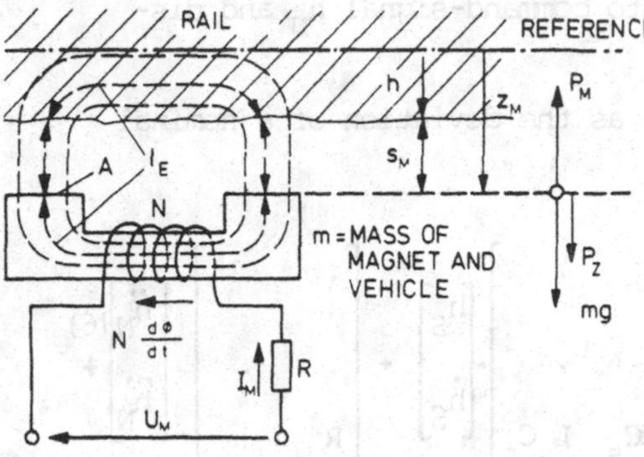

Fig.4: Single magnet model coordinates

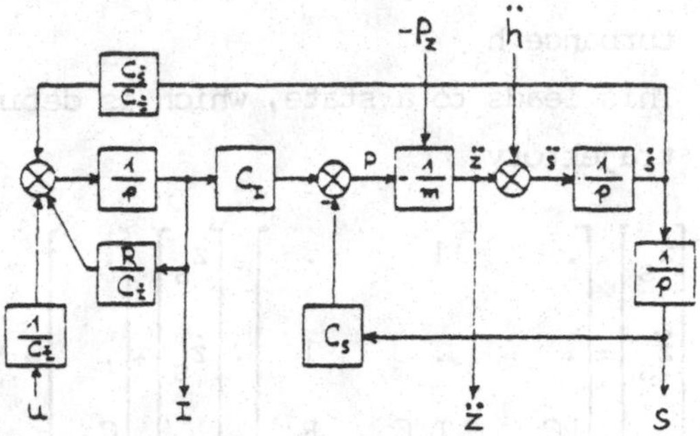

Fig.5: Single magnet model with system disturbances

For control system synthesis the nonlinear magnet equations must be linearized about the nominal gap

$$P = C_I I - C_s s; \quad U = C_{\dot I}\dot I - C_{\dot s}\dot s + R\,I \tag{3}$$

$$C_s = \frac{2P_o}{(s_o+\varepsilon)}\ ; \quad C_I = C_{\dot s} = \frac{2P_o}{I_o}\ ; \quad C_I = \frac{2P_o(s_o+\varepsilon)}{I_o^{\,2}} + L_s \tag{4}$$

$$U_M = U_o + U; \quad I_M = I_o + I; \quad s_M = s_o + s; \quad P_M = P_o + P$$

and as a linearized system the single magnet may be described by a state vector $(s, \dot s, \dddot z)$.

$$\begin{bmatrix} \dot s \\ \ddot s \\ \dddot z \end{bmatrix} = \begin{bmatrix} \cdot & 1 & \cdot \\ \cdot & \cdot & 1 \\ \frac{RC_s}{mC_{\dot I}} & \frac{L_sC_s}{mC_{\dot I}} & -\frac{R}{C_{\dot I}} \end{bmatrix} \cdot \begin{bmatrix} s \\ \dot s \\ \ddot z \end{bmatrix} + \begin{bmatrix} \cdot \\ \cdot \\ -\frac{C_I}{mC_{\dot I}} \end{bmatrix} U + \begin{bmatrix} \cdot \\ -1 \\ \cdot \end{bmatrix} \ddot h + \begin{bmatrix} \cdot & \cdot \\ \cdot & \cdot \\ \frac{R}{mC_{\dot I}} & \frac{1}{m} \end{bmatrix} \cdot \begin{bmatrix} P_z \\ \dot P_z \end{bmatrix} \tag{5}$$

4.2 System with disturbances an command-signals

For a correct definition of the state "to be controlled" now the rail-

input to the system has to be divided into command-signal h_N and disturbance h_S.
This leads to a state, which is defined, as the deviation of a nominal trajektory.

$$\begin{bmatrix} \dot z_S \\ \ddot z_S \\ \dddot z_S \end{bmatrix} = \begin{bmatrix} \cdot & 1 & \cdot \\ \cdot & \cdot & 1 \\ \frac{RC_S}{m\dot C_I} & \frac{L_S C_S}{m\dot C_I} & -\frac{R}{\dot C_I} \end{bmatrix} \cdot \begin{bmatrix} z_S \\ \dot z_S \\ \ddot z_S \end{bmatrix} + \begin{bmatrix} \cdot \\ \cdot \\ -\frac{C_I}{m\dot C_I} \end{bmatrix} u + \begin{bmatrix} \cdot & \cdot \\ \cdot & \cdot \\ -\frac{RC_S}{m\dot C_I} & -\frac{L_S C_S}{m\dot C_I} \end{bmatrix} \cdot \begin{bmatrix} h_S \\ \dot h_S \end{bmatrix} + \begin{bmatrix} \cdot & \cdot \\ \cdot & \cdot \\ -\frac{R}{\dot C_I} & -1 \end{bmatrix} \cdot \begin{bmatrix} \ddot h_N \\ \dddot h_N \end{bmatrix} + \begin{bmatrix} \cdot & \cdot \\ \cdot & \cdot \\ \frac{R}{m\dot C_I} & \frac{1}{m} \end{bmatrix} \cdot \begin{bmatrix} P_z \\ \dot P_z \end{bmatrix} \qquad (6)$$

Obviously the minumum effort to control this system would be the compensation of the inputs h_S, h_N with a simple compensation law,

$$U_K = -\frac{RC_S}{C_I} h_S - \frac{L_S C_S}{C_I} \dot h_S - \frac{Rm}{C_I} \ddot h_N - \frac{\dot C_I m}{C_I} \dddot h_N$$

stabilization with a minimum-energy-controller, which follows from the minimization of a quadratic cost-function, if only the manipulated variable is weighted in the criterion, and feedback of the state z_S, $\dot z_S$, $\ddot z_S$. Proceeding this way would lead to the following ideal dynamic behaviour (Fig.7).

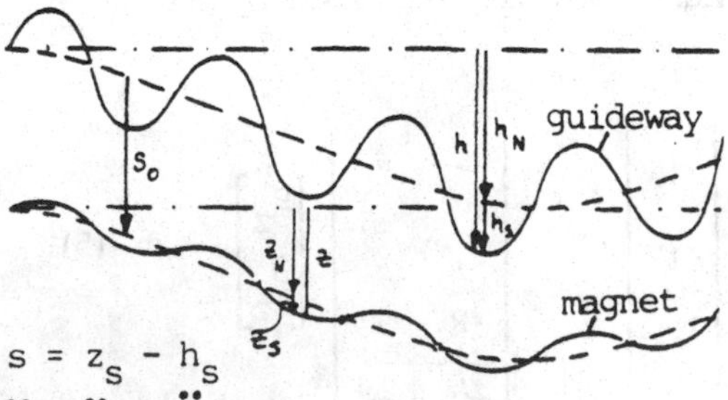

$s = z_S - h_S$

$\ddot z = \ddot z_S + \ddot h_N$

Fig.6: Guideway with command signal and disturbance

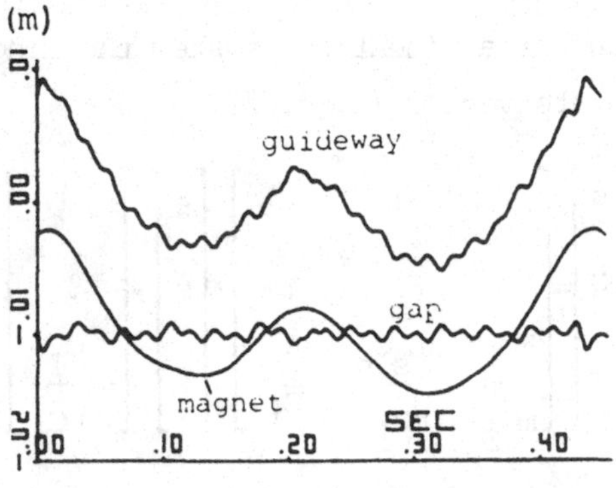

Fig.7: Single Magnet riding ideally over the actual guideway

4.3 Optimal bandwidth

For the 3-mass substitute model of the modular structure vehicle the following transfer-functions between magnet force and the motion of the magnet may be established.

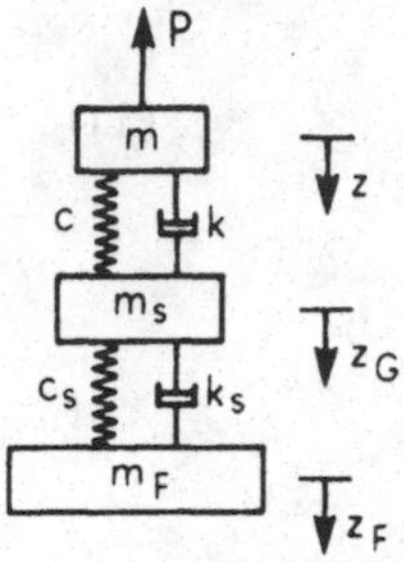

$$Z_F = \frac{k_s p + c_s}{m_F p^2 + k_s p + c_s} Z_G$$

$$\left[m_s p^2 + (k+k_s)p + (c+c_s)\right] Z_G = (k_s p + c_s) Z_F + (kp+c) Z$$

$$(mp^2 + kp + c) Z - (kp+c) Z_G = -P$$

From these equations follows

$$\ddot{U}(p) = \frac{P}{Z} = -(mp^2+kp+c) - \frac{(kp+c)^2}{\left[m_s p^2 + (k+k_s)p + c + c_s\right] - \dfrac{(k_s p + c_s)^2}{m_F p^2 + k_s p + c_s}} \qquad (8)$$

the transfer-function between magnet force and magnet movement, given a modular structure vehicle.

An elevated guideway may be assumed to have a periodic course and thus for the vehicle is a periodic disturbance, a superposition of sinus-inputs. For any of these sinus-inputs now the according partial power may be determined following two principally different objectives.

Contouring (s-control) means that the magnet follows the rail exactly and the power consumption for this case must increase with the frequency of the disturbance. From the linearized equation for the magnet force (3) follows $s = 0 \Rightarrow z = h$ and thus

$$I = \frac{C_S}{C_I} S = -\frac{C_S}{C_I} H; \quad N_1 = \frac{R}{2} |I|^2 = \frac{R\,|\ddot{U}(p)|^2}{2\,C_I^2} |H|^2 \qquad (9)$$

Platforming (z-control) means moving with constant force following none of the rail disturbances and power consumption for this case will be only dependent on the input-amplitude, not on the frequency as follows from (3) with $z = 0 \Rightarrow s = -h$

$$I = \frac{C_S}{C_I} S = - \frac{C_S}{C_I} H; \quad N_2 = \frac{R}{2} |I|^2 = \frac{RC_S^2}{2C_I^2} |H|^2 \tag{10}$$

The optimal bandwidth must follow from $N_1 = N_2$, i.e. contouring for $\omega < \omega_{opt}$ and platforming for $\omega > \omega_{opt}$.

From $N_1 = N_2$ follows $C_s = |\ddot{U}(j\omega)| \rightarrow \omega_{opt}$ (11)

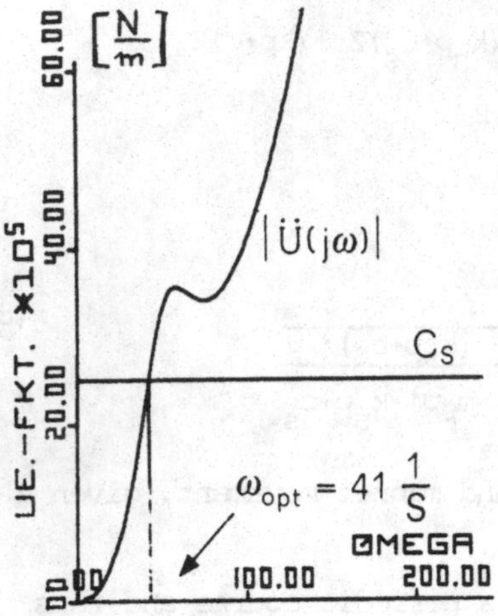

Fig.8: Optimal bandwidth for the 3-mass-substitute-model of the TR06-vehicle

4.4 Physical boundaries

For the transfer-functions S/H, U/H, I/H thus the following ideal demands can be established

$$\frac{S}{H} = 0; \quad \omega < \omega_{opt} \quad \text{contouring}$$
$$\frac{S}{H} = -1; \quad \omega > \omega_{opt} \quad \text{platforming} \tag{12}$$

and from the linearized magnet equations (3) follows

$$\frac{I}{H} = \frac{\ddot{U}(p)}{C_I}; \quad \frac{U}{H} = (C_I p + R) \frac{\ddot{U}(p)}{C_I}; \quad \omega < \omega_{opt} \tag{13}$$

$$\frac{I}{H} = -\frac{C_s}{C_I} \; ; \quad \frac{U}{H} = -(C_I p + R)\,\frac{C_s}{C_I} + C_s p \; ; \qquad \omega > \omega_{opt} \tag{13}$$

Any realization of the control concept can only result in an approximation to these demands. From the comparison of ideal and real amplitude-responses for the TR06-vehicle-structure (Fig.9) the degree of approximation can be seen.

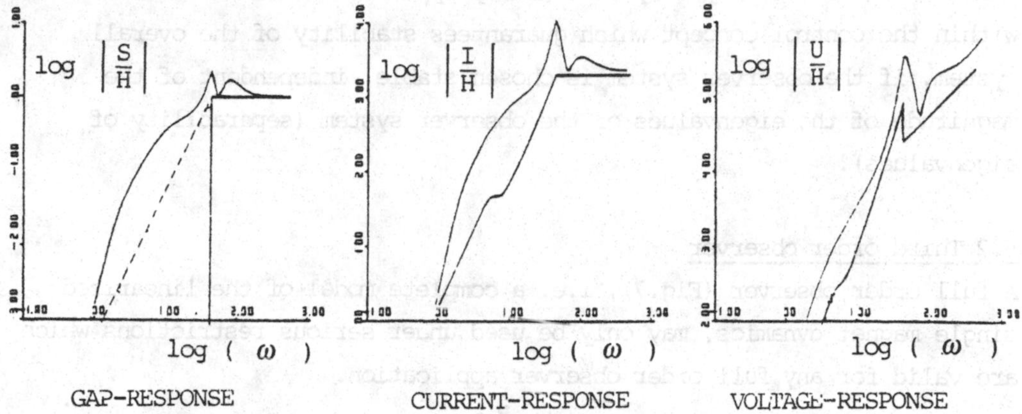

Fig.9: Comparison of ideal and real amplitude responses for the TR06-vehicle

5. CONTROL SYSTEM SYNTHESIS

Possible measurements at the magnet-level are the magnet gap "s", the magnet-acceleration "z" and the magnet-current "I", where I is dependent from s and z and additionally biased by the disturbing force "P_z". So for control system design the measurement equation.

$$\begin{bmatrix} s \\ \ddot{z} \end{bmatrix} = \begin{bmatrix} 1 & \cdot & \cdot \\ \cdot & \cdot & 1 \end{bmatrix} \begin{bmatrix} s \\ \dot{s} \\ \ddot{z} \end{bmatrix} \tag{14}$$

is to be preferred.

To stabilize the system (5) feedback of s is necessary, which cannot be measured and thus must be estimated by suitable means.

5.1 Observer concepts

A solution, satisfying with respect to stability and amplification of measurement errors, may only be found by application of an observer within the control concept which guarantees stability of the overall system, if the observer system is chosen stable, independent of the magnitude of the eigenvalues of the observer system (separability of eigenvalues).

5.2 Third order observer

A full order observer (Fig.7), i.e. a complete model of the linearized single magnet dynamics, may only be used under serious restrictions which are valid for any full order observer application.

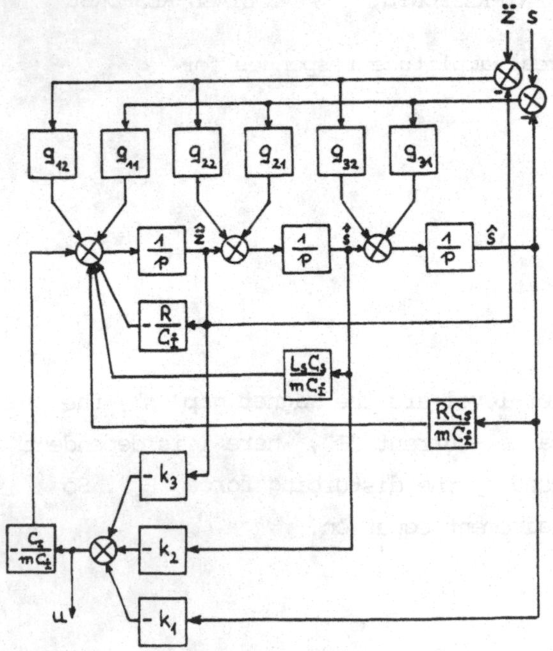

Fig.10: Third order observer for the single magnet

The magnetically levitated single mass without disturbances has a system description

$\dot{x} = A\,x + B\,u\ ;\ y = C\,x$ (15), the control law is $u = -K\,\hat{x}$ (16)

and the observer system with a model different from the real system is

$$\dot{\hat{x}} = A_M \hat{x} + B_M u + G(y - C_M \hat{x}) \tag{17}$$

For a system as described above the use of controller and observer leads to a decoupled overall system (state x and estimation error $\tilde{x}$); if the observer model is correct in its structure and parameters.

$$\begin{bmatrix} \dot{x} \\ \dot{\tilde{x}} \end{bmatrix} = \begin{bmatrix} A-BK & -BK \\ 0 & A-GC \end{bmatrix} \cdot \begin{bmatrix} x \\ \tilde{x} \end{bmatrix} \qquad \tilde{x} = \hat{x} - x \tag{18}$$

The overall system will be stable, if the subsystems A-BK and A-GC are stable. The critical point is the stability of the overall system, if the observer model $A_M = A + \Delta A$; $B_M = B + \Delta B$; $C_M = C + \Delta C$ does not agree with the plant A,B,C. The overall system then has the form

$$\begin{bmatrix} \dot{x} \\ \dot{\tilde{x}} \end{bmatrix} = \begin{bmatrix} A-BK & -BK \\ \Delta A - \quad BK - G\Delta C & A_M - GC_M - \Delta BK \end{bmatrix} \cdot \begin{bmatrix} x \\ \tilde{x} \end{bmatrix} \tag{19}$$

and stability of the subsystems does no more guarantee the stability of the overall system. Separability gets lost, because the structure of the overall system changes. Thus, depending on the system even small parameter variations may lead to instability and especially for errors in the input matrix B or output matrix C there is generally no possibility to design appropriate controllers and observers, to make the overall system insensitive, as errors ΔB, ΔC are amplified by the controller and observer gains.

The application of a single magnet model within the observer system, to stabilize a MAGLEV-vehicle, thus must be eliminated for reasons of parameter sensitivity and more serious, within a modular structure vehicle, because observer structure and dimension do not agree with the real system.

A third order ovserver designed by the minimization of a quadratic cost function, applied to a single magnet on a rigid rail with parameters corresponing to the real magnet of course gives satisfying stability. The same observer, applied to a single magnet on elastic rail gives stability only, for a rather high gain and a onedimensional representation of a modular structure vehicle (magnet, magnet frame and cabin), which has been proven to be a good substitute model for the extended vehicle on elastic rail, cannot be made stable for any gain.

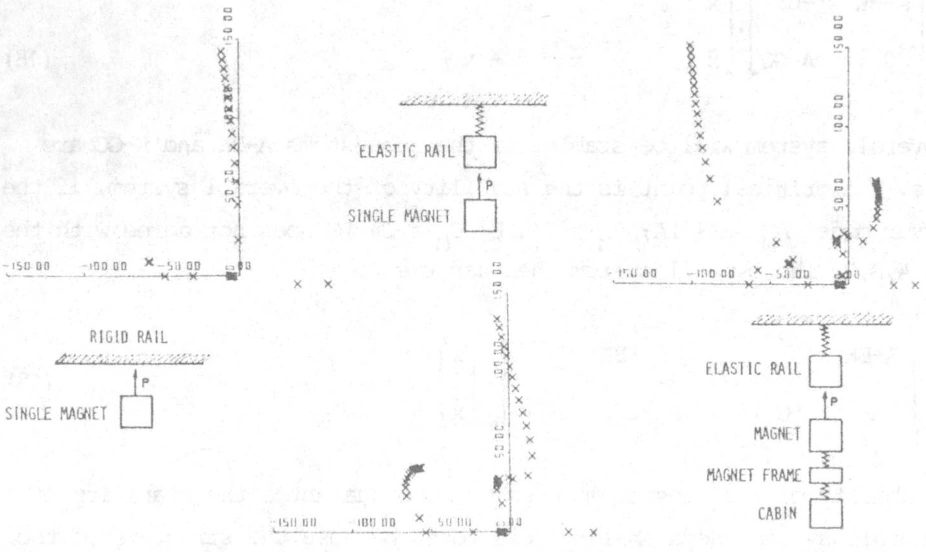

Fig. 11: Root-loci: third order observer
gain variation (0 - 1.5 nominal value)

The same difficulties arise for variations of the model parameters. For variation of the nominal gap between 5 mm and 20 mm an observer of third order, designed for a nominal gap of 10 mm yields an unstable overall system for increasing gap, even for the design-case single magnet-rigid rail.

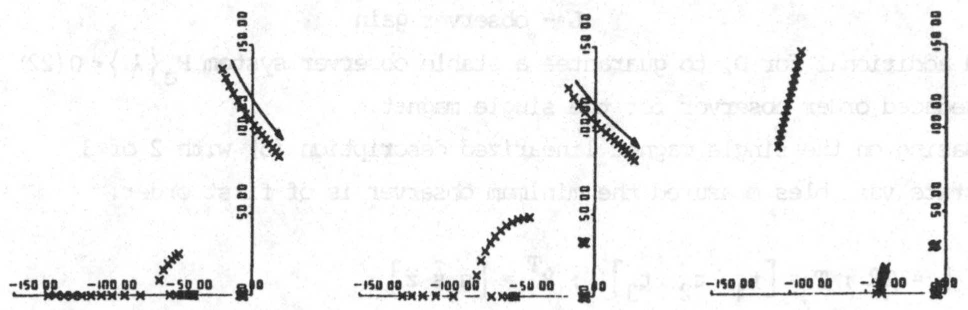

Fig.12: Root-loci: third order observer

parameter variation (magnet gap 5-20 mm)

Synthesis of an insensitive observer thus is the problem to solve and for MAGLEV-vehicles the solution from the very beginning was a reduced order observer, designed insensitive against parameter variations and structure of the model.

5.3 Reduced order observer synthesis at the example of the magnetically levitated single mass

- Theory:

$$\dot{x} = A\,x + B\,u \qquad \text{system description}$$
$$y = C\,x \qquad \text{measuring equation} \qquad (20)$$
$$u = -K\hat{x} \qquad \text{control equation}$$

observer equations

full order	reduced order
$T\,\big|\;\dot{\hat{x}} = A\,\hat{x} + B\,u + G\,(y - C\hat{x})$;	$\dot{\xi} = D\,\xi + T\,B\,u + L\,y$

(with ξ a subspace $T\cdot x$ of the state x shall be estimated)

$$\dot{\xi} = \underbrace{(TA - \underbrace{TGC}_{L})}\hat{x} + TBu + \underbrace{TGy}_{L}; \qquad \dot{\xi} = \underbrace{DT}\hat{x} + TBu + Ly$$

(the two braced terms $TA - LC$ and DT are linked by "=")

From this follows, that the matrices D, T, L have to fulfil the condition DT = TA - LC with: D - observer system (21)

L - observer gain

and additional for D, to guarantee a stable observer system $R_e\{\lambda_i\} < 0$ (22)

- Reduced order observer for the single magnet.

  Basing on the single magnet linearized description (5) with 2 of 3 state variables measured the minimum observer is of first order.

$$\xi = T\hat{x} \; ; \; T = \begin{bmatrix} t_1 & t_2 & t_3 \end{bmatrix} \; ; \; \hat{x}^T = \begin{bmatrix} s & \hat{s} & \ddot{z} \end{bmatrix} \; ;$$

$$A = \begin{bmatrix} \cdot & 1 & \cdot \\ \cdot & \cdot & 1 \\ a_1 & a_2 & a_3 \end{bmatrix} ; \qquad D = -d, \quad L = \begin{bmatrix} l_1 & l_2 \end{bmatrix}$$

From DT = TA-LC three equations for the determination of the parameters to choose follow.

- $dt_1 = t_3 a_1 - l_1$
- $dt_2 = t_1 - t_3 a_2$
- $dt_3 = t_2 + t_3 a_3 - l_2$

At this point some constraints have to be introduced for reasons of independence from system parameters, i.e. to guarantee an insensitive observer system. The necessary choice is $t_3 = 0$. Otherwise the reduced order observer concept would suffer from the same sensitivity - disadvantages as a full order concept.

$dt_1 = l_1$; $dt_2 = -t_1$; $t_2 = l_2$ and with l_2 choosen 1 $t_2 = 1; t_1 = -d; l_1 = -d^2$.

The reduced order observer system thus is:

$$\dot{\xi} = -d\,\xi + \begin{bmatrix} -d^2 & 1 \end{bmatrix} \cdot \begin{bmatrix} s \\ \ddot{z} \end{bmatrix} ; \qquad \xi = T\hat{x} = -ds + \hat{s}, \quad \hat{s} = \xi + ds \tag{23}$$

and the control law becomes $u = -(k_1 + k_2 d)\, s - k_2 \xi - k_3 \ddot{z}$ (24)

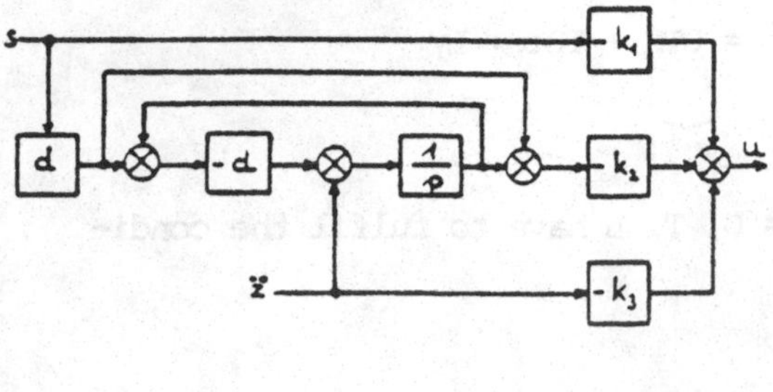

Fig.13: First order observer for the single magnet

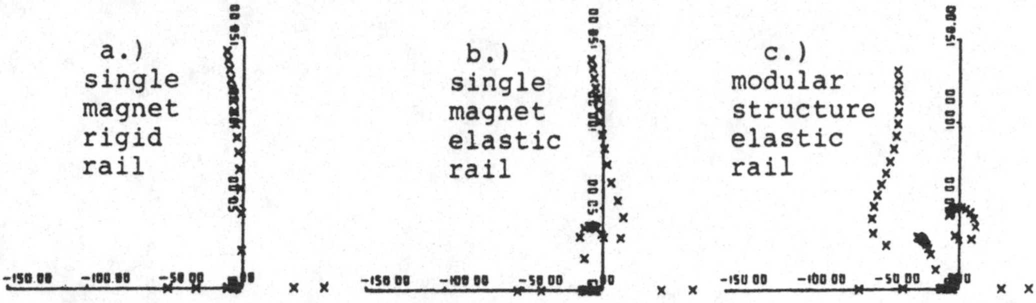

Fig.14: Root-loci: first order reduced observer
gain variation (0 - 1.5 nominal value)

The application of this reduced order observer, designed for the single magnet on rigid rail obviously still gives satisfying stability for the one-dimensional representation of the modular structure vehicle on elastic rail at nominal gain.

The problem of conditional stability, introduced by the eigenvalues of the elastic rail (Fig.14 b,c), which may cause some difficulties, hovering at zero velocity, can be governed by adding feedback of the magnet flux to the basic control law, i.e. a constant force component is added to supress stimulation of the rail.

Insensitivity against variations of the model parameters also is guaranteed, applying the reduced order observer for the single magnet as well as for the 4-mass representation (Fig.15).

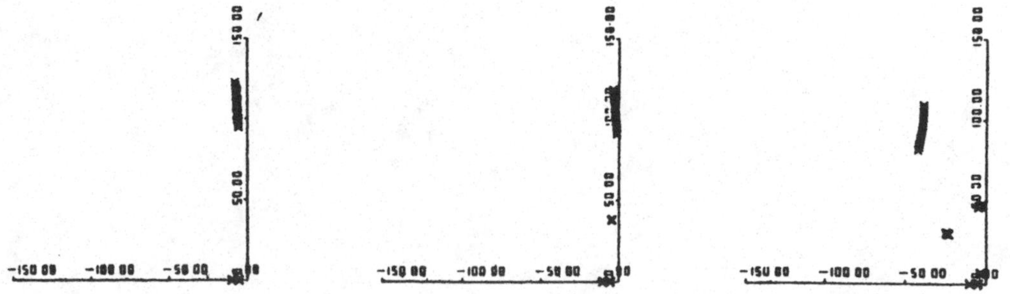

Fig.15: Root-loci: first order reduced observer
parameter variation (magnet gap 5-20mm)

MAGNETICS AND EXPERIMENTAL RESULTS OF MAGLEV-VEHICLES

Dr. Gerhard Bohn
Magnetic Levitation and Guidance Systems
department, Spaceflight Division, Messerschmitt-Bölkow-Blohm GmbH, Munich, FRG

1. Introduction

The beginning of the attractive magnetic levitation technic was in the year 1935 when H. Kemper [1] succeeded in his first private experiment on electromagnetic levitation. Further systematic research and development work he did at the "Areodynamische Versuchsanstalt" in Göttingen until 1943. [2]

The modern development of magnetic levitation started in 1968, when the companies Messerschmitt-Bölkow-Blohm (MBB) and Krauss-Maffei (KM) investigated intensely the transportation volume of the last 20 years of this century and the worldwide standard of transportation technology. The result was the necessity of a magnetically levitated and guided vehicle with a maximum speed of 500 km/h [3] . In the course of developments this velocity is reduced to 400 km/h because of vehicle weight and energy costs.

Within the electromagnetic research program five large scale vehicles have been built and succesfully tested. The first four vehicles were constructed in competition between the two companies MBB and KM:

May 1971: MBB demonstrated a 5 passenger-carrying vehicle on a track of 700 m length. Maximum velocity was 90 km/h.

October 1971: KM demonstrated the 10 passenger vehicle TRO2 with a maximum speed of 164 km/h on a 1000m long guideway.

The task for the follwing years was the development of components and to reach the maximum speed of 400 km/h.

January 1974: KM began experiments with the 20 passenger carrying vehicle TRO4 which reached a maximum speed of 253 km/h on an elevated 2,5 km long guideway.

January 1975: MBB began experiments with the unmanned test sled KOMET which reached 401,3 km/h on 1300 m long guideway. This test sled was accelerated by steam rockets.

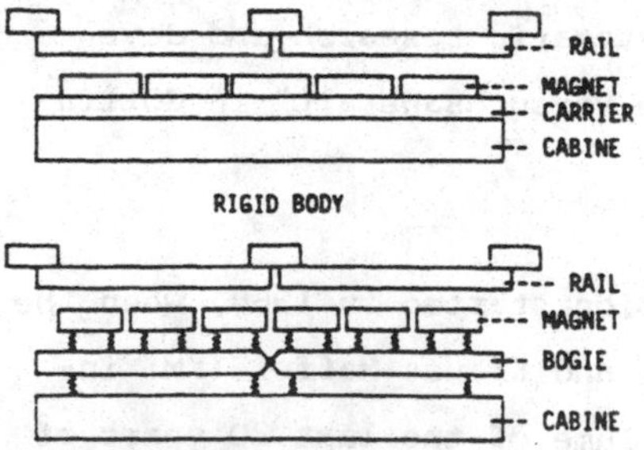

Fig. 1 Vehicle Structures

In the year 1975 the german government demanded a joint venture group of MBB and KM for the further developments. The next step was the modification of the mechanical structure of the KOMET vehicle. This is seen in fig. 1

The first four vehicles were built as rigid bodies, where the magnets were directly fixed to the vehicle body as it is shown in the upper plot. Further developments must aim to reduce the electrical losses, to raise the reliability and to diminish the demands on the guideway. These ideas led to the modified vehicle KOMET M. The aim can be reached by a modular structure where the magnets are suspended by springs and damping elements to the bogies an the passenger cabin is suspended by springs and damping elements to the bogies. This is shown schematically in the lower plot. Every magnet has its own controll system and its own current driver. With this concept we got very good experimental results as will be shown later.

The fifth vehicle TRO5 built according to this mechanical structure was the first electromagnetically levitated and guided vehicle authorized for public transportation. The vehicle is shown in fig. 2.

It consisted of 2 sections, each carrying 32 persons. This facility was built under the responsibilty of 3 companies. MBB and KM constructed the vehicle, and Thyssen the stationary parts guideway, propulsion and stations. It served as a transportation and demonstration vehicle on the International Exhibition in Hamburg 1979.

Fig. 2 IVA-vehicle TRO5

The present task is to construct a closed guideway with the length of 31,5 km and a revenue vehicle TR06 for long time tests. This facility is beeingt built in Emsland in northern Germany. The experiments will begin in 1983.

2. Designing and Modelling of Magnets

In order to give an impression of the vehicle structure fig. 3 shows a cross of the vehicle and the guideway. Magnets and rails are arranged to levitate and guide the vehicle. The magnet types which are used up to now are the U-shaped magnet with a solid iron core and rail (fig. 4)

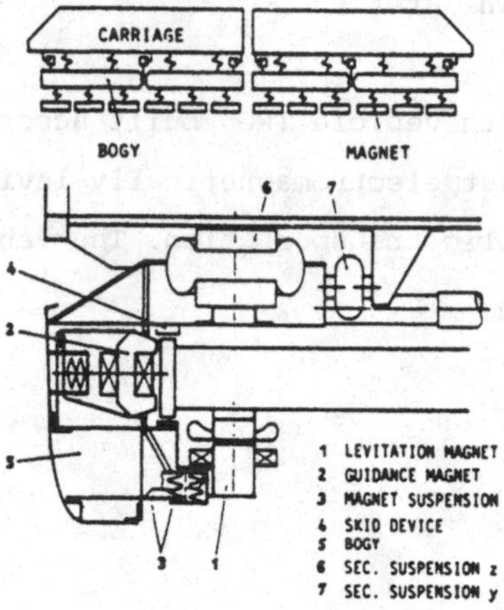

Fig. 3 Mechanical Structure

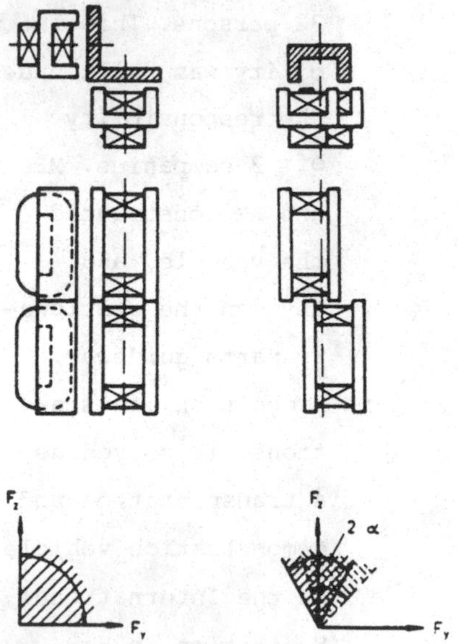

Fig. 4 Arrangements of Levitation and Guidance Magnets

In the TR05 and TR06 vehicle long-stator magnets as shown in fig. 5 are used with laminated cores and rails, which integrate the function levitation, propulsion and inductive power transmission into the vehicle. This power transmission is done by an inductive winding which is located in the magnet poles. The inductive effect is generated by the oscillating flux density in the air gap due to the teeth in the stator rail.

This kind of power transmission was not yet applicated in the vehicle TRO5. There we had a 440 V onborad battery.

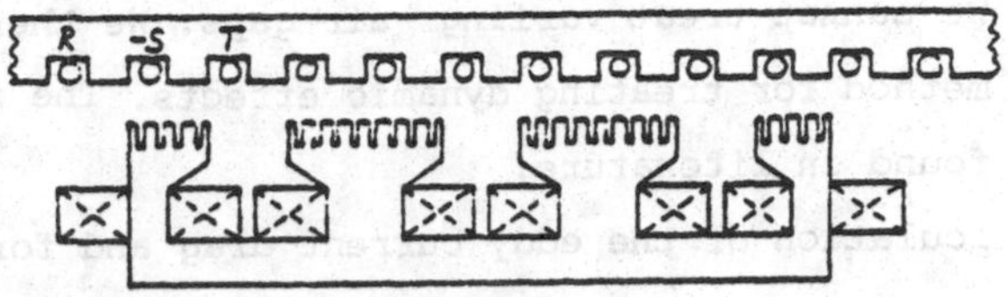

Fig. 5 Long Stator System with Linear Generator

In order to be able to construct a good and eco-nocmic levitation and guidance system we have to establisch a satisfactory procedure for designing and modelling magnets. The criteria are

temperature stability
minimum weight
minimum energy consumption

given forces for several operating points (air gap width) and a given maximim time variation of the magnet force $\mathring{F} = dF/dt$ in the nominal oprating point. The criteria can only be fulfilled by a compromise, what means that we have to find an optimal magnet design. In order to reach this aim we need a systematic and economic design program. Designing the magnet and calculating all necessary qualities with a three dimensional and if necessary time depended method is found to be to time consuming.

We therefore employ the following six step procedure:

(1) Calculation of the geometrical and electrical data of the magnet by means of a simple but non-linear (saturation effects) analog network [7] for the magnetic circuit using a cost functional as constraint which includes costs for the total magnetic levitation system.

(2) Calculation of force-and flux-characteristics by a finite difference method [8]. The calculation of force and voltage for the superposition of a constant and a harmonically oscillating current component is possible with our program, but this has several disadvantages:

a) Permeability is constant in time, we therefore cannot include dynamic saturation effects.

b) The computation time is very long as we had only good success if we switched on the frequency step by step.

c) We cannot treat variing air gaps. We therefore established a new method for treating dynamic effects. The foregoing methods can be found in literature.

3) Calculation of the eddy current drag and force loss if the iron backing of the magnets is a solid iron rail.

4) Calculation of the dynamic properties by frequency responses $\partial U/\partial I$, $\partial U/\partial s$, $\partial F/\partial I$, $\partial F/\partial s$, where U is voltage, I current, F the magnetic force and s the air gap width between magnet and rail.

5) Correction of the theoretical results by experiments if necessary.

6) The total information of the preceeding steps is integrated into an electromechanical mathematical model where the parameter of the model structure are adjusted to the foregoing calculated characteristics.

In order to be able to calculate the eddy current effects what means drag and force decrease caused by the solid armature rail when the magnet moves along its rail we deform the magnet as shown in fig. 6. The upper half space is filled with a ferromagnetic medium which carries on the surface a periodic arrangement of current layers which simulate periodic arrangement of coils. The same thing we do also in the direction of motion x. Region I is the airgap and region II the spread out armature rail. We assume the permeability as constant. Now we can solve these field problem by fourier series technique and cal-

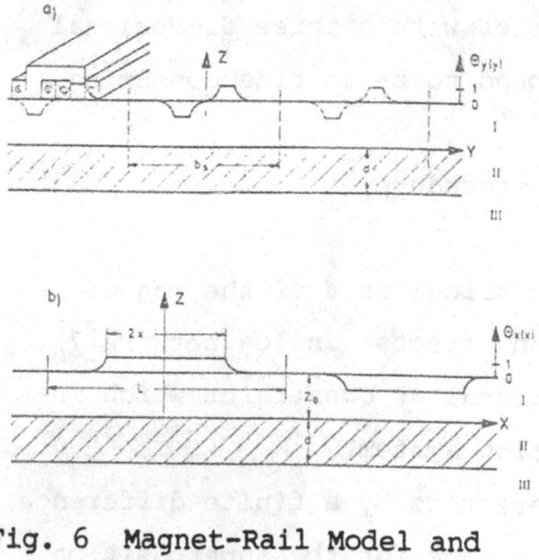

Fig. 6 Magnet-Rail Model and MMF-Distribution

culate the forces by Maxwell's stress tensor. As we have solved the problem only for constant permeability and as saturation effects play an essential role we have to make experiments with a magnet and an armature rail of the same material which is used for the final guideway. From these experiments we get an effective permeability $\mu_{(v)}$ which depends on the magnets velocity. Some results are shown in fig. 7 and fig. 8.

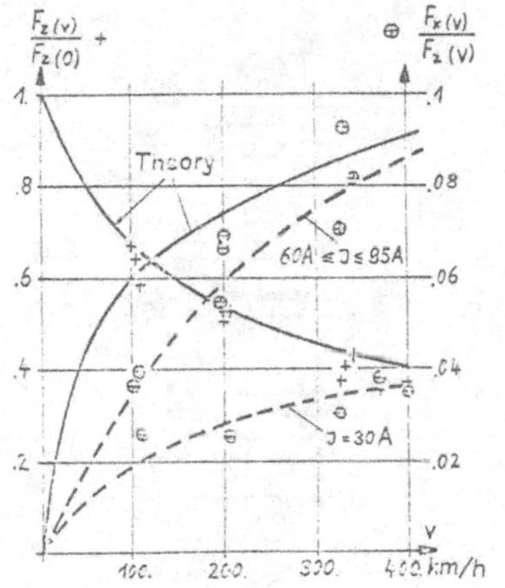

Fig. 7 Eddy Current Effects
1 Magnet, Gap: 15mm

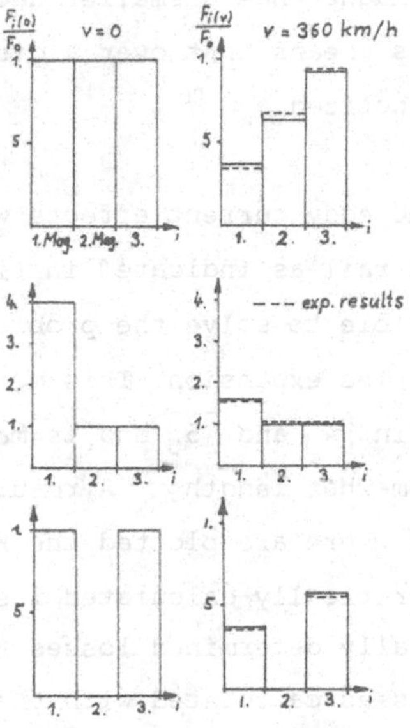

Fig. 8 Eddy Current Effects
Different Excitations

The experimental results we got from measurements with the KOMET vehicle. The magnets were fixed at the bottom of the vehicle with its force downward to a test rail. The magnets were fed by constant current. The vertical and horizontal forces were measured by a quarz facility. Fig. 7 shows the results for one magnet together with theoretical results. As can be seen, the curve for the attraction force F_z (v) / F_z (O) is practically independent of the coil current whereas the breaking force in the low current region shows a dependence on current. The attraction force decreases to 40% of the value at rest while the breaking force F_x in the nominal working point reaches 9% of the attraction force at 400 km/h. The adjustment of $\mu_{(v)}$ was done for the nominal force with I = 60 A. Fig. 8

shows the magnet forces for three different excitation configurations of three magnets. The upper plot shows that along three equally excited magnets the force increases rapidly at the velocity of 360 km/h. If the first magnet of the row is excited to four times of its nominal force when stationary the following magnets do not show any decrease in their attraction force. If the second magnet fails the lower plot shows that the third magnet has a smaller decrease in its attraction force than the first. This means that over a distance of one meter the rail remains partly magnetized.

For dynamic eddy current effects we deform magnet and rail as indicated in fig. 9. Then it is possible to solve the problem by a fourier series expansion. This method is described in [4] and [5] and is mathematically somewhat lengthy. A result is shown in fig. 10 where are plotted the ratios between theoretically calculated losses and experimentally determined losses P^{exp}. P^{mod} are the losses calculated with the sketched theory and P^{id} are calculated by considdering only an ideal inductance in connection with an ohmic resistance.

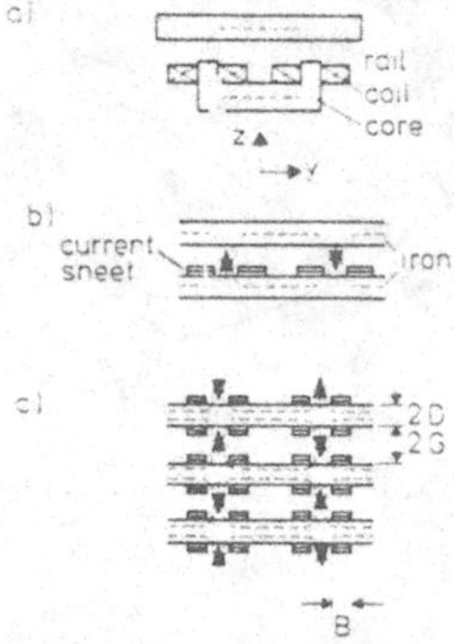

Fig. 9 Model for the Calculation of Dynamic Effects

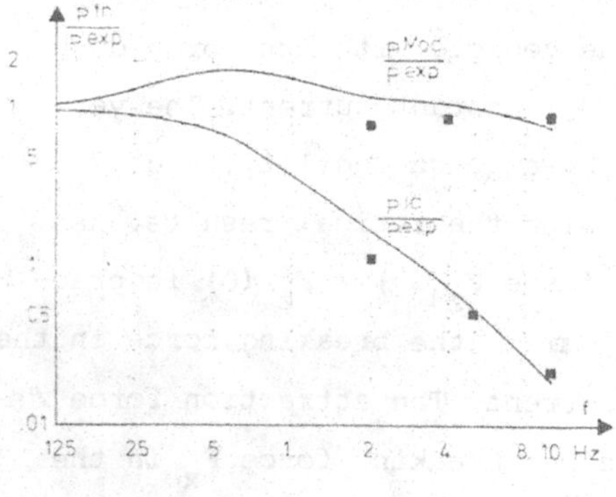

Fig. 10 Comparison of Calculated and Experimental Losses

The square dots show values from newer experiments.

We integrate these effects in a mathematical model for the total flux. For simplicity I will neglect the eddy current effects. The equation for the coil current is

$$U = IR + \mathring{\psi}$$

Where U is voltage, I the coil current, R the ohmic coil resistance and the flux coupled with all windings. We approximate this flux by the mathematical structure

$$\psi = \frac{1}{\alpha_{(s)}} \cdot \mathrm{th}(\alpha_{(s)} \cdot L_{o(s)} \cdot J) + L_s \cdot J$$

where

$$\alpha_{(s)} = \alpha_o - \frac{\alpha_1}{s + \varepsilon_1} \;, \quad L_{o(s)} = \frac{c}{s + \varepsilon_o} + d$$

With this analytic expression we calculate the magnetic energy

$$W = -\int \psi \, dJ$$

and by differentiating this with respect to the airgap s we get the attraction force

$$F = \frac{\partial W}{\partial s} = -\frac{\partial}{\partial s} \int \psi \, dJ \qquad (1)$$

Now it is possible to adjust the parameters $\alpha_o, \alpha_1, \varepsilon_o, \varepsilon_1$, c and d so that (1) fits the theoretically or experimentally determined force characteristics. By a further experiment we measure the flux ψ and get the last parameter L_s. This analytical model serves as a magnet model for dynamic simulations.

If stationary eddy current effects must be taken into account (the leading magnet is mostly affected) one has to adjust (1) to the corresponding lowered force characteristics.

3. Experimental results from the vehicle KOMET, KOMET M and TR05

As already mentioned the original vehicle KOMET was a rigid body where the magnets were directly fixed to the vehicle body. The further development led to a modular structure. This concept has the advantage that above a certain perturbation frequency from the track only the magnet mass has to follow exactly. In addition we have the possibility to reduce the air gap from 14 mm to 10 mm which reduces the static losses. The decentralized concept exhibits a well-natured fallout behaviour.

Based on these ideas the vehicle KOMET was modified to vehicle KOMET M. The electrical losses of a levitation system can be devided into five parts

- static losses (0 Hz)
- dynamic losses ($0 < f \leqslant 15$ Hz)
- losses from pulse width modulation ($f \geqslant 200$ Hz)
- losses in cables and current drivers
- losses from battery resistance

The static losses depend primarily on the air gap width, the dynamic losses on the guideway quality and on aerodynamic forces and on the masses which have to follow the rail. Losses from pulse width modulation are primarily eddy current losses in the magnet core and the armature rail.

In comparison with these three contributions to the total losses the losses in cables and current drivers are very low, approximately 3 %. We did not measure the losses due to the battery resistance, but this resistance influences the dynamic behaviour of the vehicle as a high current consumption lowers the battery voltage what causes still higher a. c. losses. Fig. 11 shows the losses from the first four points of the above table per vehicle ton depending on the vehicle velocity for the two levitation concepts.

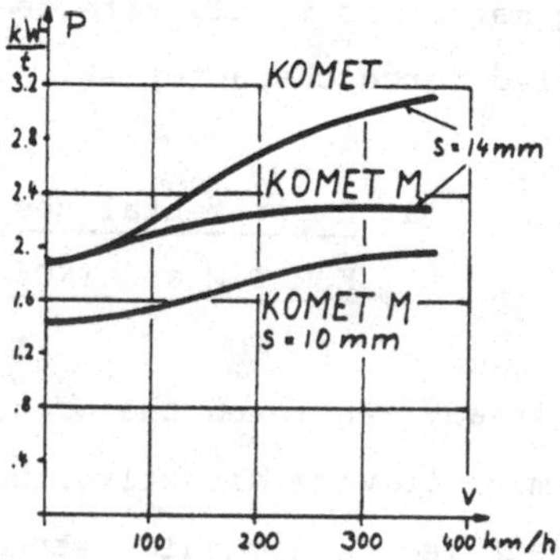

Fig. 11 Total Losses for Different System Concepts

Only the losses of the levitation systems were taken into consideration. The dynamic losses of the KOMET version are 1.2 kW/t for 360 km/h where as the losses of the version KOMET M are only 0.4 kW/t. This version enabled us also to ride with an air gap of 10mm which reduces the losses once more by nearly o.4 kW/t. In order to see the influence of the guideway quality we divided the guideway into three parts of different qualities (fig. 12). 108 m were very exact and stiff (normal guideway), 48 m were very stiff with a sinusoidal perturbation of 12m wavelength an 2mm amplitude (distorted guideway), and 48 m were elastic. Here the distance of the pillars was 6m and the static bending nearly 2 mm (elastic guideway).

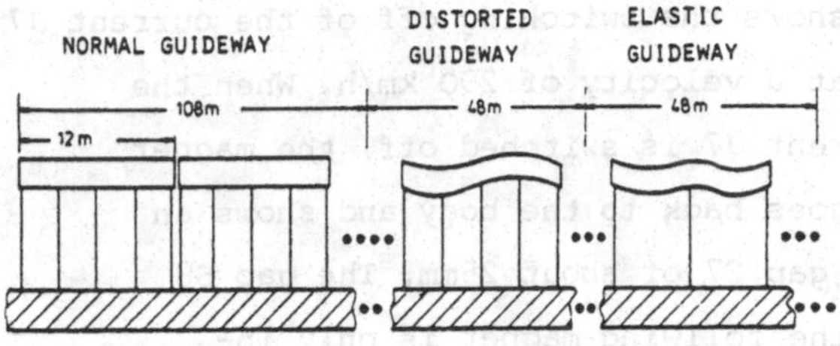

Fig. 12 Guideway for Experiments

The following plots show results only form KOMET M.

The air gap response on guideway irregularities shows the histogramm of fig. 13 in dependence on the reduced order oberserver frequency ω_s [6].

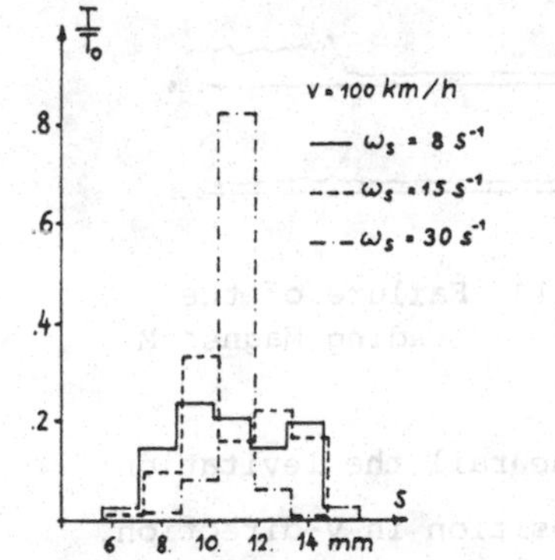

Fig. 13 Response of the Air Gap on Guideway Irregularities (distorted)

High values of ω_s mean a strong coupling. The measurements were done within the distorted section of the guideway. For ω_s = 8 s^{-1} the magnet follows with some overshoot. Within an interval of 5mm the probability is nearly constant which means that the magnet does not follow the rail. For ω_s = 15 s^{-1} we can already see a maximum which raises to a distinct peak for ω_s = 30 s^{-1}. Within 85% of the riding time the air gap is in an interval of 1.5 mm. The flanks of this distribution come from the gap sensors which measured the air gap very locally. Between the guideway beams we had gaps of 5 to 10mm which cause a signal of $\Delta s \approx 2$mm. Therefore the signals had to be filtered for the control system. This is also the reason why we could not go to higher frequences of

ω_s. Otherwise we would get stability problems.

The decentralized structure of KOMET M admits fallout of a magnet or current driver or control system. The worst case of fallout is that of a leading magnet because it has to magnetize the rail and, based on eddy currents in the armatrue rail, has the highest static coil current. Fig. 14 shows the switching off of the current J7 of the leading front magnet M7 at a velocity of 290 km/h. When the current J7 is switched off, the magnet M7 goes back to the bogy and shows an air gap S7 of about 25mm. The gap S8 of the follwing magnet is only influenced within a half second, while the current J8 rises for a short time to 100 A and then evens out at 80A. The third magnet M9 is influenced only very slightly as can be seen from the gap and current signals S9 and J9.

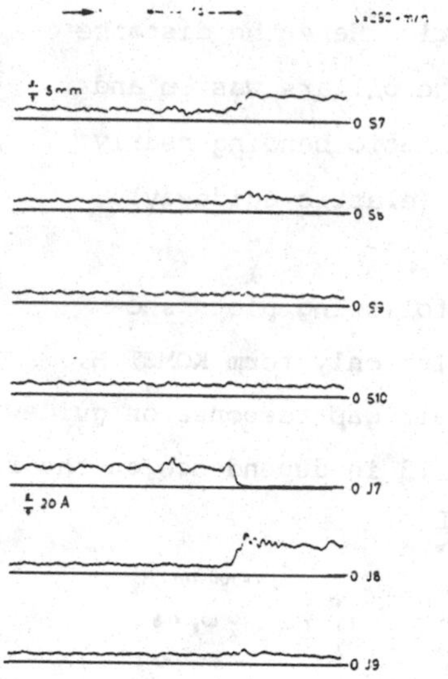

Fig. 14 Failure of the Leading Magnet M7

Now I wll show some experimental results from the TR05 vehicle. Here the magnets serve as levitation and propulsion elements as it is shown in fig. 5. Because of the slots and the teeth in the rail the levitation force depends in a weak manner on the magnets position in y-direction. This is no severe problem in the static case but for velocitiy greater than zero this perturbation can excite vibrations and as all magnets must have the same position to the teeth all magnets would vibrate synchronously.

In order to eleminate the fundamental frequency of this effect we detuned the pole distance of magnet and rail, so that the alternating parts of the forces on the seven poles of one magnet sum up to zero. On the other side we now get moments on the magnet as every magnet pole has another slot configuration from the armature rail. As the magnets are not ideally firmly fixed to the bogy for this rotational degree of freedom but have a eigenfrequency of nearly 70 Hz and as the magnet force depends also on the rotational angel we get a synchronous combined vertical and rotational mode of all levitation magnets. The vertical acceleration amplitude was about 11 m/s2. Now the slot distance is 8cm which means that at velocity of 20 km/h we get a very high a.c. current amplitude in the battery current. This can be seen by the following table.

Table 1: Battery current

velocity	V	0	20	30	60	75	km/h
mean battery current	J_B°	75	90	75	90	90	A
amplitude of a.c. part	$\hat{J}_B$	35	240	100	150	150	A
mean losses	$\overline{P}_B$	33	34	35	40	40	kW

The high mean current value J_B at 20 km/h comes from the effect that the control system of the levitation magnets partly worked in its voltage limitation which causes a 1 - 2 mm larger air gap as the nominal air gap of 15 mm. This resonance state lasted for about 3 - 5 s and therefore did not influence the ride quality noticably.

From the IVA project we also got information on the eddy current losses due to the pulse width modulation frequency of the current drivers. The KOMET-vehicle current driver had a fundamental frequency of 200 Hz and the magnet cores and armature rails were solid. The IVA current drivers had a fundamental frequency of 400 Hz. The IVA guidance magnet cores and the guidance rail were solid whereas the cores of the levitation magnets and its rail were laminated. The results for the mean eddy current losses are given in table 2.

Table 2: Losses due to pulse width modulation

KOMET levitation magnet	IVA guidance magnet	IVA levitation magnet
0,4 kW	0,150 kW	0.006 kW

As the nominal losses of electro magnets lay in the order of 1,5 kW the higher frequency causes a noticably decrease of the total losses for the levitation and guidance system. The current drivers for the next vehicle TVE will at least have a frequency of 1 kHz, which is a demand from the control system.

References:

[1] H. Kemper, ETZ Heft 15, p.391, (1938)

[2] H. Kemper, ETZ-A, Heft 1, p.11, (1953)

[3] HSB Studiengesellschaft mbH, HSB Studie über ein Schnellverkehrs-system, München (1971)

[4] G. Bohn, J. Langerholc, J. Appl, Phys. 48, No.7, p.3o93 (1977)

[5] G. Bohn, J. Langerholc, J. Appl, Phys. 52, No.2, p.542 (1981)

[6] E. Gottzein, R. Meisinger, L. Miller, JEEEE Transactions on Vehicular Technology, Vol. VT-29, No.1, p.17, (1980)

[7] E. Kallenbach: Der Gleichstrommagnet, Akad. Verlagsgesellschaft Geest & Portig K.-G., Leipzig 1969

[8] G. Bohn, P. Romstedt, W. Rothmayr, P. Schwärzler, Proc. Fourth International Cryogenics Engineering Congress 1972 (Eindhoven)

in the United States
masters